Ind a
Communications
Fifth Edition

Industrial Data Communications

Fifth Edition

By Lawrence M. Thompson
and Tim Shaw

Notice

The information presented in this publication is for the general education of the reader. Because neither the author nor the publisher has any control over the use of the information by the reader, both the author and the publisher disclaim any and all liability of any kind arising out of such use. The reader is expected to exercise sound professional judgment in using any of the information presented in a particular application.

Additionally, neither the author nor the publisher has investigated or considered the effect of any patents on the ability of the reader to use any of the information in a particular application. The reader is responsible for reviewing any possible patents that may affect any particular use of the information presented.

Any references to commercial products in the work are cited as examples only. Neither the author nor the publisher endorses any referenced commercial product. Any trademarks or tradenames referenced belong to the respective owner of the mark or name. Neither the author nor the publisher makes any representation regarding the availability of any referenced commercial product at any time. The manufacturer's instructions on the use of any commercial product must be followed at all times, even if in conflict with the information in this publication.

Printed in the United States of America.
10 9 8 7 6 5 4 3 2

ISBN: 978-0-87664-095-1

ISA
67 T. W. Alexander Drive
P.O. Box 12277
Research Triangle Park, NC 27709

Library of Congress Cataloging-in-Publication Data in process

Contents

Prologue

Larry Thompson: In editions one and two, I spent a bit of time prophesizing over the state of industrial networking. Although I have found that I have been right more often than wrong, I have also found that my being right or wrong doesn't really make any difference – as I stated in the Prologue to the third edition. In the fourth edition, I stated that there would be, and there now is, a fieldbus standard: IEC 61158. That it contains eight non-interoperable standards apparently makes good sense, at least to those who wrote it. I also stated that the marketplace will decide who gets the lion's share of the fieldbus market. However, I did predict that whoever makes the best use of object-oriented programming, makes the user interface easy, and makes the system easy to purchase, install, and maintain will be the winner.

All that I stated in the previous editions was predictable enough. What has amazed me (in the time from 2005 to 2014) is the quick and thorough embracing of Internet/intranet technologies, even in the control areas, and the use of increasingly sophisticated hardware and software.

Since this book is now co-authored, I will no longer predict what I think the technological trends may be, rather, I will defer to my previous practice and simply say, "Let the market sort it out." This book is a text on how things are now and, to stay relevant, that is what it should and will be. To this end, as time goes on by, there will be further editions whenever there are major or significant cumulative technological changes in industrial data communications.

Tim Shaw: As Larry's co-author, I second his remarks. All of my prior efforts at prognostication have come to naught, so I won't attempt any here, other than to say that I believe wireless communications will continue to become more prevalent and that cybersecurity will continue to be a challenge. I hope to see more industrial communications technologies adopting and incorporating mechanisms for cybersecurity and more vendors building such capabilities into their products, rather than forcing customers to "bolt them on" afterwards.

For additional information or answers to questions on this text, including where it can be researched, contact the authors at larrymthompson@hotmail.com (or www.snowflakecomputers.com) and timshaw4@verizon.net.

Preface to the Fifth Edition

Rationale

In the 24 or so years since the first printing of this book, there have been changes in nearly every aspect of data communications. This is particularly true in industrial applications. As stated in earlier prefaces, the rationale for the original text was then, and is still now, that many people are forced to learn about data communications because the processes aren't as transparent and as "plug-and-play" as they should be. While these individuals' original intent was not to become experts in data communications, they are nonetheless now forced to learn some specific detailed facts just to accomplish their primary job functions.

This is also a fact of life for the book's original author. Industrial data communications has become quite complex and has so intertwined itself with commercial systems and IT practices that it is difficult for one person to have expertise in all of its aspects. Enter Tim Shaw (William T. Shaw), the co-author. Tim has (refer to his biography) a wealth of experience in SCADA, DCS, PLC, electrical distribution, and substation systems, as well as in cybersecurity. His additions to this book, as originator and writer, will immensely increase its value to the reader. As partners in this creation, we hope that the reader will not be able to tell where one of us hands over the baton to the other; we desire that it be a seamless document and, hopefully, one of great worth.

As stated in the four previous prefaces, there is a need to understand the technical jargon and semantics. While the reader will not find it difficult or tedious to acquire the necessary knowledge of data communications, the learning process requires the material to be organized, so as to keep focused on particular learning points, and we have indeed tried to do so. In addition, due to technological changes and, indeed, to changes in the direction and focus of industrial applications, this fifth edition contains a great deal more material than the previous editions, expanding on almost all topic areas, particularly with respect to the wireless technologies and the security considerations that have become mandatory within industrial settings.

Much as it was with the first, second, and third editions, the need to upgrade the fourth edition became apparent as soon as it was published. The field of data communications is dynamic and is continuously moving toward different end points. While it is true that the fundamentals have not changed (or have changed very little), the industrial applications of 2014 are changing at a quicker pace and, in many cases, more quickly than the revision cycles of this text or any texts that cover the same topic areas, making it difficult to maintain currency without frequent revisions (a rather costly and time-consuming process for print materials). Much of the information, particularly the principles in the previous four editions, is still valid; however, more than minor revisions were required for the fifth edition and we, the authors, hope we have accomplished these rather major revisions accurately and objectively.

Objectives

The objectives of this edition are exactly the same as those of the previous ones: to introduce the principles and applications of industrial data communications and to bring the reader to the level where he or she can communicate with other professionals on this topic.

Audience

The intended audience for this edition is the same as for previous editions: a person with some general technical education who is somewhat literate with computers, the Internet, and search engine use. Knowledge in the electrical, electronic, and computer disciplines will be of value, as this field is predominantly electrical and computational in nature, but it is not an absolute prerequisite for understanding. A familiarity with basic number systems, along with hexadecimal representation, is required; however, there is an explanation of these topics in the appendices. As in editions three and four, background history and basic media are also discussed in the appendices to the levels necessary to understand the text. For more information, there are many reference books, study materials, and computer-assisted training courses on these subjects. As stated previously, a willingness to understand new concepts and a sense of historical perspective will help. As always when reading this text, patience and a sense of humor are required.

Overall

As in all previous versions, the text ranges from basic principles to complex applications. This book *is not a design or engineering document* but a text designed to bring the reader's knowledge up to the current practice. It is not assumed that the reader already knows, or is even familiar with, industrial data communications. The concepts and applications will be discussed only to the extent necessary to grasp general concepts and/or applications. This text is written in the same less formal, more conversational style as its predecessors. This style has been the choice of most readers, as opposed to a more formal style, and, as with previous editions, this text is written (hopefully) to maximize communication with the reader.

Larry Thompson
Owner/General Manager
ESdatCo
707 Coleman St.
Marlin, TX 76661
larrymthompson@hotmail.com

William T. (Tim) Shaw
Lead Consultant/Principle
Cyber SECurity Consulting
2318 Monkton Rd.
Monkton, MD 21111
timshaw4@verizon.net

1 Communication Concepts

This first chapter deals with the fundamentals of data communications. We primarily probe and discuss the factors that affect all communications from a "big picture" perspective, so that when the details are presented in later chapters, you will know which niche is being filled. A portion of this chapter—the discussion of data organization sometimes called a *coding* (ASCII is an example)—is almost a technical history in itself (more on this subject is provided in appendix B).

Elements

Communication has particular elements. In any communication there must be a source (in data communications this is often called the *transmitter*) and one or more destinations (typically called the *receivers*). The purpose of communication is to transmit data from the source to the destination(s). The data is transmitted through a medium of one kind or another, varying according to the technology used. When we speak, we use audio waves to project sound data through the medium of air. In data communications, the media we use are: electrical conductors (usually referred to as *copper* connections), light pipes (often referred to as *fiber-optic cabling*), and electromagnetic transmission (usually referred to as *wireless* or *radio*).

This book is all about how we *move* and *organize* data. (Data itself is useless unless organized, at which time it becomes *information*.) Data communications topologies are organized as one-to-one (point-to-point), one-to-many (multi-drop), or many-to-many (networked). Figure 1-1 illustrates these three forms of data communications organization.

Note that while data communications terms can be couched in technical symbology, the concepts behind them are relatively simple. Members of a network may have defined relationships: they may have no ranking (be peers, all have the same communications value) or they may be ranked (such as master/slave). Point-to-point means just that, from one point to another, or directly from source to destination (whether workstation-to-server or peer-to-peer). Multi-drop topology more closely resembles a network than it does point-to-point topology. In general, multi-drop involves a master station of some kind with slave stations, as opposed to peer stations. (According to the following definition of a network, a multi-drop system can also be classified a network.) For now, we will define a network simply as three or more stations (whether peers, master/slave, or some other ranking) connected by a common medium through which they may share data. Later in this book we will tighten our definition, dividing networks into wide area or local area, and so on.

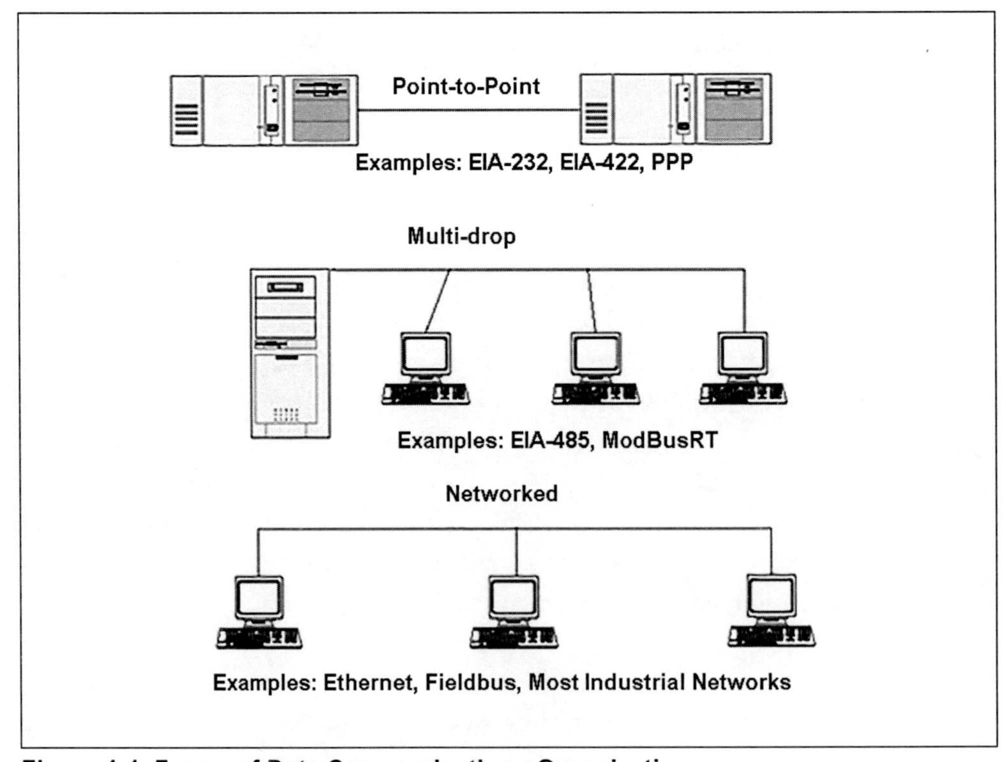

Figure 1-1. Forms of Data Communications Organization

Modes

In data communications, there are three modes of transmission: simplex, half-duplex, and duplex (see figure 1-2). These terms may be used to describe all types of communications circuitry or modes of transmission, whether they are point-to-point, multi-drop, or networked. It is important to understand these three terms because almost all descriptive language pertaining to data communications uses them.

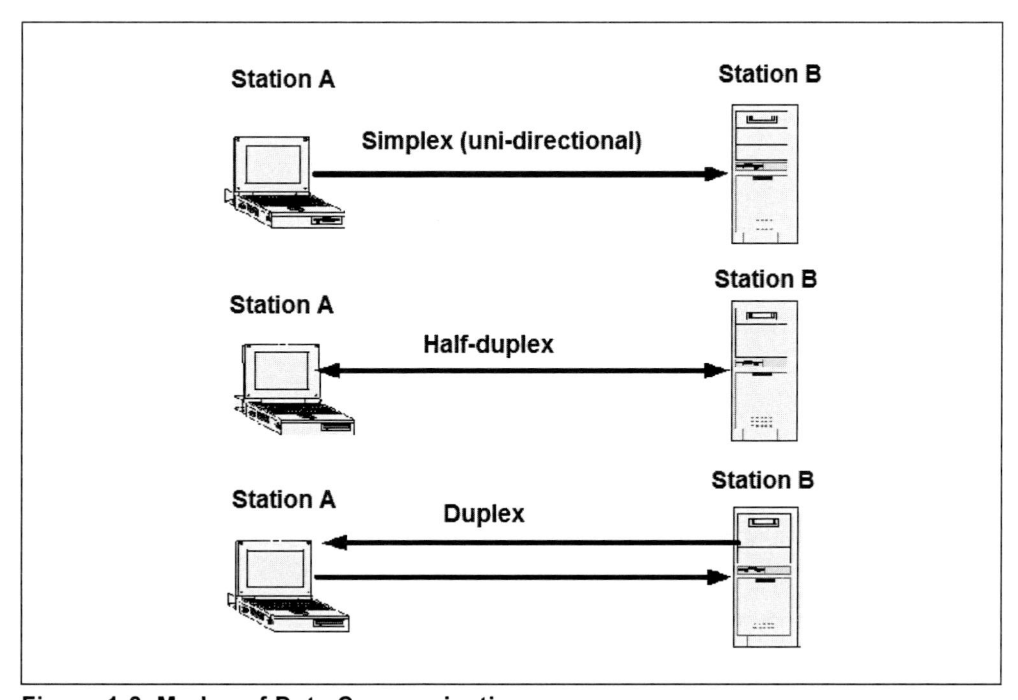

Figure 1-2. Modes of Data Communications

A communications channel may consist of a *communications circuit*, which can be either a hardware configuration (consisting of hardware components and wiring) or a "virtual" circuit (a channel that consists of software programming for communication channels that are not physically connected). "Virtual circuit" refers more to the process of communication than to the hardware configuration.

NOTES

1. The differences between a "mode" and a "circuit" are rather arbitrary. However, the reader needs to be aware that not all channels are hardware, and even though a channel is physically capable of

operating in a certain mode, it does not mean that the channel is being utilized in that mode. As an example, a duplex channel could be operated in half-duplex mode.

2. In many cases, the literature still uses the term "full-duplex" when referring to duplex mode.

You should be aware that constraints on the mode the communication channel is capable of using may be due to hardware or software. For example, if the hardware is duplex, the software may constrain the hardware to half-duplex. However, if the hardware cannot support a mode, no amount of software will cause it to support that mode, although it may appear to do so to human observation. An example is a half-duplex system that appears (due to the speed and message attributes) to be duplex to the human user.

The three data communications modes are as follows:

- **Simplex or Unidirectional Mode.** In this mode, communication occurs only in one direction—never in the opposite direction; in figure 1-2 it is from Station A to Station B. The circuit that provided this mode of operation was originally called *simplex* (in the 1960's telephone industry), but this led to confusion with more current telephony terminology. "Unidirectional" is a more appropriate name for this mode of transmission and using the name for this circuit would be much more descriptive; however, old habits (and names) are hard to change, therefore we will use the term *simplex* in this text (even though we would prefer to use *unidirectional*) so the reader will not be confused when referencing technical data.
- **Half-Duplex Mode.** In this mode, communication may travel in either direction, from A to B or from B to A, but not at the same time. Half-duplex communication functions much like meaningful human conversation does, that is, one speaker at a time and one (or more) listener(s).
- **Duplex Mode.** In duplex mode, communication can travel in both directions simultaneously: from A to B and from B to A at the same time.

Serial and Parallel Transmission

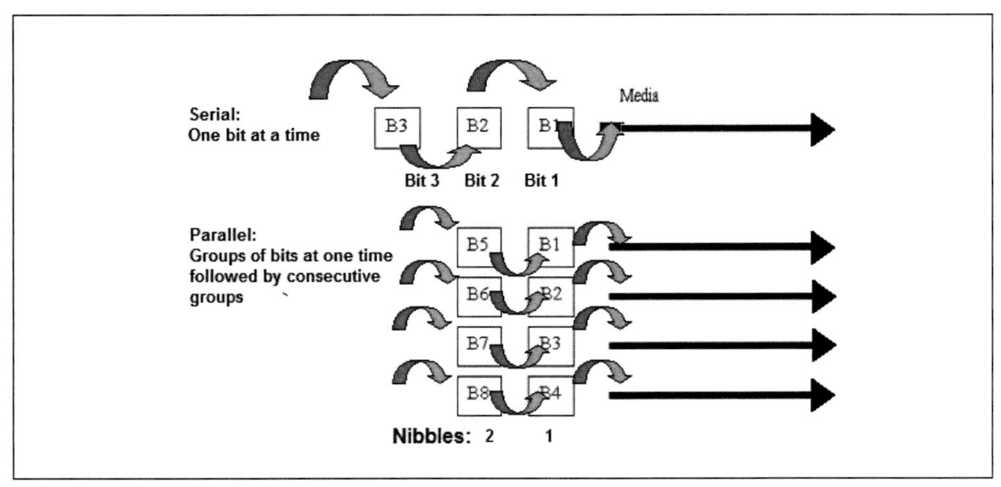

Figure 1-3. Serial and Parallel Transmission Concepts

Serial transmission (see figure 1-3) uses one channel (one medium of transmission) and every bit (binary digit, defined below) follows one after the other, much like a group of people marching in single file. Because there is only one channel, the user has to send bits one after the other at a much higher speed in order to achieve the same throughput as parallel transmission.

NOTE A nibble is 4 bits (a small byte, a byte is 8 bits).

In parallel transmission (regardless of the media employed), the signal must traverse more than one transmission channel, much like a group of people marching in four or more columns abreast. For a given message, four parallel channels can transmit four times as much data as a serial channel running at the same data rate (bits per second). However, parallel transmission running any appreciable distance (based on data rate—the faster the data rate, the more the effects for a given distance) encounters two serious problems: First, the logistics of having parallel media is sure to increase equipment costs. Second, ensuring the data's simultaneous reception over some distance (based on data rate—the higher the data rate, the shorter the distance) is technically quite difficult, along with ensuring that cross-talk (a signal from one transmission line being coupled—electrostatically or electromagnetically—onto another) is kept low. Cross-talk increases with signaling rate, so attempting to obtain a faster data rate by using additional parallel conductors becomes increasingly difficult.

Figure 1-4 illustrates what the signals would look like in serial and parallel transmission. Note that for the two 4-bit combinations, it took only two timing periods (t0-t2) to transmit all 8 bits by parallel transmission, whereas the serial transmission took eight timing periods (t0-t8). The reason is that the serial transmission used one media channel, while the parallel transmission used four media channels (Channel 1 for bits A–E; Channel 2 for bits B–F; Channel 3 for bits C–G; and Channel 4 for bits D–H).

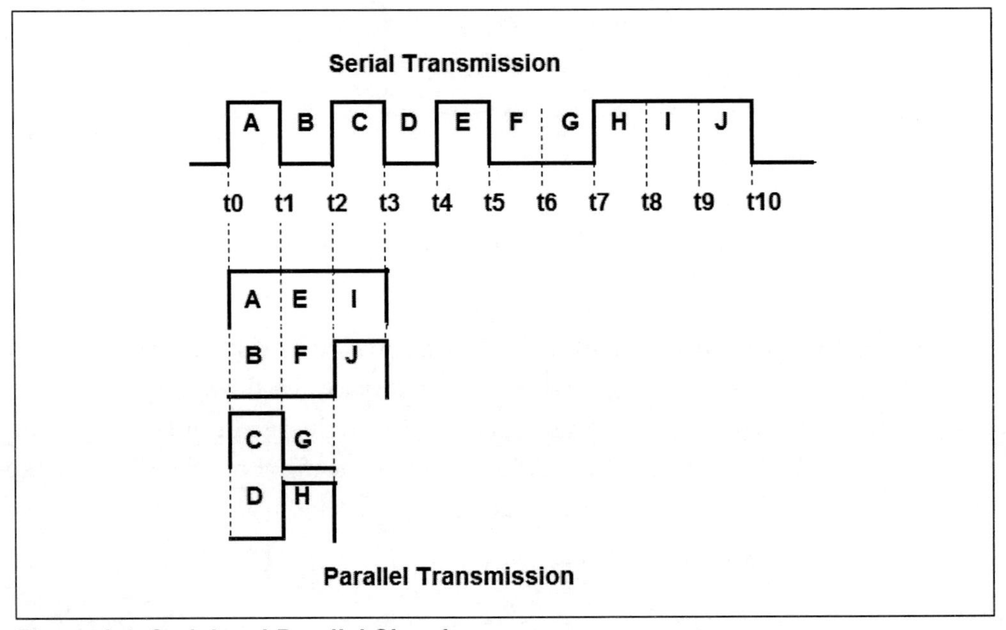

Figure 1-4. Serial and Parallel Signals

For these reasons, most data transmissions outside of the computer cabinet, with the exception of those at very low speeds, are by serial transmission. In the past, inside the computer cabinet (and indeed up to about 1 meter outside it), the buses used parallel transmission since the necessity for high-speed data transfer outweighed the cost. However, as bus speeds (the bus data rate) continue to increase, newer technologies, such as Peripheral Component Interconnect Express (PCIe) and Serial AT Attachment (SATA, a form of Integrated Drive Electronics [IDE]), are using serial transmission. This is because the problem of maintaining transition synchrony over a parallel bus increases drastically as speed increases. Over the past 2 decades, PC printers have been parallel (they started out as serial) because the signal bit lengths are long enough in duration (the signal is slow enough) to permit parallel transmission over a limited distance. Parallel transmission in is now being replaced by USB and other technological advances in serial transmission.

Data Organization: Signals

Digital Signals

A digital signal is defined as one with defined discrete states; no other state can exist. Binary digital signals have only two states: a binary digit is either a "1" or a "0." Do not confuse a "0" with nothing. Zero conveys half of the information in a binary signal; think of it more as "1" being "true" or "on" and "0" being "false" or "off," with no allowance for "maybe." All number systems (by definition) are digital, and we use not only binary (Base 2) and decimal (Base 10) but octal (Base 8) and hexadecimal (Base 16) as well. Octal and hexadecimal (hex) are used mostly to present numbers in a human readable form (humans apparently dislike long rows of 1s and 0s when coding or performing data analysis). Binary is used in contemporary data transmission systems as the signaling means because it can be represented by a simple on-off. (A binary digit is contracted to the term "bit".)

Now that we have defined channels and bits, what are we going to send as a signal over our channel? It is usually a pattern of bits. Bits alone are just data. We must organize our data into some form so that it becomes information. When this is done at higher levels of organization, we may call this organization "protocols" or even "application programming interfaces."

One of the bit patterns to be organized has to represent text, for in order for data organization to yield anything usable to humans, it is necessary to store and present information in a human-readable form. The patterns used for this form of information are called *codings*.

A coding is a generally understood "shorthand" representation of a signal. Codings are not used for secrecy—ciphers are. American Standard Code for Information Interchange (ASCII) is an example of a text coding. A "standard signal" could be called a *coding* and in fact is referred to as such in the daily work of industry. A standard signal is one that has the approval of the users and/or a standardization agency; it *specifies a way to organize data*. This book focuses, in many respects, on how data is organized for different functional tasks. There are many different organizations of digital signals in use, classified as standard (approved by standards organizations and may be open or proprietary, depending on the standardizing agency), open (in general use), and proprietary (limited to a specific organization; i.e., owned wholly by a specific organization). In this book, we will only touch on a few.

Analog Standard Signals

In all areas of data communications, a number of standard signals exist. An analog (or analogue) signal is any continuous signal for which the time varying feature (variable) of the signal is a representation of some other time varying quantity, i.e., analogous to another time varying signal (Wikipedia). In short, an analog signal is a model (usually electrical or pneumatic) representing quantities at any value between a specified set of upper and lower limits (whether temperature, pressure, level flow, etc.). This contrasts with a binary digital signal, which only has one of two values: 1 or 0. Perhaps the easiest of these standard signals to visualize is the standard 4–20 mA (milliamp) current loop signal, long used in process measurement and control (see figure 1-5).

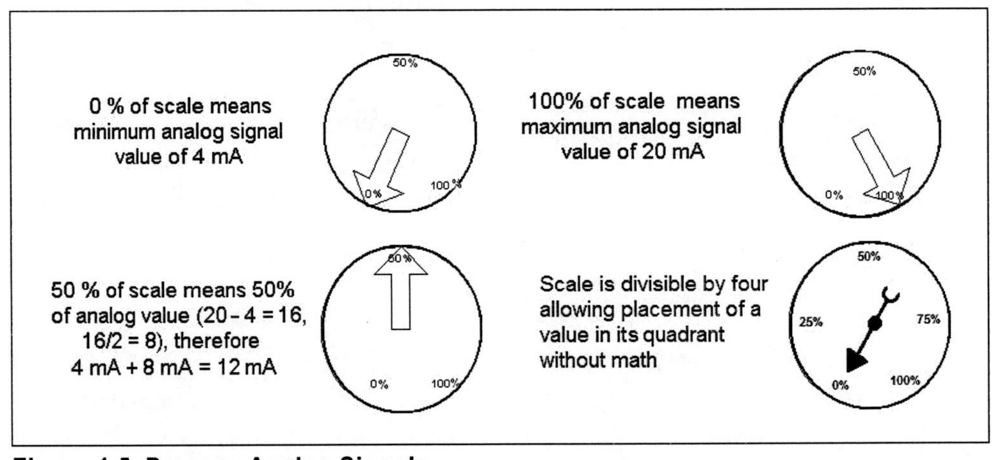

Figure 1-5. Process Analog Signals

These signals represent 0% to 100% of the range (full scale) specified. All instrumentation readings could be likened to a meter face, in which any value is allowed between 0 and 100%, perhaps specified in engineering value measurements. An example is a range of 200°C to 500°C. The 4–20 mA standard instrument signal would represent 200°C as 4 mA and 500°C as 20 mA. These are electrical values for the quantity of energy used in signaling. If you are not familiar with these units, you may not fully understand some of the signal standards and their implications. This text will try to draw the conclusions for you. However, any basic electrical text, particularly one for non-technical people, will provide more than enough background to help you understand the terms used in this book.

The 4–20 mA signal is one of the most commonly used industrial communications standards and it is used in the two-wire current loop that connects devices in an instrument circuit. The reader needs to become familiar with the terms: voltage, current, resistance, impedance, capacitive reactance, and inductive reactance. While not an electrical manual, these terms are further defined in the appendices. What follows is an abbreviated glossary:

Definitions

Voltage (measured in volts) is the difference in electrical potential (charge) that is necessary prior to any work being performed.

Current (measured in amperes or amps, nominally thousandths of an amp or milliamps) is what performs the work, as long as there is a difference in potential.

Resistance (measured in ohms) is the opposition a conductor offers to current.

NOTE The relationship between voltage, current, and resistance is known as *Ohm's Law* (E (volts) = I (amps) X R (ohms).

Capacitive and Inductive reactance are opposite in action and dependent upon frequency of change.

Impedance (measured in Ohms) is the opposition a conductor offers to a changing current value, as well as the steady state (static) resistance. It is represented as follows:

$$\text{Impedance Z (ohms)} = \sqrt{R^2 + (X_L - X_C)^2}$$

The ISA50 committee (ISA-50.01 standard) established a current loop as a standard because it has low impedance, it has a greater immunity to noise (electrical signals that are unwanted) than a high-impedance voltage circuit, and it can power the loop instruments. The receiving devices themselves are high impedance (voltage input) and acquire their input across a 250-ohm resistor (1–5VDC). This allows a known loading for each receiver in the loop. Conversion between current and voltage is provided by a 250-ohm resistor in the current loop. This simple arrangement allows the two-wire loop to both power the loop instrument and to provide the measurement signal. Standard signals allow users to pick and choose among vendors with the confidence that the inputs and outputs will be compatible. The standard signals also allow a manufacturer to build instruments for a larger group of users than if each instrument had its own defined set of signals (non-standard).

Digital Standard Signals

There have been, and still are, both open and proprietary standard digital signals in the telephone and digital data communications area (we will cover these as we go along). However, in the measurement and control areas, no open, all-digital standard signal has been established as an alternative to the analog 4–20 mA signal described in the last section. The international standard for an industrial fieldbus (IEC 61158) defines eight different fieldbuses, four of which—including PROFIBUS PA and FOUNDATION Fieldbus—are primarily used in process control. (A fieldbus is generally a local area network with its defined signaling protocols and field instruments—control being accomplished in the field instruments rather than a control room cabinet.)

Although all these fieldbuses are set as standards, they are not necessarily compatible with each other. While it appears that some of the fieldbuses have gained a majority of sales in certain niches, the marketplace has yet to determine which one will become the de facto fieldbus standard throughout automation. While not a fieldbus itself, but a physical and signal standard, Ethernet continues to gain market share and acceptance in automation areas; indeed, almost all fieldbuses and even proprietary systems now provide networking services via Ethernet.

Process measurement and control has used and still uses several digital communications standards—such as EIA/TIA 232(F) or EIA/TIA 485(A)—that detail placing data on (and removing it from) the media, which is a requirement if one is going to signal via an electrical channel. We will discuss these in chapter 3, "Serial Communications."

NOTE The Electronic Industries Alliance (formerly Electronic Association, formerly Radio & Television Manufacturers Association, formerly ... all the way back to 1924) is an association of manufacturers that develops standards. The TIA(Telecommunications Industry Association) is a subdivision of the EIA.

Data Organization: Communications Codes

Communications codes are the components of data organization that are designed for interface with humans. These codes represent letters, numerals, and controls actions that are stored and recalled, printed, sorted, and, in general, processed. IBM designed one of the first digital communications codes: the IBM 4 of 8 Code (circa 1947) which allowed error detection on a character-by-character basis. This section presents a cursory review of this and other communications codes.

The IBM 4 of 8 Code was a proprietary code. Other manufacturers typically had their own proprietary codes, all of which were incompatible with each other. In 1963, the U.S. government released the American Standard Code for Information Interchange (ASCII). Several months later, IBM released its own code, an extended version of the 4 of 8 Code: EBCDIC, which is short for "Extended Binary Coded Decimal Interchange Code" (more on this later). EBCDIC is mathematically related to the 4 of 8 Code. The 4 of 8 Code was uppercase and allowed only 64 of the 256 bit combinations available with an 8-bit code. EBCDIC had all combinations available to it because it used a cyclic redundancy check character (CRCC) for error detection instead of the parity bit (bit 7) in the 8-bit (ASCII) character. The U.S. government was (and is) a large buyer of data equipment, thus, ASCII gradually gained acceptance with many vendors (other than IBM), who relinquished their proprietary codings and adopted ASCII.

The IBM 4 of 8 Code

The IBM 4 of 8 Code is illustrated in table 1-1. In this code, there are four 1s (ones) and four 0s (zeros)—that is 8 bits, or an octet—for each character. This arrangement was used to detect character errors, yet it carried a large overhead; that is, the ratio of bit patterns used for error detection to those for data transmission is high. About 70 characters were used for error detection out of the 256 characters available.

Table 1-1. IBM 4 of 8 Coding

Character	Bits	Character	Bits	Character	Bits	Character	Bits
0	01011001	A	10001110	K	01001011	U	00101101
1	10000111	B	01001110	L	11000011	V	10100101
2	01000111	C	11000110	M	00101011	W	01100101
3	11001001	D	00101110	N	10100011	X	11100100
4	00100111	E	10100110	O	01100011	Y	00011101
5	10101001	F	01100110	P	11100010	Z	10010101
6	01101001	G	11100001	Q	00011011		
7	11101000	H	00011110	R	10010011		
8	00010111	I	10010110	S	01001101		
9	10011001	J	10001011	T	11000101		

Example 1-1. Using IBM 4 of 8 coding

Problem: The following signal is received. Decode the signal (using table 1-1) and determine which character has an error.

00001111010101001101011000011101100011110011000111

The transmission scheme uses 1 start bit and 1 stop bit. In order to tell when a transmission line is idle, a continuous 1 is placed on the line. To indicate a start, the first bit is always a 0. Nine bits (the start bit plus the eight character bits) later, the signal places a 1 bit on the line to indicate that it has stopped transmitting a character and it remains in that condition until the next start bit.

Solution: First break the code into 10-bit groups (one start bit, one stop bit, and eight information bits):

0000111101 0101001101 0110000111 0110001111 0011000111

Second, strip off the start and stop bits:

00011110 10100110 11000011 11000111 01100011

Third, convert the bits into alpha characters:

H E L ? O

The second L has five 1s and three 0s; it is a character error. All characters can only have four 1s. When a character error occurs, the entire text would be transmitted and not acknowledged as received until it was correct.

International Telegraphic Alphabet #5

Also known as ASCII, the International Telegraphic Alphabet (ITA) #5 (see table 1-2) does not make use of letter frequencies, as did the Morse code, the ITA2 Telegraph Code, and the IBM 4 of 8 Code. It does, however, use a numerically arranged code, in which the alphabet is in ascending order, which permits sorting operations and the like. ASCII is a 7-bit code. Its arrangement is well thought out; particularly in light of the technology available at the time it was designed. It was usually (and is now) transmitted as an 8-bit signal with the "Most Significant Bit" (B7) reserved for parity—an error detection scheme. The parity bit was used for a number of years; however, the technological advantages to using an 8-bit data byte (particularly for sending program code) eliminated the need to use the B7 bit for parity. Today when parity is used, another bit is added to the 8 bits as the parity bit, making the signal 9 bits in length (11 bits when the start and stop are added). For our present purposes, we will set B7 to 0 and concern ourselves only with B0 through B6.

Table 1-2. ASCII (ITA5) Alphabet

				B6-	0	0	0	0	1	1	1	1
				B5-	0	0	1	1	0	0	1	1
				B4-	0	1	0	1	0	1	0	1
B3	**B2**	**B1**	**B0**									
0	0	0	0		NUL	DLE	SP	0	@	P	`	p
0	0	0	1		SOH	DC1	!	1	A	Q	a	q
0	0	1	0		STX	DC2	“	2	B	R	b	r
0	0	1	1		ETX	DC3	#	3	C	S	c	s
0	1	0	0		EOT	DC4	$	4	D	T	d	t
0	1	0	1		ENQ	NAK	%	5	E	U	e	u
0	1	1	0		ACK	SYN	&	6	F	V	f	v
0	1	1	1		BEL	ETB	‘	7	G	W	g	w
1	0	0	0		BS	CAN	(	8	H	X	h	x
1	0	0	1		HT	EM	)	9	I	Y	I	y
1	0	1	0		LF	SS	*	0	J	Z	j	z
1	0	1	1		VT	ESC	+	:	K	[	k	{
1	1	0	0		FF	FS	,	;	L	\	l	|
1	1	0	1		CR	GS	-	<	M	]	m	}
1	1	1	0		SO	RS	.	=	N	^	n	~
1	1	1	1		SI	US	/	>	O	_	o	DEL

Reading table 1-2 is simple; though, it takes some getting used to. First, notice the representations of the bit order: B4, B5, and B6 are grouped as three bits and so have a decimal value between 0 through 7 (same as hex); the lower four bits (B0 through B3) are a group of 4 bits and have a decimal value of 0 to15 (or hex 0 through F). (For a thorough discussion of number systems, including hexadecimal, refer to appendix A.)

The most significant bit (MSB) is B6. B6, B5, and B4 are read vertically. Bits 3 through 0 are read horizontally. As an example, what is the coding for the uppercase C? Locate the uppercase C; trace the column vertically and find that the first three bits are (B6) 1, (B5) 0, and (B4) 0: 100. Locate the uppercase C again and travel horizontally to obtain the value of bits 3 through 0, or (B3) 0, (B2) 0, (B1) 1, and (B0) 1: 0011. Put the bits together (B6 through B0) and you have 1000011 for an uppercase C (hex 43).

So, what would a lower case c be? Follow the same procedure and you will determine that it is 1100011 (hex 63). The difference between uppercase and lowercase is bit 5. If B5 is a 0, the character will be an uppercase letter; if B5 is a 1, the character will be lowercase. When B6 is a 0 and B5 is a 1, the characters are numerals and punctuation. Hex uses a different convention: to obtain the lower case letter from an upper case value, add hex 20; to obtain an upper case letter from a lower case value, subtract hex 20.

If both B6 and B5 are 0, then it is a non-printing (control) character. Control characters were based on the technology of the time. Table 1-3 gives the assigned meaning of these characters.

The one exception to these rules is DEL (delete), which is comprised of all 1s. The reason for using all 1s can be traced back to when the main storage medium in ASCII's beginnings was paper tape. If you made a mistake punching a paper tape (a 1 was a hole in the tape, a 0 was no hole), you then backed it up to the offending character and used the delete key, which punched all holes in the tape, masking whatever punching had been there.

Table 1-3. ASCII (ITA5) Control Characters

Mnemonic	Meaning	Mnemonic	Meaning
NUL	Null	DLE	Data Link Escape
SOH	Start of Header	DC1	Device Control 1
STX	Start of Text	DC2	Device Control 2
ETX	End of Text	DC3	Device Control 3
EOT	End of Transmission	DC4	Device Control 4
ENQ	Enquiry	NAK	Negative Acknowledge
ACK	Acknowledge	SYN	Synchronous Idle
BEL	Bell	ETB	End of Transmitted Block
BS	Back Space	CAN	Cancel
HT	Horizontal Tabulation	EM	End of Medium
LF	Line Feed	SUB	Substitute
VT	Vertical Tabulation	ESC	Escape
FF	Form Feed	FS	File Separator
CR	Carriage Return	GS	Group Separator
SO	Shift Out	RS	Record Separator
SI	Shift In	US	Unit Separator
		DEL	Delete

Example 1-2

Problem: Convert the following data into an ASCII-coded string. Note the use of hex notation. (If you do not feel comfortable with hex, you may use the binary representation for each character—but it will take up a lot more space.)

The quick brown fox. THE QUICK BROWN FOX.

Solution: Using table 1-2, we get:

T h e q u i c k b r o w n f o x.

54 68 65 20 71 75 69 63 6B 20 62 72 6F 77 6E 20 66 6F 78 2E

T H E Q U I C K B R O W N F O X.

54 48 45 20 51 55 49 43 4B 20 42 52 4F 57 4E 20 46 4F 58 2E

Extended Binary Coded Decimal Interchange Code

As described previously, the Extended Binary Coded Decimal Interchange Code (EBCDIC) was developed in the early 1960s by IBM from its 4 of 8 Code and it is proprietary to them. ASCII has only seven bits, the eighth bit being reserved for parity. One problem in using only 7 bits plus 1 bit for parity arises when computers transmit program instruction coding. Computers normally operate using an 8-bit octet or a multiple of 8 (i.e., a "word" that is 16 bits, or a "double word" of 32 bits). All 256 possible 8-bit combinations may not be used but it is likely that the computer's instruction set would use the eighth bit. With the "7 for information + 1 for parity" bit scheme, the eighth bit isn't available. Most computers using ASCII transmit 8 bits and use a ninth bit for parity, if parity is used. In EBCDIC, the blank spaces are used for special or graphic characters particular to the device using them. If required, this code can transmit "object" code—that is, the combinations of 1s and 0s used to program a computer in 8-bit increments—with little difficulty. EBCDIC did not require a parity bit for error detection but instead used a different error-detection scheme—CRCC.

One of the first interface problems of the PC age was how to perform PC-to-mainframe communications. By and large, Personal Computers (PCs) and other devices transmitted in ASCII, while many of IBM's minicomputers and mainframe computers used EBCDIC. This meant that they could not talk directly without some form of translator, generally a software program or firmware in a protocol converter. While such a conversion was not difficult to perform, it was another step that consumed both memory and CPU cycles.

Unicode

Character coding by electrical means originally became established with the Morse code, in which letter frequencies and a form of compression were used for efficiency. Fixed-length character representations, of which ASCII and EBCDIC are examples, came about in the early 1960s and are still in wide use today. However ASCII was designed for English (American English at that, the term *American* being the definitive) and, as communications are now global, there are many languages that require a larger number than 128 patterns (7 bits). Today, Unicode is an industry standard for encoding, representing, and handling text as expressed in the majority of writing systems. The latest version of Unicode has more than 100,000 characters, covering multiple symbol sets. As of June 2015, the most recent version is *Unicode 8.0*, and the standard is maintained by the Unicode Consortium (Wikipedia). Instead of representing a character with 7 or 8 bits, Unicode uses 16 bits. For users of ASCII, this is not a problem (as long as the upper 8, more significant, bits of the 16 total are accounted for) because the Unicode conversion for ASCII is hex 00 + ASCII. In other words, the upper eight bits are set to 0 and the ASCII character set follows. The following is excerpted from Microsoft Corporation's Help file for Visual Studio 6 (MSDN Library, April 2000):

"Code elements are grouped logically throughout the range of code values, which is called the *codespace*. The coding begins at U+0000 with standard ASCII characters, and then continues with Greek, Cyrillic, Hebrew, Arabic, Indic, and other scripts. Then symbols and punctuation are inserted, followed by Hiragana, Katakana, and Bopomofo. The complete set of modern Hangul appears next, followed by the unified ideographs. The end of the codespace contains code values that are reserved for further expansion, private use, and a range of compatibility characters."

This language is disseminated under the auspices of the Unicode Consortium. It does not dictate how the character should appear but merely that it should be processed as this character. The printing is up to the software and hardware. Most modern office suites (Microsoft Office 2000—2013 and Libre Office 2.0 and up) are set up to handle Unicode. One must remember that this is an international world, made much smaller by data communications and, therefore, having a common language standard (for processing purposes) is a good starting point for world-wide communication.

Data Organization: Error Coding

How we organize our data has a lot to do with how we recognize and correct transmission errors. These are alterations to our intended signal and are caused by noise or other media properties. For example: when listening to a music selection on a DVD, if there is a scratch on the DVD surface (greater than the DVD player's error correcting abilities), it will change the music (not necessarily improving it). For the purposes of this discussion, we will assume that the data to be transmitted is correct; the matter of erroneous data is not within the scope of this book.

Anytime you place data on a medium and take it off, there is a probability of error. Whether the probability of an error is high or low depends on the media and signal properties. After the data has been taken off of the medium, we can only detect errors if the data was organized in a way to detect errors prior to transmission. The IBM 4 of 8 Code used an early form of error detection. If any character had more or less than four 1s in 8 bits, it was in error. The entire block (84 characters) would then be retransmitted. Parity is an extension of this concept.

Parity

Parity is a means of error detection. Using 7 bits of an 8-bit structure, as ASCII does, will leave 1 bit extra (in most cases B7 is set to 0) and the parity bit is added to the 8-bit character, making it 9 bits. Parity (in the telecommunications sense of the word) means to count the number of 1s in a character (the 0s could be counted, but traditionally only the 1s were counted). The agency determining the system's operating specifications will have decided on which parity to use: "odd" or "even." If "odd" parity is used (see table 1-4), then the parity bit will be whatever value is required to ensure an odd number of 1s in the character (including the parity bit).

Table 1-4. Character Parity–Odd Parity Selected

	Char. A	Char. B	Char. C	Char. D
Parity	1	0	0	1
B7	0	0	0	0
B6	0	1	0	1
B5	1	0	0	0
B4	1	1	1	1
B3	0	0	1	0
B2	1	0	1	0
B1	1	0	0	0
B0	0	1	0	0

Table 1-5 illustrates how a character's parity is determined. Note that the characters are arranged vertically. This form of parity is called *vertical parity*, for reasons that will become evident momentarily.

Example 1-3

Problem: Using odd parity, add the correct parity bit to the following ASCII string:

What is your name?

Solution: Use ASCII (from table 1-2) as shown in table 1-5 to determine the bit combinations of each character. Add a 1 as the parity bit to ensure that the total of 8 bits (vertically) has an odd number of 1s.

Table 1-5. Determination of Block Parity Bit

BIT	W	h	a	t		i	s		y	o	u	r		n	a	m	e	?
Parity	0	0	0	1	0	1	0	0	0	1	0	1	0	0	0	0	1	1
B7	0	0	0	0	0	0	0	0	0	0	0	0	0	0	0	0	0	0
B6	1	1	1	1	0	1	1	0	1	1	1	1	0	1	1	1	1	0
B5	0	1	1	1	1	1	1	1	1	1	1	1	1	1	1	1	1	1
B4	1	0	0	1	0	0	1	0	1	0	1	1	0	0	0	0	0	1
B3	0	1	0	0	0	1	0	0	1	1	0	0	0	1	0	1	0	1
B2	1	0	0	1	0	0	0	0	0	1	1	0	0	1	0	1	1	1
B1	1	0	0	0	0	0	1	0	0	1	0	1	0	1	0	0	0	1
B0	1	0	1	0	0	1	1	0	1	1	1	0	0	0	1	1	1	1

Block Parity

The vertical character format mentioned previously is based on the punched cards used in early data entry and programming. Because punched cards (at least IBM's cards) contained 80 columns and historically most data was on these cards, transmissions were in an 80-character "block." (The 80 columns of data were known as a record.) Two framing characters were added at the start of the data block and one framing character was added at the end of the data block. An 84th character, called the *block parity character*, was added to the end of the transmitted block. It was computed from the other 83 characters. Table 1-6 illustrates the 84-column vertical format.

Table 1-6. Block Parity

	SOH, DLE, or STX	SOH or STX	1st Char.	Information of 78 characters	Last Char.	ETB or ETX	Block Parity (84th column)
Parity	0	0	1		1	0	1
B6	0	0	1		1	0	1
B5	0	0	0		0	0	1
B4	0	1	0		1	0	1
B3	0	0	0		1	0	0
B2	0	0	0		0	1	0
B1	1	0	0		1	0	1
B0	0	0	1		0	0	0

Parity was first determined vertically for each of the block's 83 characters. Next, parity was determined horizontally for each bit by counting the number of 1s in each row for all 83 counted columns (the 84th column is not counted). The horizontal parity bits were placed in the 84th column, thus creating the block parity character. The 1s in the 84th column were counted to determine the vertical parity bit for the block parity character, the same process as determining parity for all other characters. To determine if parity was met for all the characters in the block, parity was determined horizontally for the Parity row by counting the 1s in the 83 counted columns and the result was compared to the vertical parity bit for the block parity character. The vertical parity bit for the block parity character must match the horizontal parity result for the Parity row, if not, there was an error. This scheme is known as *vertical and horizontal parity checking*, or *block parity*.

With block parity, fewer than one in 10,000 errors remain undetected. However, this scheme exacts a heavy penalty in overhead. One parity bit for each character adds up to 84 bits, or more than 10 8-bit characters.

Error Correction

In addition to having bits devoted to assisting with error detection, what means could be used to correct an error, once one was detected? The answer is normally an automatic retransmission (repeat) query (reQuest), known as an ARQ (explained later in this chapter). This method had significant ramifications in many applications since the transmission device had to store at least the last transmitted block. (When using horizontal and vertical parity, data has to be transmitted in blocks.) There also had to be some scheme to notify the transmitter that the receiver had successfully received the transmission or that it was necessary to retransmit the block due to error. In most cases, half-duplex would be too inefficient in terms of *transmission time versus line-setup time* (i.e., the time it takes for the transmitting device and the receiving device to establish connection and synchronization).

Over time (less than 60 years), several schemes have been devised to increase the "throughput" (defined here as the number of bits correctly received at the end device). Today, many schemes use a block that varies in length, depending on the number of errors detected. In transmissions that have few errors, such as those over optical fiber or a local area network (LAN), the block could be made longer. In media that has frequent errors, such as short wave radio, the block could be made shorter.

Blocks are generally of a fixed length, or a multiple thereof. Packets, on the other hand, have a fixed minimum and maximum length but can vary in length between these two limits. We normally speak of packet transmission in modern data communications, although the difference between a block and a packet is now more one of semantics than of practice. (It is worth noting that the longer the block/packet, the greater the chance of an undetected error and that parity and block parity check schemes were far less effective with multiple-errored bits than with single-bit errors. This is the reason the CRCC method is superior in terms of efficient use of bits for error detection, particularly for multiple errored bits.)

Cyclic Redundancy Checks

Any time data is placed on a medium (e.g., wireless, unshielded twisted pair [UTP], magnetic or optical rotating media, and fiber cable), there is a probability of error; thus, an error detection scheme must be used. Most modern devices use a far more efficient scheme of error detection than parity checking. Though ASCII was designed for a vertical and horizontal (at times called *longitudinal*) block-checking code, it can also be used with a cyclic redundancy check (CRC) scheme that creates a check character.

A CRC character (CRCC) is the number 8-bit check character developed by the CRC. A cyclic code divides the text bits with a binary polynomial (the CRC), resulting in a check character (today, nominally 16, 32, or 64 bits; that is 2, 4, or 8 characters). This is done by checking every bit in a serial bit stream (the bits that make up a packet, also called a *bit stream*) and combining data from selected (and fixed) bit positions. Figure 1-6 illustrates the generation of a simple CRCC, which is sometimes (incorrectly) called a *checksum*. A number of different CRCCs are used today. The primary difference between them is the "pick-off point" (the power of X in the representative equation, such as the CRCC-CCITT [International Telegraph and Telephone Consultative Committee or Comité Consultatif International Téléphonique et Télégraphique] that uses $G(x) = X^{16}+X^{12}+X^5+1$). The advantages of one pick-off point or another depends on the application. For example, a communications channel may have different error behavior than a magnetic medium, such as a hard drive. In any event, the transmission protocols used in industrial data communications will probably use the CRCC-32 scheme (a local area network CRC of 32 bits).

Within a given size block (packet or frame), a certain size character or characters will be generated. A common (CRC-CCITT) type uses two 8-bit octets, forming a 16-bit check character. The CRCC characters are generated during transmission and added to the transmitted packet. The receiving end receives the bit stream and computes its own CRCCs. These must match the transmitted CRCCs or there is a detected error. Note that even on a block of 80 characters, this scheme only uses 16 bits, in comparison to the 88 bits used with the vertical and horizontal parity check. The CRCC method is used for writing to a disk drive or to almost any magnetic medium and is the method most often used (whether the CRC is 16-bit or 32-bit) in modern data communications.

Checksum

Any of a number of error-detection codes may be called a *checksum*. Many times the CRC-16 or CRC-CCITT is called a *checksum*; however, they are actually cyclic codes, whereas a checksum was originally designed as a linear code. For instance, all of the 1 states in a block may be totaled to obtain a checksum. (This usually happens through a process known as modulo (mod) addition, where a number is divided by the modulo number but only the remainder is used. An example is 11 mod 4: 11 divided by 4 equals 2 with a remainder of 3, so the answer is 3. This checksum is tacked on as a block check character or characters. The efficiency of a checksum in detecting errors is not as high as with a cyclic code, yet the circuitry to produce the checksum is less complex.

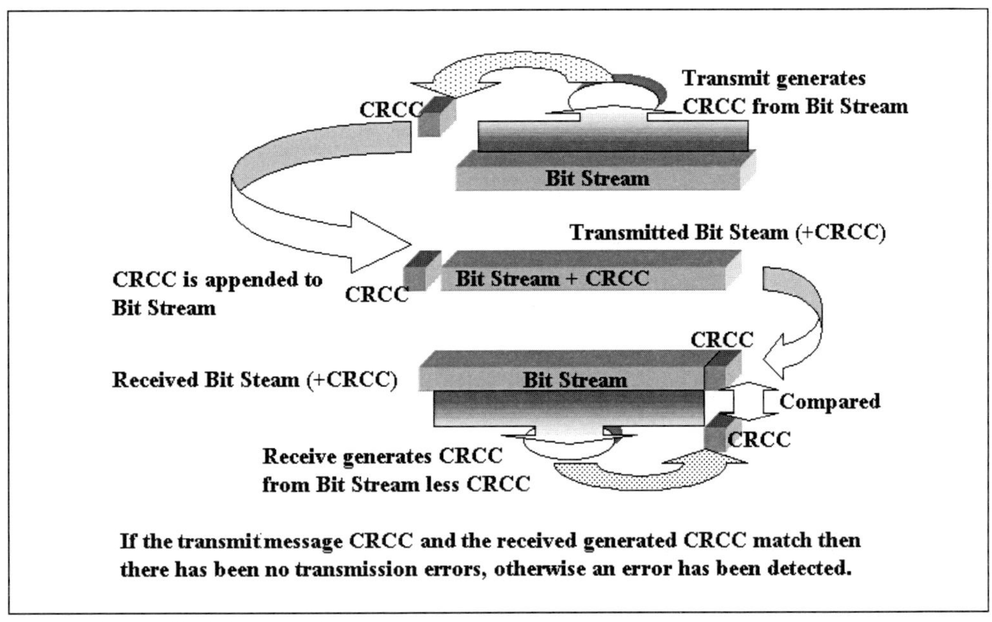

Figure 1-6. CRCC Concept

NOTE The abbreviation for modulo varies depending on the programming language. In COBOL and older programming languages, modulo is abbreviated MOD (11 MOD 4). In C-style languages, the abbreviation is the percent (%) symbol (11%4).

There are other error detecting codes besides the checksum and the CRCC. These codes exist specifically for detecting and (sometimes) correcting errors. Some of these codes are cyclic, some are linear, some are quasi-cyclic, and some are polynomial. Some of these codes are sufficiently long in pattern length to be used for error detection and correction.

ARQ

What hasn't yet been explained is what happens when an error is detected (at least in general terms). The following is a simplified discussion of a generalized automatic retransmission (repeat) query (reQuest), the ARQ method of error correction.

Before beginning our discussion, it is important to note the change in terminology as multiple terms are used to refer to the data being transmitted. Prior to HDLC, data sent (typically asynchronously) was defined by control characters and was referred to as *data blocks* (usually an 80-character [IBM punched card] record). There would be a set of delimiters at the start (DLE or STX) and an ETX and block parity character at the end of the data for a total of 84 characters. As time went by and data began to be sent by bit-oriented protocols (HDLC); these blocks of data were called *packets* (about 1970). Since 1990, the terminology has been changed again (but certainly not by everybody); the Layer 2 octets are called *frames* (framed by the protocol standard) and the arrangement of data in Layers 3 and above is referred to as *packet data units (PDU)*. Of course, some still refer to the Layer 2 octets as "packets" when they mean "frames."

In normal operation, the transmitter sends a frame (a multiple set of octets) to the receiver. The frame is received (we are using a CRCC in this example as the error detection method; methods may vary but the ARQ system is basically the same). If the CRCCs match, the receiver sends an ACK (Acknowledgment). If they do not match, the receiver sends a NAK (Negative Acknowledgment) and the transmitter then retransmits the frame. This is illustrated in figure 1-7.

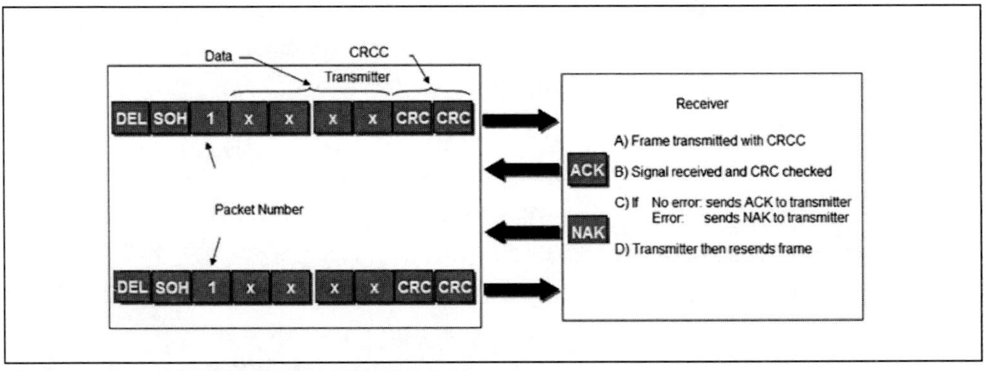

Figure 1-7. ARQ Normal Error Detection

If an occasion should occur where the frame is so badly corrupted that the receiver does not recognize it as a frame, then the receiver will send nothing. After sending a frame, the transmitter waits for a response (in many systems, this is an adjustable [by software] period of time). If a response does not arrive within the expected time, the transmitter assumes the worst and resends the frame. This condition (not getting a response in time) is called a *timeout* and is illustrated in figure 1-8.

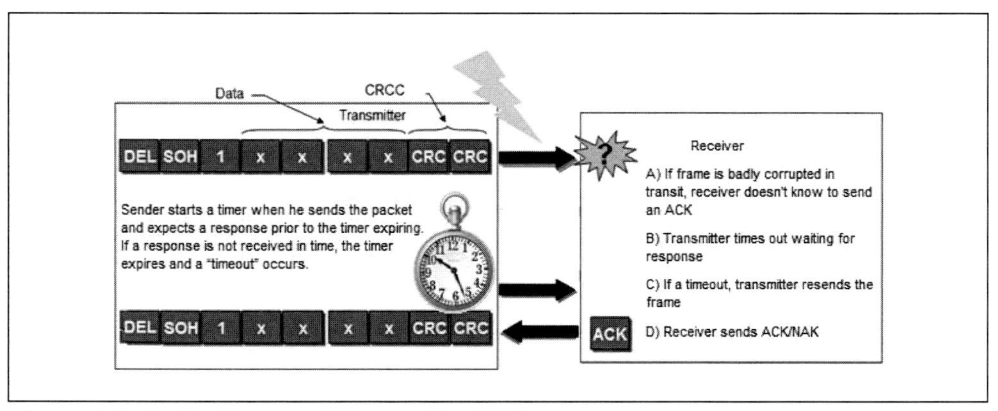

Figure 1-8. Badly Damaged Frame, No ACK

On the other hand, suppose that the frame is received correctly and an ACK is returned. What if the ACK is "bombed" (or lost)? If this happens, the transmitter will time out and will resend the frame. The receiver will detect that it has two frames with the same frame number and that both were received correctly. It will then drop one of the frames and send another ACK, as shown in figure 1-9.

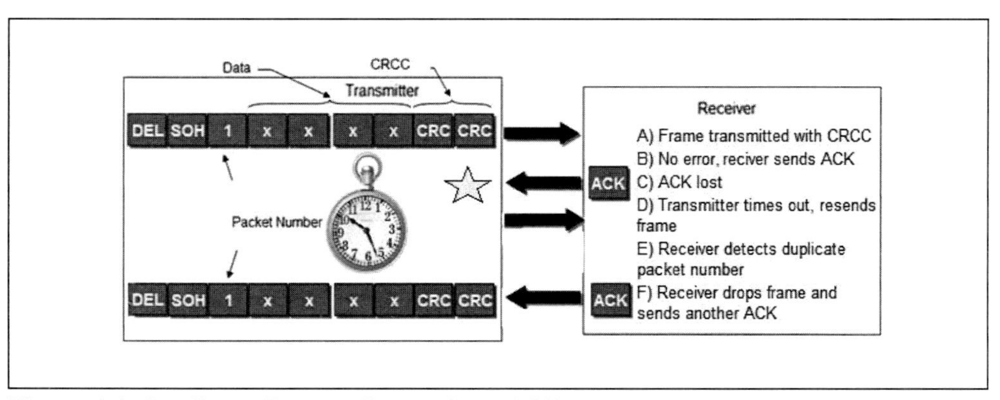

Figure 1-9. Duplicate Frames Due to Lost ACK

There are many ways to enhance error correction via ARQ and they are accomplished in individual protocols. Determining how many frames can be transmitted prior to receiving an ACK/NAK has ramifications for throughput. The more frames transmitted prior to an ACK/NACK, the greater the throughput; however, more frames must be stored after being received, so the memory requirements are greater.

Forward Error Correction

Although the ARQ method of error correction discussed above detects an error (by whatever means) and then retransmits the portion of the message that was in error, what happens when there is simplex (unidirectional) transmission? The ARQ method will not receive an ACK/NAK because the circuit is unidirectional, so ARQ cannot be used on a simplex type circuit. In this case, or in cases where the medium is too noisy to allow for any significant throughput, forward error correction (FEC) may be employed.

FEC requires that the communications network's pattern of errors be analyzed and an error-correcting algorithm be designed to fit the identified error patterns. This can be done for most real-world circuits without extraordinary difficulty. The main problem is that error-detection and location overhead can be as much as 1 parity bit (or error-detecting symbol) for each data bit. Using this approach would immediately cut throughput in half. In some cases, however, the data must be 100% error-free (without regard to capital expense or throughput) because this transmission may be the only meaningful transmission sent or the only time given to the circuit for any throughput to occur. (Think real-time process data, satellite tracking data, nuclear plant critical monitoring, and so on.) For these applications, FEC is appropriately employed.

A thorough discussion of error-correcting algorithmic codes and associated theory is beyond the scope of this text. However, many good references, at many different knowledge levels, are available on this subject. Error correction is integral to data communications. Knowing that an error exists is not the end goal, the goal is to detect an error and determine how to correct it while ensuring data integrity.

Data Organization: Protocol Concepts

A protocol is an accepted procedure. We use protocols in personal interactions all the time. When you are introduced to someone in the United States, the protocol is to shake hands, unless there is a ceremony requiring another form of protocol or you are being introduced to a large number of people. In a class, or throughout a conversation, there is a protocol that tells you to wait until another person is finished speaking before speaking yourself. There is a physiological reason for this protocol: humans are effectively half-duplex—they typically cannot talk and listen very well at the same time.

Terminologies

Asynchronous generally means that something may occur at any time and its occurrence is not tied to any other event. The old "start-stop" teletypewriter signal (see appendix B), with its 1 start bit and 1 stop bit, is a good example of an asynchronous transmission method. The teletypewriter signal started out using motor speed (reference appendix B) as the main means of synchronizing the start-stop bits within each character. In today's vernacular, any start-stop signal (which is seldom used in today's world) is assumed to be asynchronous. In actuality, a start-stop signal may be transmitted synchronously or asynchronously.

Synchronous generally means "tied to a common clock." The clock signal is typically transmitted along with the data, usually in the transition of data from one state to another. Since synchronous transmission uses bit timing (the device clock [receive and transmit] is in synchrony with the change in bit value [from 0 to 1, or 1 to 0]), each bit of data must be accounted for. Almost all modern data communications are synchronous transmissions.

The terms *baud rate* and *bits per second* are often used interchangeably: this is incorrect. Baud rate is the line modulation rate; that is, the number of signal changes (decisions such as: one to a zero, zero to a one, or number of degrees leading or lagging the reference phase) per second that are placed on the communication medium in a given time (the maximum value of the number of changes supported by the medium is called the *bandwidth of the channel*). Bits per second (bps), on the other hand, is the data transmission rate of the device that is transmitting or that a device is capable of receiving. Depending on the type of signaling and encoding used, the baud rate (the line modulation rate) and the bit rate (generally the throughput speed) may be quite different.

Baud per second is a term that is used to describe a change (like in acceleration or de-acceleration) in a transmission medium signaling state, not a line speed. Ads for a "33.6 Kbaud modem" are in error; such a modem would require a line with a bandwidth great enough for 1/33600-second rectangular pulses (not possible with a voice-grade wireline). What this specification means in reality: these are 33,600-bit-per-second (33.6 Kbps, the data bit rate) modems. Such modems use a 1200 baud line and achieve their higher data rate by sending a signal state representing 56 bits of data per baud. That is, they make one line state change (baud) for each 56 bits of data (one change of phase or amplitude for each 56-bit combination). This is accomplished through a high-level

> **Example 1-4**
>
> **Problem:** Determine the required line modulation rate if a device transmits a rectangular pulse that is 1/1200 of a second long but the transmission occurs only once per hour (1 bit per hour).
>
> **Solution:** The bit rate is one pulse per hour. However, the medium must be capable of passing a rectangular pulse of 1/1200-second duration. To determine the required baud rate, divide the *shortest information element* (usually a bit time) for non-coded signals *into 1*. The interval for 1/1200 of a second is 0.0008333; dividing this into 1 gives, of course, 1200. The baud rate required is 1200.

coding technique called *trellis encoding*. Modems running at 33.6 Kbps (kilobits per second) require a 600 baud line in each direction (a total of 1200), rather than one that supports 33,600 signal changes per second in each direction. A standard dial-up telephone line can support a total of 1200 signal changes per second (about 3.3 KHz bandwidth). Some refer to the baud rate as symbols per second (see appendix C).

Protocols

As electronic message handling became the norm, it became imperative to build as much of the communications link control into the terminals as possible. To this end, "communications protocols" arose. Most protocols were developed by individual vendors for their own systems. Since these protocols are generally incompatible with the protocols and equipment of other vendors, the customer is generally locked into one manufacturer, promoting incompatibility.

All data communications efforts involve protocols of one kind or another. A protocol is a different organization of data, as opposed to shaping groups of 1s and 0s into characters. Communications protocols are either *character-based* or *bit-oriented*. This means that the information we are looking for will take the form either of characters telling us information or of bit patterns (other than characters) telling us information.

Character-Based Protocol

One example of a character-based protocol is Binary Synchronous Communication (Bi-Sync) (another is the teletypewriter system; see appendix B). The

IBM Bi-Sync protocol, one of the first and most widely used of the proprietary protocols, was developed to link the IBM 3270 line of terminals to IBM computers in a synchronous manner. (This protocol may also be used in a system with asynchronous signaling; that is, with start-stop characters, provided the text mode is used.) The Bi-Sync protocol is character-based; control depends on certain character combinations, rather than on bit patterns. It also requires that the transmitting terminal be able to store at least one block of data while transmitting another.

When transmitting under Bi-Sync, the hardware is responsible for avoiding long strings of 1s or 0s. In a synchronous system, if a significant length of time is used to transmit only one state or the other, the receiver loses its bit synchrony. As a result, communication is disrupted or, at least, it is in error. Most modern hardware uses a scrambler to ensure data state transitions. The scrambler contribution is removed at the receiver, since the scrambler pattern is performed on a scheduled basis. In addition, most phase continuous systems use Manchester encoding, which moves the state of the bit (1 or 0) into the transition of data (a transition from 1 to 0 indicates a 1 state, while a 0 to 1 state transition indicates a 0, rather than the level—a voltage state representing 1 or 0). Manchester encoding is explained in detail later. Table 1-7 lists some of the Bi-Sync control characters. Note that the ASCII control characters are used if Bi-Sync is used in an ASCII-based system.

Table 1-7. Bi-Sync Control Characters

SYN	Synchronous Idle	Used to synchronize receivers
SOH	Start of Header	Indicates routing information
STX	Start of Text	Indicates message text starts
ETX	End of Text	Indicates end of transmitted text
ITB	End of Intermediate Block	More blocks coming
ETB	End of Transmission Block	Do error count, block over
ACK	Acknowledgment	Received block OK
ACK1	(Same as ACK)	Received odd block OK
ACK2	Acknowledgment 2	Received even block OK
NAK	No Acknowledgment	Bad block, retransmit
ENQ	Enquiry	Go ahead and send
DLE	Data Link Escape	Pay no attention to control characters until a DLE pair appears again
EOT	End of Transmitted Text	This transmission has ended

Example 1-5

Problem: Show the Bi-Sync sequences required to transmit the following message: "This is a short block." Just indicate their location in the message stream. To produce a minimum block size, no pad characters (characters used to allow a block, packet, or frame to meet minimum length requirements) are included. ASCII in hex notation is used.

Solution:

16 16 01 41 02 54 68 69 73 20 69 73 20 61 20 73 68 6F 72 74

SYN SYN SOH A STX T h i s i s a s h o r t

20 62 6C 6F 63 6B 2D 04 17 BCC1 BCC2 16 16

b l o c k . EOT ETB CRC1 CRC2 SYN SYN

The receive portion of the circuit that just transmitted will now wait for an ACK (06) before proceeding. If a NAK (15) occurs, the circuit will retransmit this block of text.

Bi-Sync normally uses duplex transmission (both directions simultaneously). Although it could use half-duplex, it was not primarily intended for that type of operation. Before duplex Bi-Sync was used, a block (packet or frame) would be transmitted, the line would be turned around (the transmit and receive modems would swap functions and be resynchronized), and then the original transmitter (now a receiver) would wait for the original receiver (now a transmitter) to send either an ACK or a NAK. If a NAK was received, the line would be turned around again (swapping the transmit and receive functions) and the original transmitter would retransmit the block. Duplex operation would not have been faster, except for eliminating the turnaround time, because the transmitter could do nothing else until it received a response to its transmitted block. To speed things up a bit, the Bi-Sync protocol had the transmitter store two blocks, so it could wait for an ACK while transmitting the second block. ACK1 and ACK2 signals were used to differentiate between ACKs. The primary benefit in this case was that transmission could take place without line turnaround (duplex) or without halting transmission until an ACK was received. It should not be forgotten that when Bi-Sync was developed (1967), the cost of data storage was approximately one US dollar per bit in 1980 dollars—a significant expense.

Because the control characters are intertwined with the text, they must be sent in pairs to ensure that they are identified. How would you transmit a machine or computer program in object (machine-executable) form, if that code were composed of 8-bit octets, some of which may very well be the same as the control codes? To make this possible, Bi-Sync allows a transparent mode, in which control characters are ignored until the receiver detects several data link escape (DLE) characters.

Bi-Sync is dependent upon character-oriented codes. In modern communications, there is a need to make the protocol independent of the transmitted message type. In other words, it should make no difference to the protocol what bit patterns the message consists of or even in what language it is composed, as long as it is in 8-bit octets. Link Access Procedure-Balanced (LAP-B) protocol (described later in this chapter) and other bit-oriented protocols provide that functionality rather gracefully.

Bit-Oriented Protocol

Bit-oriented protocols use a concept called *framing*, where there are bit patterns before and (in some schemes) after the information frame (which is why it is called a frame now, rather than packet or block). In framing, there is a binary pattern (which the protocol ensures cannot occur in the bit stream) that is the start delimiter (starting point). There are also binary patterns that indicate the addressing and what type of frame it is (e.g., the frame contains information or is of a supervisory nature), as well as some method of sequence numbering followed by the user data. A frame check sequence (FCS), normally a CRCC, follows the user data, which is typically a variable number of octets. The user data is surrounded by the protocol; that is, the protocol "frames" the user data (see figure 1-10). There may also be a stop delimiter or the frame may use the CRCC as the delimiter.

Link Access Protocol-Balanced (LAP-B) is a bit-oriented protocol. It is very similar in both structure and format to other bit-oriented protocols: High Level Data Link Control (ISO HDLC), Advanced Data Communications Control Procedure (ANSI ADCCP), and IBM's Synchronous Data Link Control (SDLC). IBM uses SDLC, which is a subset of HDLC, in its Synchronous Net-

Start
Delimiter
Stop
Delimiter
Address & Packet
Control Information
Your data
goes here
CRCC
Ones and zeros are organized into different groups whose patterns have meaning to the protocol detecting the frame. Note that in many cases the CRCC will be the stop delimiter.

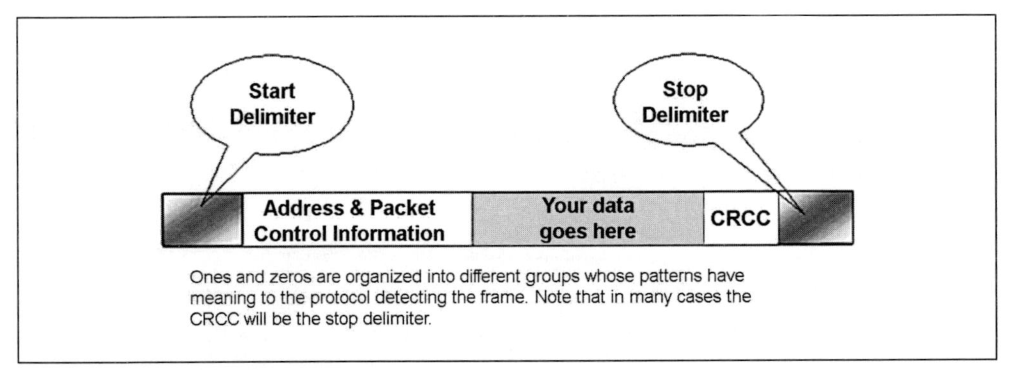

Figure 1-10. Frame Concept

work Architecture (SNA). LAP-B, ADCCP, and HDLC are quite similar. For that reason, only LAP-B will be discussed here.

Figure 1-11 illustrates an LAP-B frame. Notice that it is bounded by "flag" characters. In other words, the flags "frame" the data. The flag is an 8-bit octet, starting with a 0 followed by six 1s and ending with a 0. It is inserted at the beginning and end of each transmitted frame. The protocol only allows the frame to have this pattern at its start and end. It does so by using a technique called *zero insertion* or *bit stuffing*.

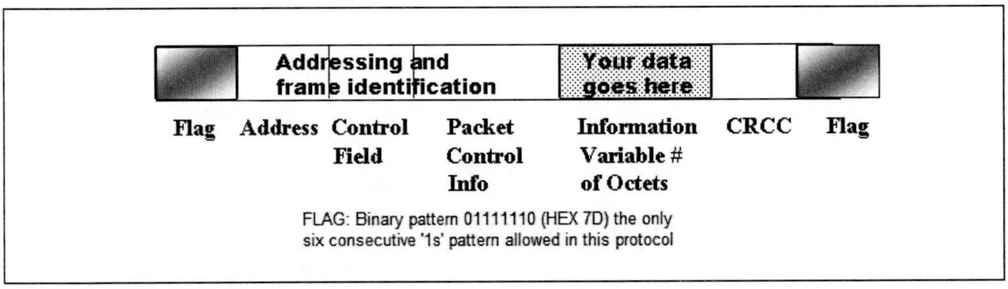

Addressing and
frame identification
Your data
goes here
Flag
Address
Control
Field
Packet
Control
Info
Information
Variable #
of Octets
CRCC
Flag
FLAG: Binary pattern 01111110 (HEX 7D) the only six consecutive '1s' pattern allowed in this protocol

Figure 1-11. LAP-B Frame

Zero Insertion. In normal transmission, any time the protocol detects five 1 bits in the data stream, the rule is to insert a 0. The receiver protocol, upon detection of five consecutive 1s, knows to remove the 0 that follows. It is that simple. Figure 1-12 illustrates 0 insertion and removal.

The LAP-B protocol will allow reception of up to 7 or 128 frames, depending on the system requirements, before it must have the first frame acknowl-

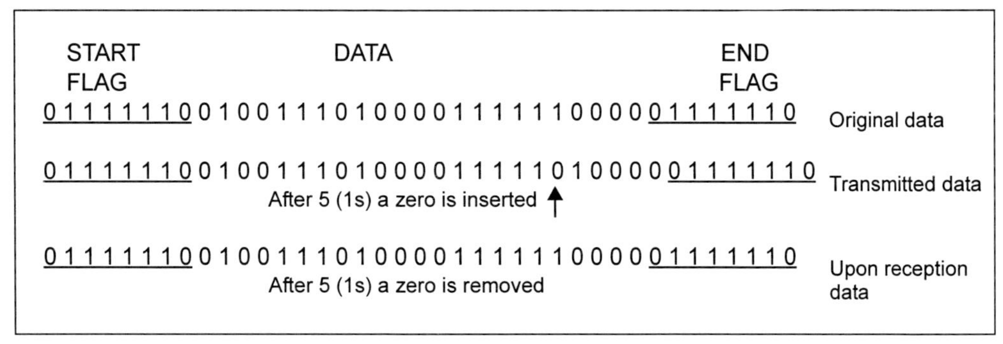

Figure 1-12. Zero Insertion and Removal

edged. Medium-induced errors are detected through the frame check sequence (FCS) and identified by frame number.

An error will destroy the frame number. The transmitter receives notice because the destination receiver ignores an errored frame and notifies the transmitter upon receiving the next good frame that it is not the one it anticipated. The receiver will request retransmission of the damaged frame and perhaps all subsequent frames, depending on what type of system is in operation. Note that the data carried in the frame has no bearing on the protocol, as it concerned only with individual bits in a frame, not what their ultimate meaning will be.

We will discuss bit-oriented protocols in more detail in chapters 2 and 7.

Protocol Summary

We have looked at two protocols, one character-based and one bit-oriented. We left out much of the inner workings and subsequent details from the discussion in order to illustrate only the salient points of some protocol concepts. As you proceed through the next chapter, you will find that the discussed protocols were specifically Data Link protocols; these bit-oriented protocols will again be discussed in chapter 2. For this chapter, it is only important that you see how 1s and 0s are organized to frame the characters you are transmitting.

Character-based protocol commands are directly related to the data within the frame. Certain characters must be sent twice or by some other method used to detect control characters as text. It is an older method, as bit-oriented protocols have generally replaced character-oriented protocols.

The bit-oriented protocols do not depend on the data stream contents; they operate independently. This means that such a protocol recognizes certain bit patterns that the protocol does not allow to occur in the data (e.g., LAP-B addresses) as it frames and error-checks packets (frames). LAP-B may be used in any system but it is primarily used in point-to-point and multi-drop systems.

Framing the data usually means having some form of start delimiter (flag), some sort of addressing and control process, the actual user data, an error check on the frame, and a stop delimiter (if the error check is not the stop delimiter itself). In lieu of a stop delimiter, some protocols count the octets in the frame and provide a frame length count (for variable length frames). Other protocols send frames with a fixed length. As the user data is book-ended by the framing, the user data is then said to be *encapsulated*. Figure 1-10 illustrated the general frame; take note of its similarity to the LAP-B frame in figure 1-11. In the end, the user data is a series of organized 1s and 0s and a bit-oriented protocol only knows how many bits it requires in a given protocol, not what they represent.

Summary

In this chapter we have reviewed the organization of data, as organized into characters (for transmission, storage, and presentation to humans) and as organized to detect errors in transmission. The first data transmission code (IBM's 4 of 8 Code) devoted more overhead to the error-detection scheme than to the data transmitted. Use of cyclic redundancy codes has minimized the necessary overhead, while enhancing the accuracy of error detection.

More than anything, this chapter has served to introduce you to the foundational concepts of data communications and how communications are organized into modes of transmission, into character codes, and into protocols. In the following chapters, we will organize data further into information (protocols and such) and implement what we have discussed in this chapter. The key point to acquire from this chapter is that data, 1s and 0s, means nothing unless it is organized into some accepted structure that then enables it to become useful information.

Bibliography

Please note that when Internet references are shown in this book, the address was valid at the time the chapter was created. Websites come and go, so occasionally one of the references will no longer work. It is then best to use a search engine to locate the topic. The web addresses are shown to provide credit for the information referenced.

Keogh, J. *Essential Guide to Networking*. Upper Saddle River: Prentice Hall, 2001.

Microsoft Corp. *Unicode*. MSDN Library, April 2000.

Peterson, W. W. and E. J. Weldon, eds. *Error Correcting Codes*. 2nd ed. Boston: MIT Press, 1988.

Sveum, M. E. *Data Communications: An Overview*. Upper Saddle River: Prentice Hall, 2001.

Thompson, L. *Electronic Controllers*. Research Triangle Park: ISA, 1989.

Thompson, L. *Industrial Data Communications*. 4th ed. Research Triangle Park: ISA, 2006.

Wikipedia. Various pages.

2 Communications Models

Modeling

Before explaining the models used in data communications, it might be useful to explain what models are and what purpose they serve. A model is a simulation of a real object. It may be a mathematical description, it may be an analogue of the original (a model of the object's physical characteristics), it may be a set of logical constructions, or it may even be a set of functions. A model can contain all of these properties. We use models to simplify explanation when the real object is too complex to visualize in human terms and to represent objects that we cannot physically capture.

In communications, we use models of different types to explain functional or circuit operation and to design communications. In this chapter we look at seven models. We first consider the International Organization for Standardization (ISO) Open Systems Interconnection (OSI) model, which is a model of the functionality required to communicate from the source end user to the destination end user. The next model we examine is the Internet model, the model that is the most widely used. Another model we look at is the Institute of Electrical and Electronic Engineers' (IEEE) 802 LAN (Local Area Network) model. We then close the chapter by discussing a set of four modern applications models and two data exchange architectures utilized by those models. To understand data communications, it is important that you grasp what these models represent, as almost all discussions of protocols and standards (both in this book and in general) are based on these models.

The question at this juncture is: how are the groups of 1s and 0s, the data's organization, transmitted from one end user to another? It is not enough that data be organized; there must be a system for moving this data from one location to another. This system is comprised of rules (provided by protocols): who has access to the medium and when (rules about medium access), who does packet (frame) error detection, who counts packets, who performs routing, who is responsible for end-user-to-end-user communication, who keeps track of the variety of traffic, who ensures that the traffic is compatible with the host or remote stations, and who interfaces with the end-user programs. All of these functions have to be performed when communicating from one end user to another.

ISO OSI Model

The OSI model describes these functions and generally specifies the order in which they take place in transmission. The OSI model is complex and understanding its functions is crucial to understanding data communications. Yet, the information needed to gain this understanding is actually quite simple. Although most readers have probably been introduced to the OSI model, it is usually explained only in brief statements or in the implementing standards themselves. One reason for this is that OSI is a model of functions, not a hardware or software specification. The OSI model goes from the concrete (the physical—you can touch it) to the abstract (the application—it is a set of virtual functions). If a system is OSI compliant, specific OSI standards are implemented in each OSI layer. If a standard other than OSI determines a particular OSI layer's functions, it does not necessarily make the implementation OSI compliant. The best way to begin an explanation of the OSI model is with an analogy.

Mail Analogy

Figure 2-1 is a simple analogy of communications functions. In the "good old days" before businesses computerized, how did you send a purchase order (PO) to a vendor? First, you researched the technical specifications of the product you wanted (if your company did not have equipment preferences) and, perhaps, the manufacturer or vendor.

Next, you filled out a standard requisition form. This means that you put the information into a standard format that was acceptable to both your company and to the vendor. This form was usually typed by a secretary. It was then determined how many copies of the PO to make and to whom the copies should be addressed within the vendor company. All the appropriate information was then attached to the PO.

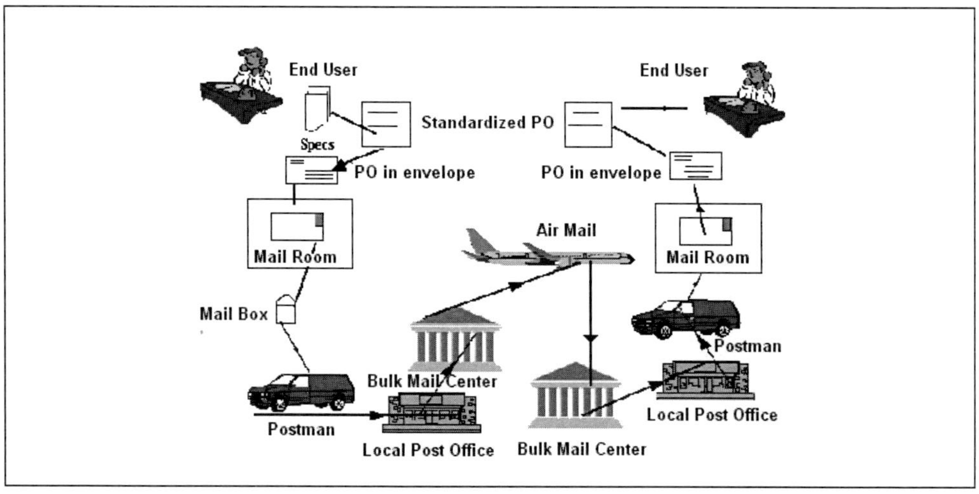

Figure 2-1. Mail Analogy

A mailing label was produced, by a secretary or perhaps in the mail room, and the PO was placed into an envelope for mailing. (At this point, the information on the PO is irrelevant; only the address on the label is important.) The mailroom placed postage and the label on the envelope and dropped it into the box outside for pickup. The mailroom attendant did not care what was in the envelope, only where it was going and how much it weighed. The mail truck picked up the envelope and took it to the local post office (indicated by the last two digits of a five-digit ZIP code in the United States).

The employees at the local post office checked the ZIP code in the recipient's address to determine if the envelope was to be delivered locally. The employees were only concerned with the first three digits in the ZIP code, as they indicate the address of the regional bulk mail center to which the envelope is being sent. If the zip code in the recipient's address is the same as the one for the local post office, the letter would be validated at the local post office and sent out for delivery the next route day. If the item was to be delivered to another zone, it went to the regional bulk mail center, typically by truck. The regional bulk mail center then determined where the item was to go next, based only on the first three digits of the ZIP code in the recipient's address. It might go by plane, train, truck, or bus to the receiving regional bulk mail center in that zone.

From the receiving regional bulk mail center, your PO went to the local post office and from there to the vendor's mail stop, to the vendor's mail room, to the mail entry (or similar position) clerk, and then finally to someone who would read the PO to determine what you required.

From the moment that you placed your vital information (the PO) into the envelope, no one cared what that information was until it was received at the vendor by the person who could act on it: the end user. Information was added (to the envelope) and the envelope was moved from place to place, without any regard as to what was in the envelope, as long as the envelope was of standard dimensions and weight. Conversely, when you filled out the PO, you did not consider how many places the envelope that contained it would be picked up at and delivered to on its way to the end user.

This analogy bears a close similarity to the OSI model for end-user-to-end-user communications, where the original sender is one end user and the person who will read the purchase order is the end user at the receiving end. One critical question should be addressed here: how did the sending user know that the letter was received? The sender doesn't; there is no response except for the receiving end user processing the PO and sending the product. This is analogous to the "connectionless-oriented" transmission or OSI Type 1 method. You send your data (encapsulated or "framed" by the necessary overhead) and regardless of the number of times it is handled (these instances are sometimes referred to as *hops* when discussing routed communications), you (the sender) receive no confirmation (acknowledgment) from the receiving end. OSI Type 1 transmission (no acknowledgment) is referred to as a *datagram*.

If you sent multiple packets at one time, they would probably arrive at different times and probably in random order. Using Type 1 transmission, there would be no response from the receiver. However, using OSI Type 3 transmission, the sender receives confirmation when the entire message has been received (that is, all the packets have been received) by way of an acknowledgment or from being handled at a different layer of functionality.

If you wanted to monitor the movement of your PO, you would send it as a registered letter. In this method of delivery, the letter must be signed for every time the letter is handled. In many ways, this is analogous to "connection-oriented" transmission or OSI Type 2. One similarity of Type 2 transmission to registered mail is that Type 2 almost always takes longer to arrive than Type 1, just as registered mail takes longer to arrive than regular mail. Another example of connection-oriented communications is a dial-up telephone call. You establish the connection before any transmission takes place and the circuit remains connected (hopefully) for the length of the transmission. (The author is indebted to George Stiefelmeyer for this postal example. Although the author has taken a few liberties with it, he has always found it to be quite effective in explaining the OSI model.)

OSI Model

The Open Systems Interconnection (OSI) reference model is an attempt to standardize the functionality of end-user-to-end-user computer communications. Some communications systems built today are OSI compliant (meet all the OSI model specifications). However, the large majority of systems are not OSI compliant, although they do implement the functionality in one or more of the OSI layers. As an example, the Internet Protocol (IP of TCP/IP fame) is not OSI compliant, yet it is described as a Layer 3 protocol.

End-to-end computer communications encompass many disparate systems. What is defined in the OSI model is not hardware but a set of functional layers. The OSI model has seven layers (see figure 2-2). It makes no attempt to specify any modem, medium, medium access, or any of the physical standards, such as connectors, coax, and so forth. Instead, it allows existing standards that meet the OSI functional requirements to fit into place.

In the OSI model, the line of demarcation between data communications and data processing exists at the border between the Transport and Session layers. However, many non-OSI-compliant systems integrate these functional areas into the communications stack and/or the operating system, so defining the layer boundary is difficult. (Note that the User Program shown in figure 2-2 is not a layer in the OSI model. It is included merely to illustrate the connectivity paths.)

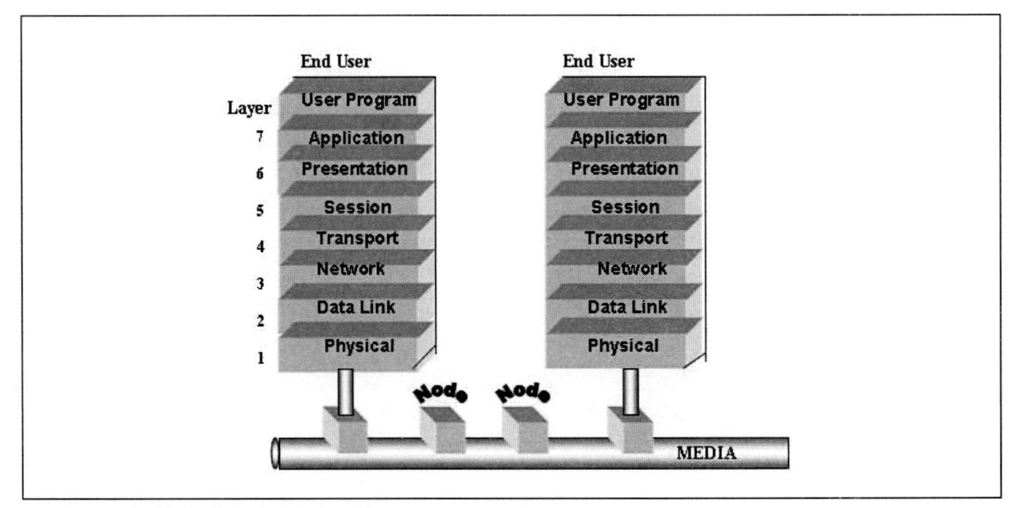

Figure 2-2. ISO-OSI Model of Interconnection

Physical Layer (1)

The Physical layer provides the mechanical and electrical means of connecting to the medium, be it copper, fiber optic, or wireless. Some examples of the Physical layer include Electronic Industries Alliance (EIA)/Telecommunications Industry Association (TIA) 232, EIA/TIA 485, or the LAN network interface card (NIC). The Physical layer provides line termination and/or impedance matching as well as synchronization of data (all of which are discussed in later chapters).

Service Access Points (SAP)

The interfaces between layers, such as between Layer 1 and 2, are known as service access points (SAPs). For every connection or task requiring communications, there will be a set of SAPs: the source service access point (SSAP) and the destination access service point (DSAP). If there is duplex operation, there will be two sets of SAPs. (SAPs are known as ports when referring to the Transmission Control Protocol (TCP, from TCP/IP) or the User Datagram Protocol [UDP]. SAPs are nothing more than addresses in memory assigned by whatever program is controlling the communications. They are well defined in the IEEE 802 series, which establishes local area network standards. If yours is a multitasking machine, more than one communications task may be in process at any time. Hence, a number of different SAPs may be active at one time. See figure 2-3 for a visual representation of SAP.

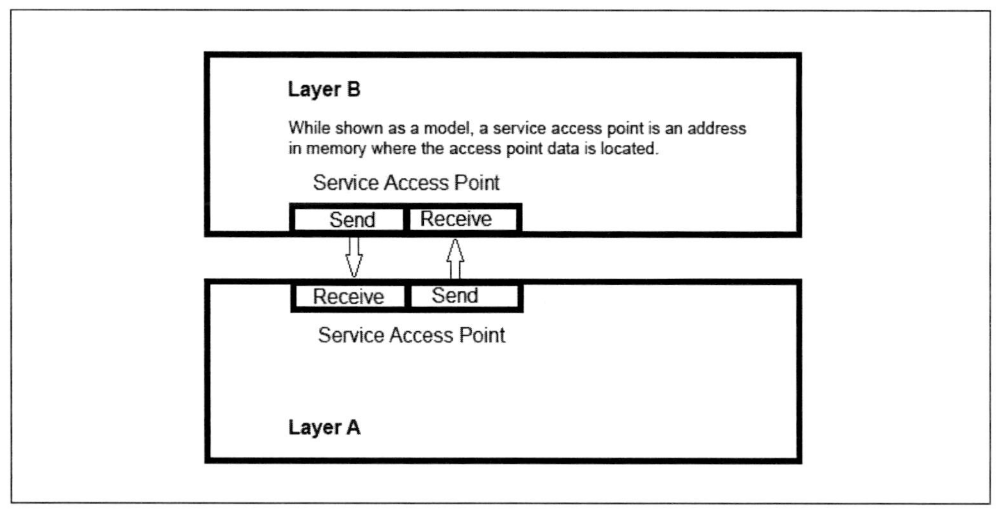

Figure 2-3. Service Access Points

Data Link Layer (2)

The Data Link layer frames (encapsulates) data and provides error-free transmission to the Network layer. This layer always ensures that frame check sequences (FCSs) check for bit errors and request retransmission of erroneous frames. For a Type 2 transmission (connection oriented), the Data Link layer also checks for missing message sequences and lost or duplicated frames. Protocols such as Bi-Sync, Synchronous Data Link Control (SDLC), Advanced Data Communications Control Protocol (ADCCP), and High-Level Data Link Control (HDLC) are Data Link layer protocols.

Network Layer (3)

The Network layer performs end-user-to-end-user routing. Specifications that fit this layer include CCITT X.25 or parts of various LAN operating systems, as well as any functions associated with data communications network routing and control. Layer 3 is almost totally focused on routing functions. The Internet Protocol (the IP of TCP/IP) and Internet Packet Exchange (IPX of Novell fame), along with the OSI Internet protocol, are all concerned with routing and the path that packets take from end user to end user. All fit into the Layer 3 functionality model.

Transport Layer (4)

The Transport layer is responsible for reliable end-user-to-end-user communications. It translates the lower-layer information into a data processing format and places the 1s and 0s into packets in the downward (toward Layer 3) direction.

The most common Transport protocols are TCP, which is connection-oriented, and UDP, which is connectionless. In a number of connection-oriented systems, the Transport layer is null (not used). Why? Because all the error checking (which includes transmission and frame assembly and disassembly), packet counting, and sequencing ("frame" and "packet" are sometimes used interchangeably) is performed in Layer 2.

"Connection-oriented" implies that a connection has been established prior to data transmission. Such is the case with many current industrial protocols using only Layers 1, 2, and 7. TCP is connection-oriented, even if the underlying system is connectionless (such as the Internet). TCP is not OSI compliant, yet it has Layer 4 functionality and is the de facto standard for data communications today.

Session Layer (5)

The Session layer is concerned with the jobs at hand and with scheduling jobs from the Applications layer (7) through the Presentation layer (6). It allocates system resources as required and communicates commands to the Transport layer. The Session layer establishes virtual connections and closes them when the communication tasks are complete.

You do much the same thing when you make installment-type payments through the mail. You mail the first payment (open a connection) and send payments until the note is paid off. If you are like many people, you make a number of these communications monthly. You don't know the actual route to the end user (note holder), nor do you know that a communication has arrived; however, you do know when your communication does not arrive because you receive a late notice. You cease communications and close your session when you pay off the note and receive the note papers. You are essentially doing what the Session layer does; only the Session layer does it in a much shorter period of time.

Presentation Layer (6)

The Presentation layer ensures that the communications data has the correct syntax for the Applications layer. Not all systems employ one type of computer architecture. Even among compatible machines some are (were) 8-bit, some 16-bit, some 32-bit, and some 64-bit. Each machine has a different set of rules. The Presentation layer ensures that communications are performed in a common language; the encryption and decryption of the data is usually performed at the Presentation layer. This layer also ensures that all the data typing and formatting will interface with the Application and Session layers.

One of the details that has to be worked out at this level is ensuring that the bit format is the same, that is, *big-endian* versus *little-endian*. If the most significant bit (from the perspective of a binary data value) is transmitted first, followed by all other bits to the least significant bit, it is called *big-endian* (used by IBM and others). In the little-endian format (used by Intel and others), the least significant bit is transmitted first, followed by all other bits to the most significant bit. It should be obvious that for any meaningful communications to occur, the bit formats must be the same. The mismatch is easy to correct, however, it must be identified first.

Application Layer (7)

The Application layer is where the system meets the end user's programs. Application layer functions are extremely high level compared to the bit functions at the Physical layer. The application in question might take the form of a process controller, a database manager, or any of the myriad of user applications. The Application layer is much like an operating system using a graphical user interface (GUI). An icon is selected that has a number of associated instructions; however, all that will be noticed by the user is the result. The Application layer takes requests and gives output to the user in the user's required form. This is where the network interfaces the user's programs when using an agreed-upon protocol (such as Hypertext Transfer Protocol [HTTP], Internet Message Access Protocol [IMAP], or even Telnet).

The Tortuous Path

Nothing is as simple as it would appear. During the transmission process, data is encapsulated in each layer with information required by the next layer. For example: What happens when a user program makes a simple file query, "Where is this file name located?" Application utilities (such as File Transfer, and Access Control and Management [FTAM]) are all programs that reside in the Application layer. If an external program wants to retrieve a file located on some distant (but accessible) point on the network and it has the file name (but usually not the file actual location), the program passes the file name to the Application layer (Layer 7). The utility adds header data to the file name indicating that this is a file query regarding location and path, and then routes the query to the Presentation layer.

The Presentation layer (Layer 6) provides additional information and directions about translating the Application data into a standard format. The new information is inserted in front of the Application data, and then the query is passed to the Sessions layer. (This additional information will be needed by the *receiving* Session layer before it passes the request to the Applications data.)

The Session layer (Layer 5) will queue the file query for execution and establish a virtual circuit (that will be disconnected when the transmission is completed). This additional transmission information will be tacked in front of the added Presentation layer information (which is in front of the Application data, which is in front of the actual file query request).

Next, the Session layer passes the request on to the Transport layer (Layer 4), where it will be packetized (a better word to use here than "framed" because framing is actually done in the Data Link layer) into a packet data unit (PDU). Depending on system constraints, there will be a minimum and a maximum packet size in bytes or octets (either one means 8 bits). The query is now sent to the Network layer.

The Network layer ensures that both the receive and send IP addresses (for TCP/IP) are in the proper locations in the protocol header at the sending end, and that this is the correct station at the receiving end. Based on the address, the Network layer determines the routing path and inserts that information in front of the information provided by the Transport layer. The request is then passed to the Data Link layer.

The Data Link layer (Layer 2) "frames" all the data, adds start and stop delimiters, and then generates and adds the frame check sequence (FCS), normally a CRCC bit error-detection. Finally, the request is received by the Physical layer (Layer 1) which accesses the medium, as dictated by the Layer 2 Media Access Control, and dumps the data out onto the medium in accordance with established rules.

At the receiving end, the same things happen, only in reverse. At each step up, the bits added to the original data are stripped away and used to pass the data up to the next layer until, finally, only the original data is presented to the user at the receiving end. Figure 2-4 illustrates the encapsulation process at each layer as the data proceeds from the Application layer to the Physical layer.

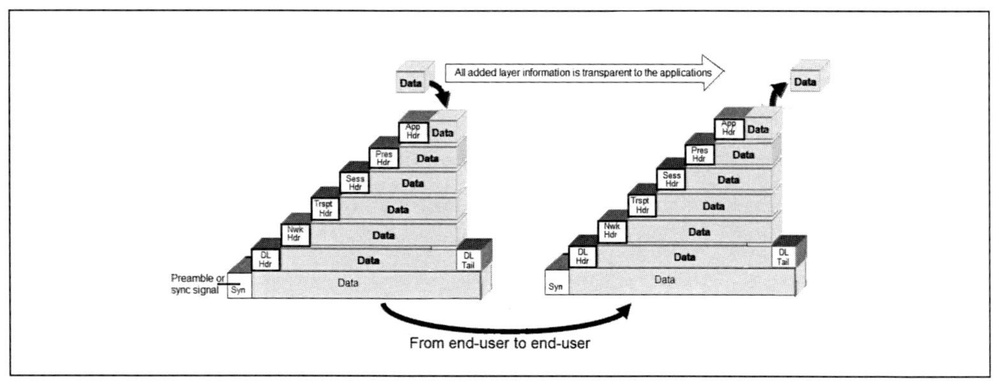

Figure 2-4. Layer Encapsulation

The Internet Model

Originating at about the same time as the OSI ISO model, the Internet model defines data communications in only five layers, as figure 2-5 illustrates.

Internet Layers
Application
Transport
Network
Data Link
Physical

Figure 2-5. Internet Layers

The Internet model is the most widely used protocol model for all data communications. The Data Link and Physical layers generally follow the IEEE 802 model (described next), but the Network layer always uses the IP (Internet Protocol) format, which is currently defined as either version 4 (IPv4) or version 6 (IPv6). These versions are defined by Internet standards that can be found at the following web pages:

IP version 4: http://www.ietf.org/rfc/rfc0791.txt

IP version 6: http://www.ietf.org/rfc/rfc2460.txt

These two standards are not directly or backward compatible. It is expected that IPv6 will phase out IPv4 within the next few years. However, since version 4 is in wide use, the two standards are both likely to co-exist for the next 10 years. In any case, there is no change to the OSI model and, indeed, the reason that functions are layered is that the only change will be in the internal workings of the Network layer (Layer 3). The Data Link (Layer 2) and Transport (Layer 4) layers can retain the same interfaces, while their internal functions do not (in theory) change at all. Early implementations used a dual stack (a complete set of Layers 2, 3, and 4—one for IPv4 and one for IPv6), while more contemporary versions use a dual IP Layer (containing a single stack with only a dual Layer 3, for IPv4 and IPv6).

The most common Transport protocols are TCP and UDP. UDP is a simple protocol that requires no acknowledgment or error checking. While TCP uses ports similar to UDP, it is different in that it overlays acknowledgment and end-to-end error checking (along with options) to guarantee that data is delivered.

The TCP/IP suite, as currently delivered, provides a number of standard applications (depending on the vendor), in addition to the Layer 3 and 4 functions. While some of these protocols implement only the Application layer functions, some also provide all or a partial implementation of the Presentation and Session layer functions. The TCP/IP suite makes no claim to be OSI compatible. Some of the listed applications and their protocols did not exist at the creation of TCP/IP, some have been greatly modified, and some have been determined to be downright security hazards and are not used. The standard utilities in the TCP/IP suite of protocols include:

- SMTP – Simple Mail Transport Protocol
- SNMP – Simple Network Management Protocol
- FTP – File Transfer Protocol
- NTP – Network Time Protocol
- DHCP – Dynamic Host Configuration Protocol
- LDAP – Lightweight Directory Access Protocol
- MIME – Multipurpose Internet Mail Extensions
- SOAP – Simple Object Access Protocol
- CIFS – Common Internet File System
- IPsec – A Layer 3 (mostly) protocol for encryption (voluntary in IPv4, mandatory in IPv6)

- ModbusTCP – A recently added data-transfer command sequence protocol, using the same protocol data units (PDU) but differing in the application data units (ADU) from the two other implementations of ModBus

The IEEE 802 Model

IEEE 802 was established as a local area network specification. This standard divides the OSI Data Link layer into two distinct sublayers: Media Access Control (MAC) and Logical Link Control (LLC) (see figure 2-6). Various SAPs are defined for the sublayers' connectivity. IEEE 802 also defines how different standard networks are to be physically connected: what form of media access they should use and how the user will interface data to the Data Link layer Logical Link Control (LLC).

The number 802 (February 1980 was the first meeting of this committee) is the designation for the main IEEE committee for LAN standardization and specification. There were, and are, various subcommittees refining portions of the 802 specification, which defines many different aspects of LAN technology. An example is 802.3, the standard for a Carrier Sense Multiple Access with Collision Detection (CSMA/CD) network and a formalization of Ethernet. The 802.3 standard originally had two physical specifications but it has many more now, as will be explained later. The IEEE 802 model is illustrated in figure 2-6.

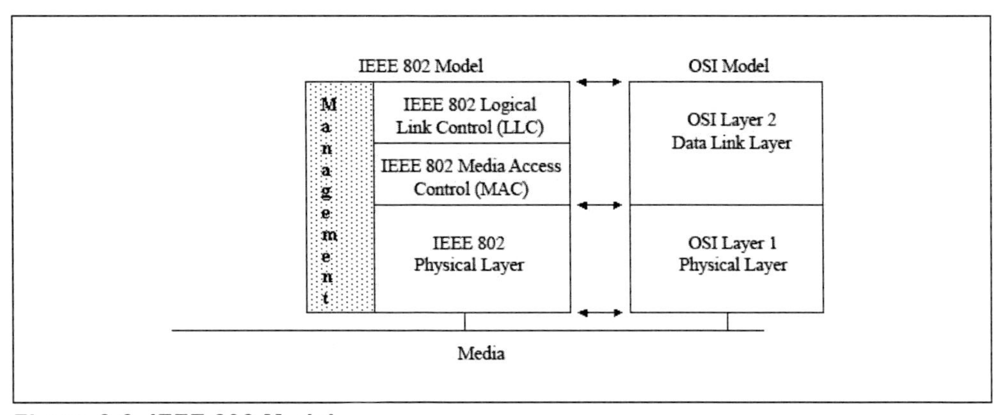

Figure 2-6. IEEE 802 Model

IEEE 802 currently specifies six different LAN technologies, many of which will be discussed at some length in later chapters:

- **IEEE 802.3:** This standard defines the CSMA/CD network (now known as Ethernet) and describes (at present) six LAN types: 10BASE5, 10BASE2, 10BASE-T, 100BASE-T, 1000BASE-T, and 10000BASE-T.
- **IEEE 802.4:** This standard defines the token-passing bus, a bus topology that uses a token-passing access. Three different types are described, two broadband and one carrier band. Token-passing bus was the basis for the now obsolete Manufacturing Automation Protocol (MAP) effort.
- **IEEE 802.5:** This standard describes a nonproprietary version of the IBM Token Ring, running at 4/16/100 Mbps (megabits per second).
- **IEEE 802.11:** This standard describes wireless local area networks.
- **IEEE 802.15:** This standard describes personal area networks.
- **IEEE 802.16:** This standard describes wireless wide area networks.

The rationale behind the IEEE 802 model and its specifications is this: if you meet the external interface requirements of the LLC layer, then your communications will work regardless of the underlying MAC technology being used.

Unless a Network layer (Layer 3) is available to the LLC, the system cannot route data outside of its own network (the system is "not routable"). The reason for this is that there is no provision or mechanism for finding addresses other than Data Link (Layer 2) addresses. Many local area network protocols (e.g., Microsoft's NetBEUI and IBM's SNA) were not designed to be routable by themselves.

It should probably be noted here that many older industrial LANs (such as the distributed control system [DCS] and the programmable logic controller [PLC] system) are "islands of automation" and were not designed to be connected to any other system. Most industrial networks did not originally employ routable protocols. This was because there was no requirement to route: the network topologies were "flat" (no computer had to send a message via an intermediate computer to a computer on another network) so that overhead could be minimized, as could all the other layers with the exception of

Layer 1, Layer 2, and Layer 7. If there is no Layer 3, then the protocol, by definition, is not routable by itself. However, a non-routable frame could (through encapsulation in a routable protocol) be made routable using some form of protocol convertor or gateway router. Figure 2-7 illustrates this, displaying two typical industrial network non-routable nodes. Note that each of these protocols contains a pattern of 1s and 0s that make up the protocol data. As stated previously, this pattern may be encapsulated in a routable protocol (as user data) and sent over many routable networks to a final destination (which will be running that non-routable protocol), with the only necessity being spoofing the time-outs used in the non-routable system to mimic acknowledgments to be made within a specified time.

Notice that FOUNDATION Fieldbus is called a *three-layer network*; that is misleading. It divides the Data Link layer into three sub-layers (remember that IEEE 802 was divided into two) and it also adds a fourth layer (not foreseen by the creators of the seven-layer model), called the *User layer*. This is where the Electronic Device Definitions (EDD) reside that make FOUNDATION Fieldbus plug-and-play.

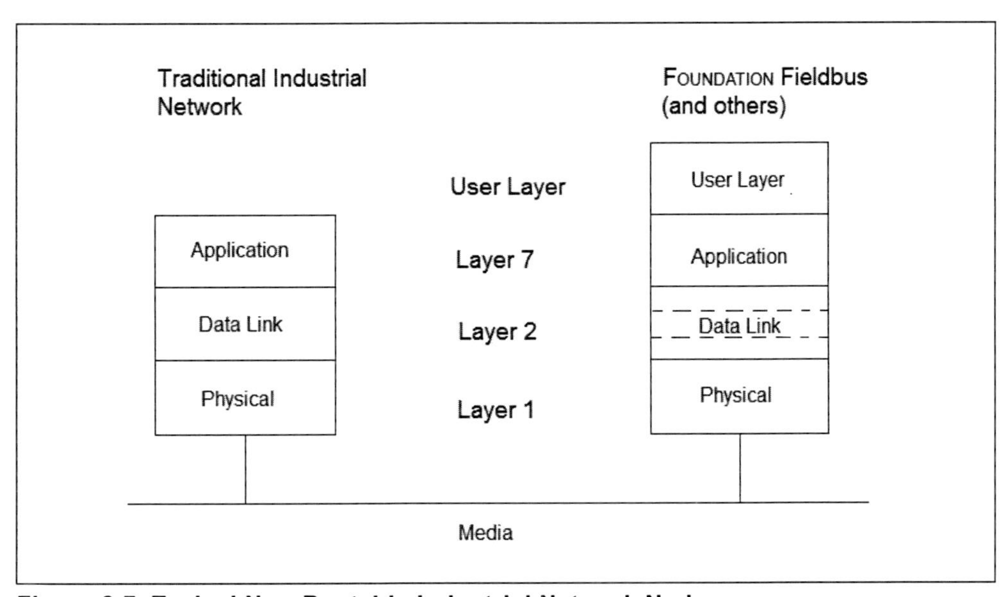

Figure 2-7. Typical Non-Routable Industrial Network Node

Application Models

One other set of communication models remains to be discussed before we move on to the details of each model. These are application models (not to be confused with the Application Layer, these are more like "apps" and should be referred to as *user applications*); they dictate how communications and networking software applications are applied. We will look at four types of application models: one-tier, two-tier, three-tier, and N-tier. However, we will begin by looking at the Application Model itself (see figure 2-8). Note that there are three services: User, Business, and Data. These will be more fully explained in later chapters.

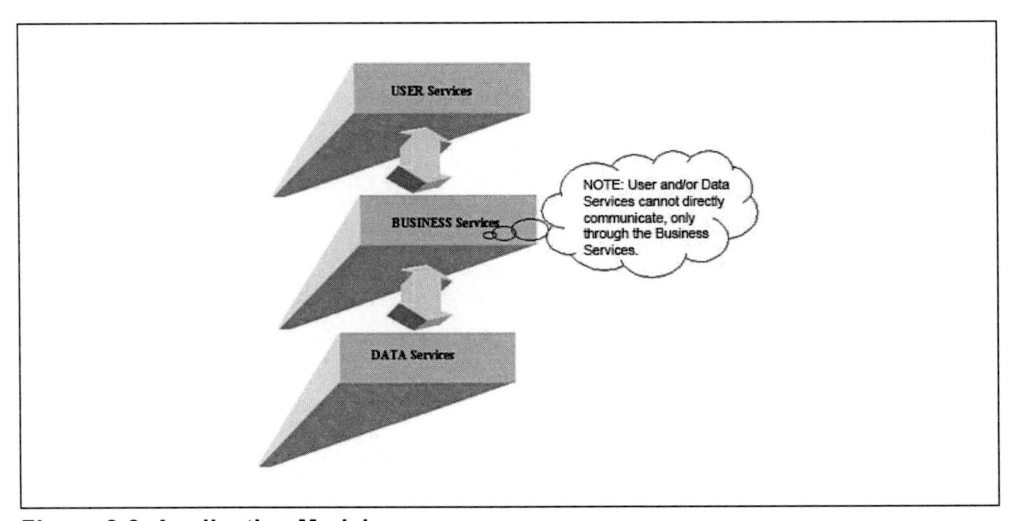

Figure 2-8. Application Model

User Services are composed of the graphical user interface and presentation logics and are sometimes referred to by those names. Data Services provide the organization of 1s and 0s. Typically data is organized in a database system. The database will have a management system, which is the name typically given to the database program. The majority of databases of any size are managed by relational database management systems (RDBMSs), such as Access, Oracle, or SQL Server. Although the data does not have to reside in a database, in most cases it does. The Business Services are an extremely important part of the system; they supply the organization, as well as the boundaries, procedures, and rules.

One-, Two-, Three-, and N-Tier Models

Where these three services reside determines which tier model (one-, two-, three-, or N-) you are using. An example of a one-tier model is a stand-alone PC running a check-balancing program in which check data is stored in a personal database, such as Access. In this case (and all one-tier models), all the services are located on one machine: the User services are your graphical user interface (GUI), the Business Services are in the check-balancing program, and the Data Services are provided by Access. A one-tier model is shown in figure 2-9.

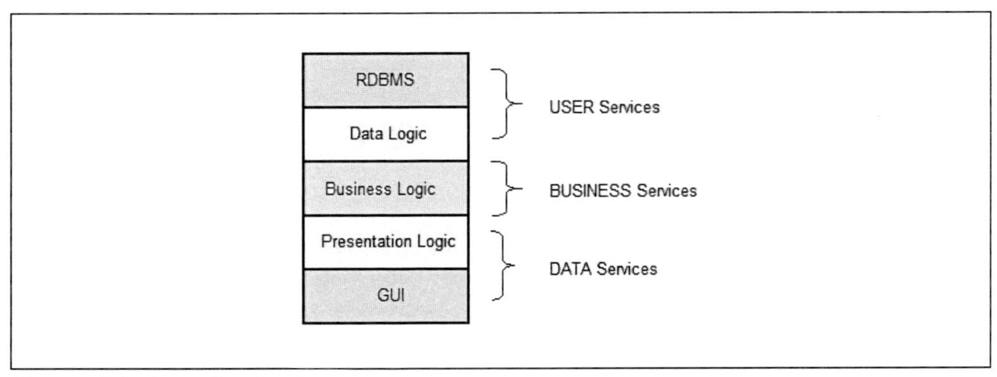

Figure 2-9. One-tier System

Two-tier systems are typically called *client-server systems*. A client is defined as a user of services and a server is defined as the resource for services. A single machine can be both a client and a server, depending on the software installed on it and its configuration, as described previously for the one-tier system.

Client-server systems were established in the 1980s, when personal computers were first networked to improve performance over simple file sharing. "Client-server" refers to the functional model for the relationship between the node (client) and the shared resource (server). The client-server software architecture is query-response based and is modular in its presentation, Business, and Data Services. It was intended to improve network performance, in comparison to the performance of a centralized server-based architecture that uses time-shared access.

Originally, PC networks were based on file sharing: the file server (the provider of resources) downloads files from a shared location to the workstation (requester of services) and the application is then run on the workstation. This arrangement works best if the number of stations requesting files and the amount of data moved are small.

Because of the performance limitations of file sharing, the client-server arrangement gained popularity as memory and data storage increased in capacity and dropped in per-bit price. The reduced system costs for microprocessor-based systems (for servers in particular) and the appearance of powerful relational database systems enhanced the popularity of the database server application and, thus, it replaced the simple file server. By using a relational database management system (RDBMS, referred to throughout this chapter as a DBMS—a database management system), user queries could be more efficiently handled. By using a query-response system, rather than file transfer, the DBMS significantly reduced network traffic.

In two-tier architecture, the interface to the presentation system (user display logic) is located in the client workstation and the DBMS is in a server that typically handles multiple requests from multiple clients. The location of the Business Services determines whether this is a "thin" or a "fat" client. If the Business Services are located in the client, it is a fat client; if they are located in the server, it is a thin client.

In summary, we have two types of clients—thin clients and fat clients—based on where the Business Services reside. You could run both the server and the workstation on one computer and have two tiers on one machine; however, a server is typically a separate machine.

The server's requirements will depend on whether a fat or thin client model is used. Fat clients require higher-power workstations and the demands on the server are only for the database application. Thin clients require a very robust server, as both the Business and Data Services for each client run on the server. Figure 2-10 illustrates a two-tier system for a fat client and for a thin client.

As well as it performs vis-à-vis the original file-sharing arrangement, the two-tier approach does have its limitations. Because each user maintains a connection while using the DBMS, the number of possible simultaneous connections is limited. The user connection is maintained even when there is no data transfer (actually until the client logs off, which in some cases could be 8 hours or more).

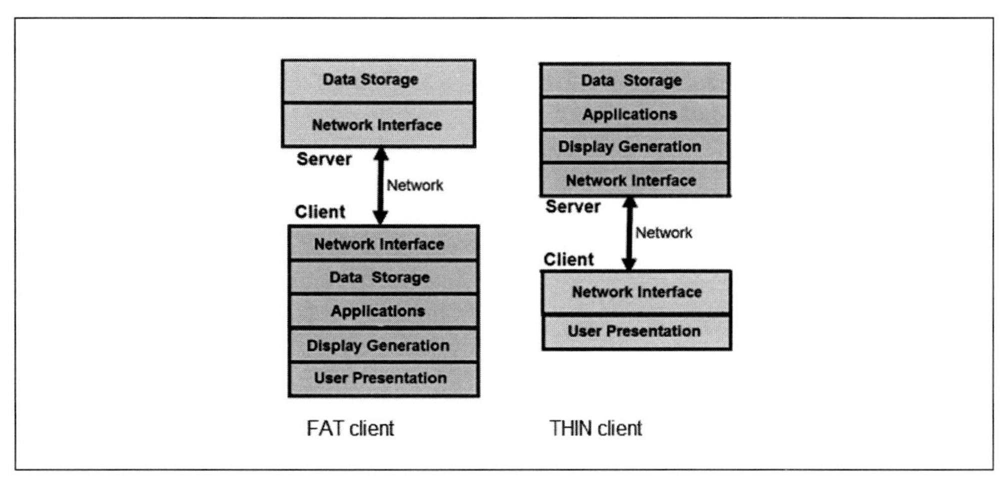

Figure 2-10. Two-tier Systems

Three-tier and N-tier systems use application servers and very thin clients. A three-tier system is typical of a distributed network (e.g., intranet and industrial networks), in which there is an application server running the Business Services. N-tier implies that the system has more than three tiers and involves Internet connectivity. Three-tier and N-tier systems are illustrated in figures 2-11 and 2-12, respectively.

The three-tier architecture was designed to avoid the limitations of the two-tier system. In the three-tier system, a middle tier is added between the presentation interface at the client and the interface to the DBMS Data Services. By placing the Business Services (or rules) in the middle tier and, generally, on its own application server, the Business Services can perform queues, run applications, and perform prioritizing, scheduling, and transaction processing. The three-tier architecture runs the Business Services application on a host (application server), rather than on either the Data Services server or the client system. This application server shares business logic, computations, and a data retrieval service (an interface to the DBMS). As a result, a number of efficiencies are realized: upgrades to the business rules need to be performed only on the server; the interface to the database server has greater integrity, and so on. In addition, because security and program integrity are on one machine (not many), administrative control is greatly facilitated.

The three-tier architecture offers significantly better performance, with a greater number of clients than is possible within the two-tier architectures. This is because the RDBMS is not held hostage to users who are logged on.

When the client typically requests data (which is located in the DBMS), the business rules determine if the client requests are allowed and then interfaces the request to the DBMS. A response to the client is made in the form (nowadays) of a page, using a browser as the presentation logic. There is no permanent connection. After the response is obtained and the information is downloaded to the presentation page, the connection is broken. It is re-established when the client makes another request.

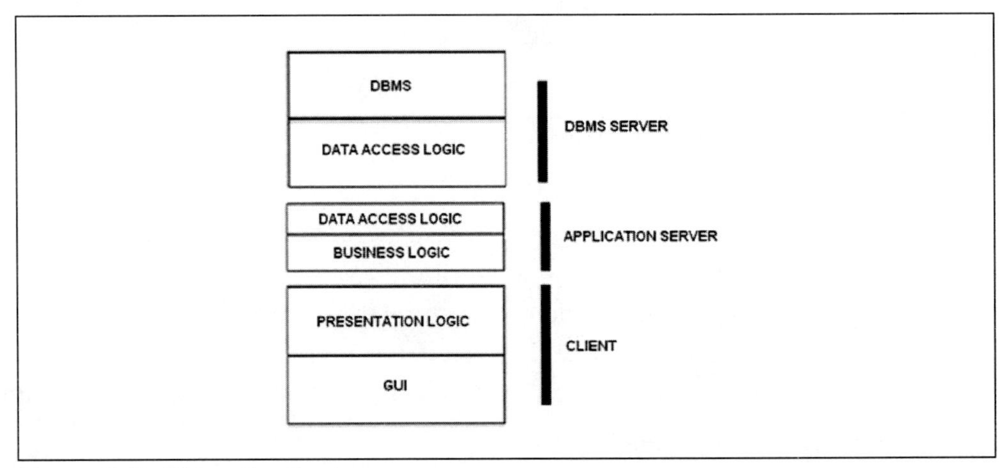

Figure 2-11. Three-tier System

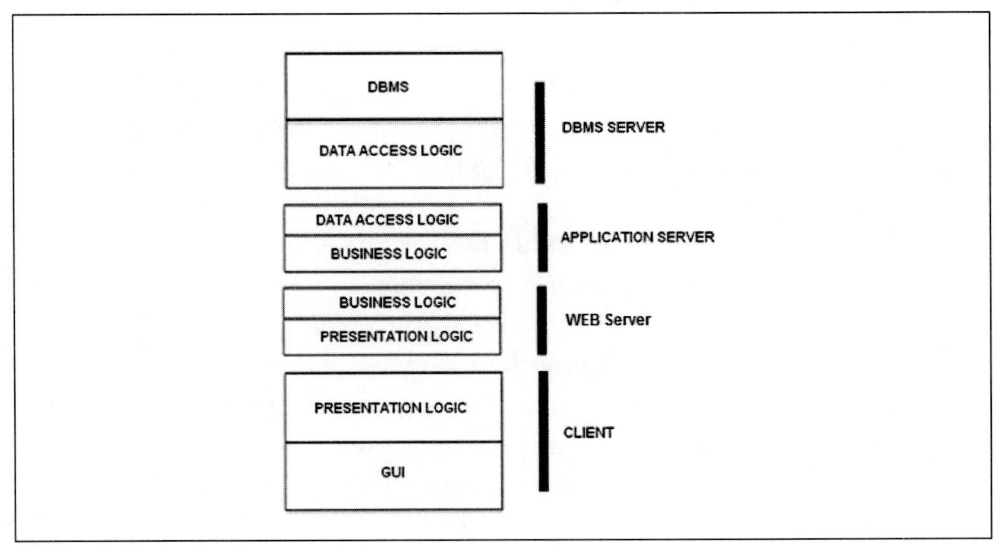

Figure 2-12. N-tier System

The sections below describe two of the methods applications use to obtain information from a process that has undergone changes. For example, when an output changes from 4 to 16 mA in a control system—how do interested applications obtain this change?

Data Exchange Architectures

Producer-Consumer

One method for exchanging data is the producer-consumer model. The server is the *producer*. It broadcasts data on the control system network or multicasts it to a multicast group. Clients are the *consumers* that listen for incoming data. Transmitting changed values without waiting for a query from the client is called a *push* method, as the server is pushing changed data to the client without data first being requested by the client. Much of the data in a control system is of a read-only nature and certain process values are of interest to many applications. If these values were supplied by a server using a traditional *pull* method, it could only supply this data to the clients in a query-response mode. The pull method wastes bandwidth on the server's network and CPU cycles because the client has to either anticipate the value changing, or run a polling routine (a scheduled query-response), or have the server raise an exception to a particular client followed by the query-response.

In many network environments, multicast is more efficient than broadcast. If multicasts are to be used throughout the control network, then all affected routers must support them. Clients that are not on this control network or are not a part of a multicast group will have to acquire the data via other means.

Publisher-Subscriber

In control systems, the publisher-subscriber approach is a better method for exchanging data. Using the publisher-subscriber method, a client application (the subscriber) communicates a request to a server (the publisher) and is put on a list to receive a response when the value changes; the subscriber expects to receive a response within a designated period. This request can be for a one-time receipt of data, or for data at regular intervals, or for data only when values change. In this model, the server maintains a list of clients and the data in which they are interested. If a large number of clients want the same data, the server must acquire this data set only once and transmit it to the clients on its list upon request. This is much more efficient than the query-response (client-server) model.

The publisher is normally a dedicated application. All client applications that depend on changes in the publisher application are called *subscribers*. As stated, the publisher maintains a list of current subscribers. When a client application wishes to become a subscriber, it must subscribe to an event controlled or reported by the publisher and it will be added to the list of current subscribers. The subscriber application is provided by the publisher (just as in everyday life). Another application is provided to unsubscribe. Whenever the state or states of the publisher application is caused to change, the publisher application will notify all current subscribers. The subscribers can then obtain the changed data at a convenient (for them) point in time. As long as a subscriber can process the available data within the constraints of process timing, the system will be real-time.

Summary

This chapter briefly discussed the seven-layer ISO OSI model for interconnection. Each of the seven layers is a set of functions that must be performed for end-user-to-end-user communication, regardless of media, system, or complexity. By defining the layers, standardization across vendors can be accomplished.

It should be reiterated that just because two systems are OSI compliant, it does not mean that they can communicate with each other. However, two or more systems that use the same standards to achieve the same layer functions can communicate.

The OSI model is used throughout this book (with explanations) to illustrate how the various layers are implemented by different standards and/or how whatever hardware or software being used is providing this function. This is, after all, a book about industrial data communications, an area that is rapidly becoming standards based. "Standards based" simply means nonproprietary or "open," a condition that should be welcomed by users and vendors alike.

The Internet model is introduced as the technology and the applications behind it are used on industrial intranets; the TCP/IP suite of protocols has become a defacto standard even on industrial (at least the Layer 3 and Layer 4 functions) systems.

We also looked at the IEEE LAN model and the main concepts behind it. The IEEE LAN model was intended to ensure communications provided the LLC interface requirements, regardless (within reason) of the underlying technology.

Finally, we discussed the differences between one-, two-, three-, and N-tier user application communications models. We also briefly touched on two alternate data exchange methods: producer-consumer and publisher-subscriber used by the application models.

With this and the preceding chapter behind us, we can now delve into the details of just how the various models, protocols, and communications are implemented.

Bibliography

Carnegie-Mellon University, Software Engineering Institute. *Client/Server Software Architectures—An Overview.* Pittsburgh: Carnegie-Mellon University, 2005.

Henshall, J. S., and S. Shaw. *OSI Explained: End-to-End Computer Communication Standards.* Chichester: Horwood Ltd., 1988.

Martin, J. *Telecommunications and the Computer.* Upper Saddle River: Prentice Hall, 1990.

Morneau, K. *MCSD Guide to Microsoft Solution Architectures.* Boston: Course Technology, 1999.

Stallings, W. *Local and Metropolitan Networks.* Upper Saddle River: Prentice Hall, 2000.

Stiefelmeyer, G. As quoted in various lectures. Great Falls: Stiefelmeyer International Limited, 1994.

Thompson, L. *Industrial Data Communications.* 3rd ed. Research Triangle Park: ISA, 2001.

Wikipedia. The publisher subscriber and producer consumer methodologies. 2013.

3 Serial Communications Standards

We have already introduced the concept of serial communications, the placing of one bit after another onto a single media channel. It has replaced almost all other forms of data communications due to the extremely high bit rates possible with today's technologies. However, there are many technical differences between downloading a configuration file onto an Ethernet switch or programmable logic controller (PLC) via a dedicated cable and sending data to a file server over a 100 Mbps (megabits per second) local area network (LAN). Both are serial communications, yet they differ in many ways. This chapter focuses on three Electronic Industries Alliance (EIA)/Telecommunications Industry Association (TIA) serial communications standards: numbers 232, 422, and 485. Since the United States now sits on the international standards committees, the TIA/EIA standards have their equivalency in ISO standards and, indeed, most have been changed to meet the ISO standards. Serial interface to the PC is being accomplished by newer and much faster technologies, so we will also focus on four PC-based standards that will impact industrial use and automation applications: USB 2.0, FireWire (IEEE 1394), SATA, and PCIe.

Basic Concepts

Digital communication is the exchange of binary data between digital devices (or more accurately, between a program running on one digital device and a program running on another digital device). These communications might be one-way or bi-directional. The two communicating devices might be physically local to each other or might be far enough apart that some form of intervening telecommunications equipment is needed to enable the communications. There must be a communications channel of some form to carry the bits from the sending device to the receiving device (and possibly for the reverse as well). The internal voltage-based representation of binary bits within a digital device may be unsuitable for interfacing with the communications channel medium or telecommunications equipment and, thus, some form of signal conversion may be required.

The transmission of bits must take into account the state of the receiving device, so that transmission doesn't occur unless the receiving device is ready and able to receive the data being transmitted. If, during the transmission of bits, some form of interference causes an error in the received data, there must be some means of detecting this condition and correcting the erroneous data.

All these issues must be addressed in order to ensure proper and accurate exchange of data between two digital devices, and communication standards must account for these issues. Figure 3-1 provides a generalized model of data communication.

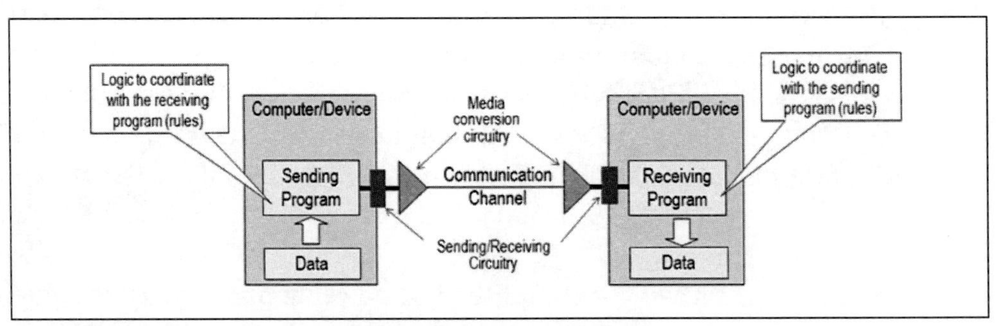

Figure 3-1. Basic Elements of Digital Communication

Definitions

Understanding how data communication equipment (DCE) and data terminal equipment (DTE) are defined is essential to any discussion of data communications. From the "dumb" hardware devices, dominant in the 1960s, to today's software interfaces, these terms are used throughout data communications (and unfortunately, not always correctly). In many early applications of digital communications, there was a need to use existing telecommunications infrastructure and/or technology (e.g., the analog telephone system or voice-grade radio) to create a communications channel when the communicating devices were physically widely separated. In those cases, there was a need for specialized equipment to interface between the devices (DTE) and the communications equipment. This was especially true when some of the communicating devices were un-intelligent (no computer/microprocessor) and were merely electronic (e.g., a "dumb" CRT terminal) or electromechanical (e.g., a teletype). The specialized interface equipment placed between the communications infrastructure and the communicating devices was designated as data communication equipment (DCE). Several electrical standards were developed for connecting DTE to DCE, particularly the RS-232 "recommended standard," which is now known as TIA/EIA-232.

Data terminal equipment (DTE) means the end device (either data source or destination) that sends (originates) data to (or receives data from) another DTE (e.g., a computer, a printer, or a PLC). The DCE interface at both ends of the channel provides the necessary connectivity for the communications infrastructure to create an end-to-end channel. Figure 3-2 (End-to-End DTE Communications) illustrates using DCE to interface DTE with available communications infrastructure. The DCE is often the point that is nearest the communications line, external to the equipment attempting to communicate. Examples are modems, line drivers, multiplexer composite output, and so on. The DCE is usually connected to a DTE, forming a complete endpoint station. The DTE is farthest from the communications line; the DCE is the closest to the communications line.

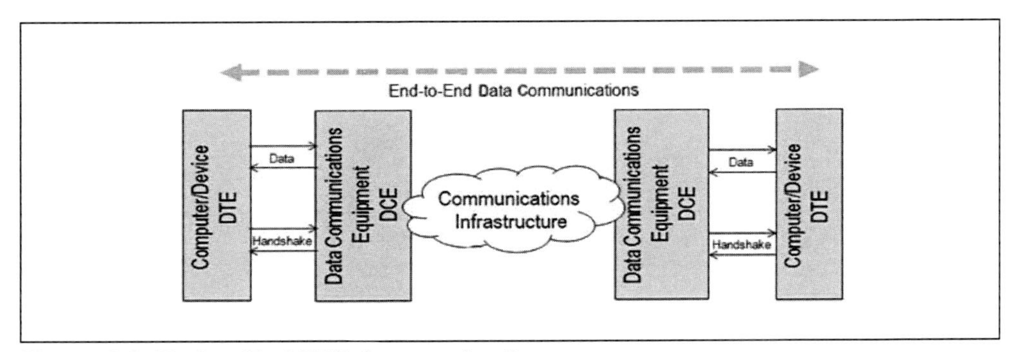

Figure 3-2. End-to-End DTE Communications

TIA/EIA Standards

In general, the TIA/EIA serial standards specify the electrical and/or mechanical interface between the DCE and the DTE. These standards are typically for the Physical layer only and make no allusion to the Data Link layer or higher layers of the OSI Model. An *analog* DCE, such as a telephone modem, is concerned with the line modulation type, media access, line speed, etc. The TIA/EIA serial standards only deal with the *digital* signals between DTEs, DCEs, and other devices.

Figure 3-3 illustrates a typical DCE/DTE setup. Any discussion of the serial transmission digital data interface between DTE and DCE starts with one of the most widely used digital standards ever developed: the Electronic Industries Alliance (EIA) 232, which is now in its F version. This standard is still frequently (but inaccurately) called by its original name: RS-232 (RS means *recommended standard*).

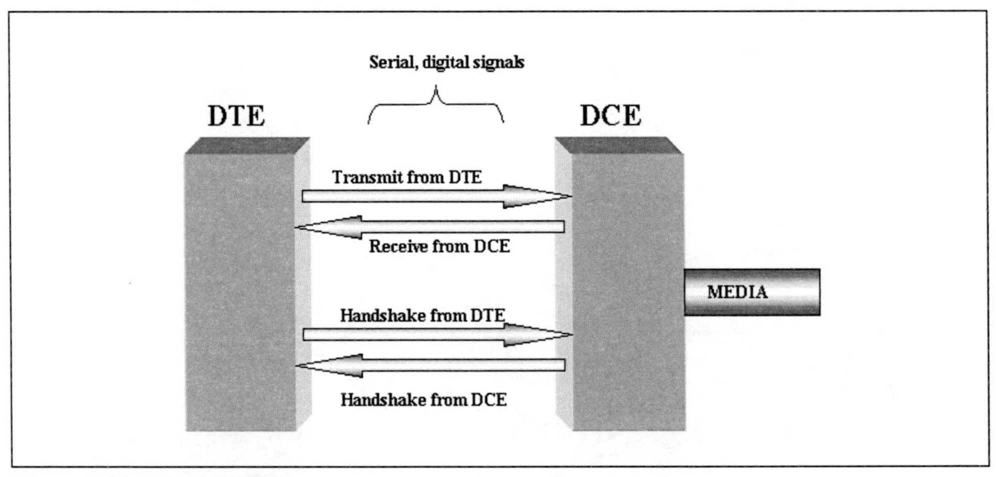

Figure 3-3. DTE to DCE

TIA/EIA-232-F

The TIA/EIA-232-F standard is for the interface that makes possible the connection between a DTE (which could be a controller, printer, computer, etc.) and a DCE (modem, etc.) by employing serial binary data interchange. When this standard was first conceived around 1962, the DTE was either a teletypewriter or a dumb terminal (a terminal with only data handling, no peripherals or microprocessor of its own) and the distant end DTE was probably a computer. At that time, the standard (then named *RS-232-C*) was originally intended to describe the interface between a modem (DCE) and a teleprinter (DTE)—nothing more. It was never anticipated that the standard would be used for everything from calculators to multiplexers.

Up through RS-232-C (1968), the standard specified only that there be a 25-pin connector (no requirement for a male, female, or subminiature type-D connector) and only described the 25 pins and their actions. The standard dictated a maximum transmission distance (cable length) of 15 m (50 ft.). A 1972 addendum to the standard stated that the 232 standard was obsolete, it should not be used in new designs, and it should be limited in existing designs to 9600 bps. The addendum further stated that EIA RS-449, accompanied by electrical standards TIA/EIA-422 or -423, should be used instead. Although TIA/EIA-422 and -423 (which will be described later in the chapter) are now in place, EIA-449 apparently never really caught on and RS-232-C continued to be used, even at 19.2 Kbps (kilobits per second; or higher for non-standard applications). With improvements in cable technology and circuit sensitivity, it has become possible to run short distances at speeds far in excess of those defined in the standard. Today it is not unusual to see two computers, or a computer and an intelligent device, communicating over a (short length) EIA-232 circuit at bit rates of 115,200 bps.

The D version of this standard was developed in1986 and published in January 1987. The E version brought TIA/EIA-232 in line with International Telecommunications Union (ITU) V.24 and V.28 and ISO 2110. It specified the 25-pin connector dimensionally and electrically. It also included several new circuits and redefined the protective ground. The following terminologies were affected:

- DCE, which was "data communication equipment" or "data set," is now "data circuit-terminating equipment."
- "Driver" is now "generator."
- "Terminator" is now "receiver."

When microcomputers began replacing dumb terminals, the connector type to be used depended upon the connector's gender and on the programming of the associated universal asynchronous receiver-transmitter (UART—an integrated circuit that performs serial/parallel conversion with associated control and logic functions).

The reason for the subminiature D-type connector (originally specified by ITT-Cannon) is simple: it was the connector used on the Bell modems in the late 1960s. The DCE had a female connector, so it (logically) was assumed that the DTE was a male connector. Because the connector type was not explicitly stated and because the DCE or DTE function could be programmed into many microcomputers, the prospective purchaser was never sure what type of cable connector was needed or how many lines would be in the cable.

An entire industry segment grew around this uncertainty, manufacturing items such as "gender menders" (two connectors, both either female or male, that were connected back to back; also called *gender changers*) which changed the gender of the cable end plug. Also produced were "null modem" crossover connectors and "breakout boxes." A null modem crossover connector consists of two connectors connected back-to-back or joined by a cable. On the 9-pin version, lines 2 and 3 were crossed; on the 25-pin version, lines 2 and 3 were crossed and several pairs (4 and 5, 6 and 20) may also be crossed. A breakout box typically had a male and a female connector with wiring between the connectors to connect all or some of the 25 lines to test points, and switches that are used to break the normal wired path between the connector pins. The user or technician would jumper different lines until the desired result was obtained. Several enterprising companies made breakout boxes complete with microprocessors that could indicate how to correctly wire the connections so the circuits would function. Figure 3-4 shows examples of gender menders, null modem connectors, and a breakout box. Example E-1 illustrates why a null modem is necessary.

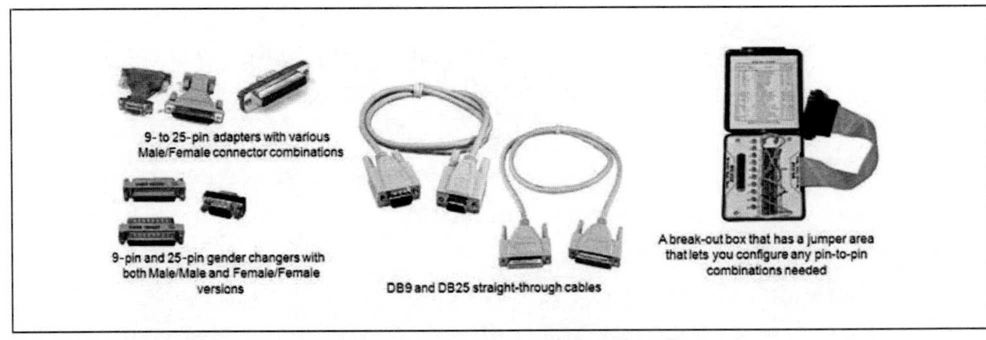

Figure 3-4. Adapters, Gender Menders, and a Breakout Box

Example 3-1

Problem: You have a PC with an external modem. The modem is the DCE and the computer is the DTE. If the computer operates a printer over a serial line and the printer is a DTE, what is the computer?

Solution: Although it would appear logical to say that the computer is the printer's DCE, it is in fact a DTE as well. To connect a serial printer to a computer, you need a null modem adapter. A null modem adapter is a male-to-female connector with the data lines crossed from pins 2 to pins 3 (and sometimes the handshake lines as well). The null modem adapter provides the solution to connecting a serial printer (DTE) to the serial port of the computer (DTE).

TIA/EIA-232-F is the current (when this book went to press) version. It is basically the same as the E version, but it has been brought into line with international standards. The main difference lies in some electrical specifications involving signal rise time. With version E, it was necessary to buy an additional standard to acquire the 9-pin specifications; however, the F version has all 9-pin specifications, but no 25- (or 26) pin connectors outlined. Table 3-1 provides a list of pin terminations and circuit numbers for the 25-pin connector (E version).

Table 3-1. TIA/EIA-232-E Pin Assignment

Equivalent. Circuit	CCITT	Pin	Function
AA	101	1	Protective Ground (Frame Ground)
AB	102	7	Signal Ground—all signals are referenced to pin 7
BA	103	2	TxD Transmitted Data, DTE to DCE
BB	104	3	RxD Received Data, DCE to DTE
CA	105	4	RTS Request to Send, DTE to DCE
CB	106	5	CTS Clear to Send, DCE to DTE
CC	107	6	DCE Ready, DCE to DTE
CD	108.2	20	DTE Ready, DTE to DCE
CE	125	22	Received Line Signal Detector (Ring Indicator DCE to DTE)
CF	109	8	Data Carrier Detect, DCE to DTE
CH	111	23	Data Rate Selector, DTE to DCEs
CI	112	11	Data Rate Selector, DCE to DTE

Table 3-1. TIA/EIA-232-E Pin Assignment

DA	113	24	External Transmitter Clock, DTE to DCE
DB	114	15	Transmitter Clock, DCE to DTE
DD	115	17	Receiver Clock, DCE to DTE
RL/CG	140	21	Remote Loop Back, DTE to DCE
LL	141	18	Local Loop Back, DTE to DCE
TM	142	25	Test Mode, DCE to DTE
			Pins of Historical Importance
SBA	118	14	Secondary Transmitted Data
SBB	119	16	Secondary Received Data
SCA	120	19	Secondary RTS
SCB	121	13	Secondary CTS
SCF	122	12	Secondary Data Carrier Detect

Since many of the signals in the original 25-pin standard were there to support dumb devices, they are generally not required when connecting intelligent devices. A subset of 9 signals (or even less) is all that is typically required for communication purposes. Table 3-2 lists the DB9 (subminiature D) 9-pin connector terminations. The DB9 is the contemporary connector for the serial interface (if any) on a modern computer.

Table 3-2. DE (DB)-9 Pin Assignment

Equivalent Circuit	CCITT	Pin	Function
CF	109	1	DCD – Data Carrier Detect
BB	104	2	RxD – Receive Data
BA	103	3	TxD – Transmit Data
CD	108	4	DTR DTE Ready
AB	102	5	Signal Ground
CC	107	6	DSR DCE Ready
CA	105	7	RTS – Request to Send
CB	106	8	CTS – Clear to Send
CE	125	9	RI – Received Line Signal Indicator

For the equivalent circuits (these are engineering models): if the first letter is an A, it is a common circuit; if it is a B, it is a signal circuit; if it is a C, it is a control circuit; and if it is a D, it is a timing circuit. An S indicates a secondary circuit. It is worth noting here that old terms are hard to change. Technically, the DTR signal is not "Data Terminal Ready" but "Data Termination Equipment Ready." Somehow it doesn't have quite the same ring to it.

Even in the 9-pin version of the standard, many of the signals are present to enable "hardware handshaking" between the DTE and the DCE. (A handshake refers to the methodology of data flow control—the request to transmit and the acknowledgment that the receiving device is ready.) As an example, DTE could originate the conversation by sending an RTS (Request to Send) to the DCE and it would reply (when ready) with the CTS (Clear to Send) signal. The DTR (Data Terminal Ready) and DSR (Data Set Ready) signals can also be used for the same purposes (generally one or the other, but not both are used; different manufacturers choose their own way). These are called *hardware handshakes* because the hardware (DCE or DTE) generates these signals to control data flow. For a direct point-to-point cable connection between a pair of PCs, it is possible to use only pins 2, 3, and 5 to create a usable circuit. In that case, the sending and receiving software needs to incorporate "rules" for coordinating who gets to transmit and when. These rules constitute "software handshaking," as no pins (connections) are used to determine data flow.

The D in the connector name (e.g., DB9, DB25) stands for the shell shape. The letters following the D specify the shell size: A is for 15 pins, B is for 25 pins, E is for 9 pins (DE), and so on. Because the 25-pin serial port connector was a DB25, many started erroneously calling the DE9 a DB9; in this text we will try to use the correct appellation (but old habits die hard). Two of the D connectors are illustrated in figure 3-5.

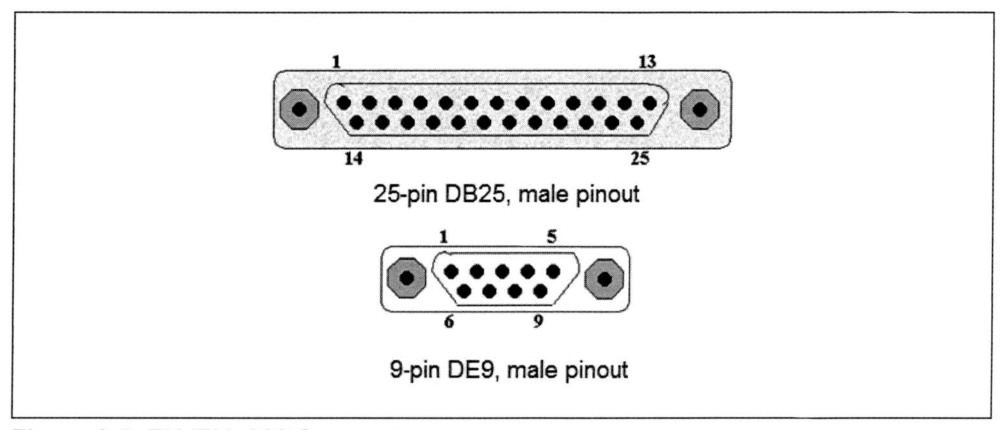

Figure 3-5. TIA/EIA-232 Connectors

The pins of historical importance (between 1962 and 1972) were concerned with the Bell 202 modem and the reverse channel, also called the *secondary channel*. This low-speed part of the bandwidth was used only for ACKs and NAKs (packet receipt responses) to keep the circuit from having to turn the line around (half duplex line, meant to exchange send and receive modems) and, thus, losing time to synchronizing the modems after the line rearrangement. (An ACK is an acknowledgment that an error-free packet has been received; a NAK is a negative response indicating that the packet contained errors.) Some modems still exist that use a secondary channel. Other pins have different uses according to who is implementing the circuitry. Only the DE9 (DB9) connectors, listed in table 3-2 by equivalent circuit and pin, still find use in asynchronous systems today.

Figure 3-6 illustrates the waveform voltage limits found in RS-232-C and EIA-232-F. The teletypewriter origin of this signal is apparent in that the most positive signal condition is a logic 0 (or SPACE) and the most negative signal condition is a logic 1 (or MARK). Teletypewriter circuits had long used an OFF condition (SPACE), no current, and the ON condition (MARK), which was a negative voltage (in relation to earth ground). This on-off arrangement is known as *neutral signaling*. Polar (for polarized) signaling does not have a legitimate no-current state (that is, none is defined). Instead, either a positive or a negative voltage was impressed across the line, so the negative condition as MARK was just carried on from the teletypewriter through EIA-232. As almost all binary logic in contemporary use has its most positive state as the TRUE or logic 1 condition (the reverse of the teletypewriter state), this tends to cause a bit of confusion, particularly when you're trying to observe EIA-232 signals.

The voltage levels defined in the standard should be considered in two ways. First, the voltage level is imposed on the wire at the transmitting end. Signal characteristics will degrade as the signal goes over a longer and longer length of wire, so the voltage seen at the receiving end will generally be lower than was applied at the transmitting end. Second, the voltage level is considered as valid at the receiving end. The current standard allows the signal to be up to ±15 volts DC, but no less than ±3 volts DC (the higher the original voltage, the longer the cable can be and still deliver an acceptable signal level at the receiving end). Any voltage between ±3 volts is considered to be noise and is never considered as a valid signal. When a device is not transmitting (or is done transmitting), the transmit line must be held at the "1/MARK" state (a -3 to

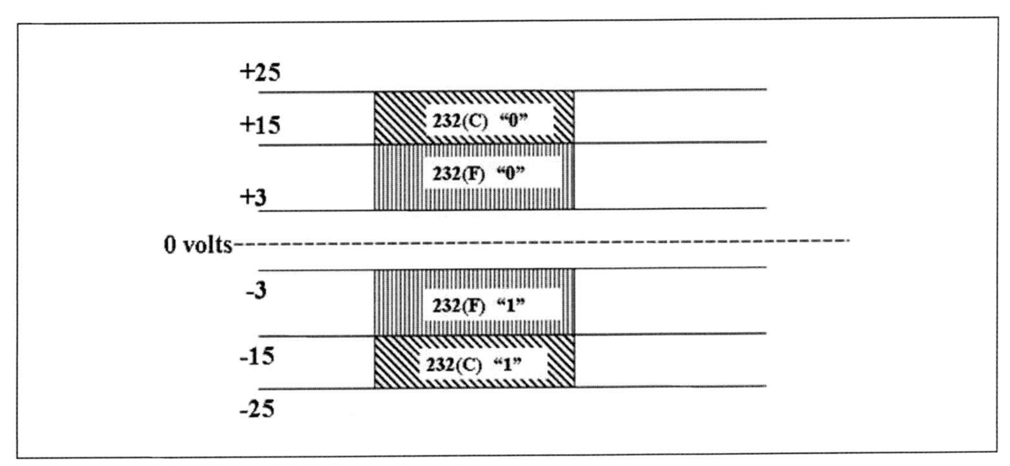

Figure 3-6. TIA/EIA-232 Voltage Levels

-15 volt level). Figure 3-7 shows what an EIA-232 signal looks like on the transmit line coming out of a COM port on a typical PC.

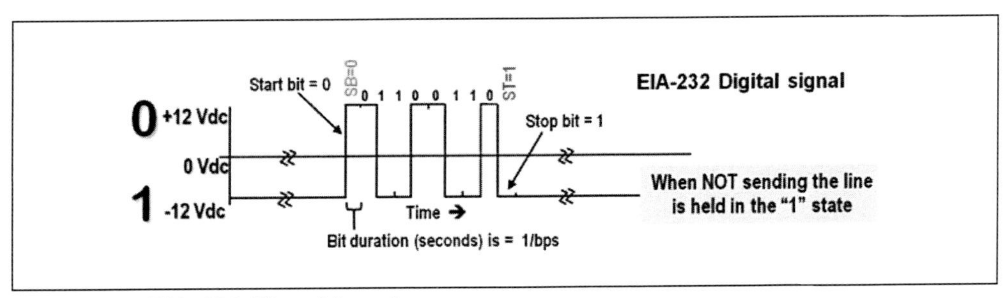

Figure 3-7. EIA-232 Signal Levels

How did such a standard become so widely used? It is because there was no alternate standard. The original standard defined a very narrow application that over time blossomed into near universality. *Remember*, when a device says it is "232 compatible," that means only that it outputs a signal somewhere between ±3 to ±15 volts DC, that certain circuit input and output impedances are within a specific range, and, with luck, that the DTE (if identified) will transmit data on pin 2 and will expect to receive data on pin 3. Whether it uses the flow control pins or software to provide a handshake is up to the manufacturer. If a device claims to meet the EIA-232 standard, it must supply the correct state on the flow control pins (4, 5, 6, and 20) for the circuit conditions. The reader should take this information to mean that there is still a great deal of variability (although nothing like in the previous 20 years) between devices that claim to be 232 compatible. Most PCs (with a serial port) expect these flow control pins to be enabled before they will communicate.

The measurement of the signal voltage at the receiving end of an EIA-232 circuit is done relative to local ground. This means that if there is a grounding problem (e.g., a "ground loop"), such that the local ground is several volts higher than at the transmit end of the circuit, the signal received may become unusable. In addition, electrical noise in the area could cause inaccurate readings at the receive end, especially at high bit rates, where bits have a very short time duration. Such problems caused the EIA to look for alternate designs that were not subject to these flaws. As a result, the EIA published the RS-449 functional and mechanical standard and electrical standards EIA-422 and 423. It also declared RS-232 obsolete and not to be used in new designs (circa 1977). RS-449 was rescinded in 1986, while TIA/EIA-232 lives on in the F version. EIA-422, 423, and 485 are discussed later in this chapter.

Desktop PCs, while certainly not the only user of 232 ports, were the majority users. It has become standard to use a male 9-pin connector and DTE configuration for the COM ports of desktop and laptop PCs, although most modern laptops (and a number of desktops) no longer offer serial ports with 25- or 9-pin connectors. This poses a problem, as you cannot directly connect a DTE to another DTE with a straight-through (pin N-to-pin N) cable. Figure 3-8 shows why.

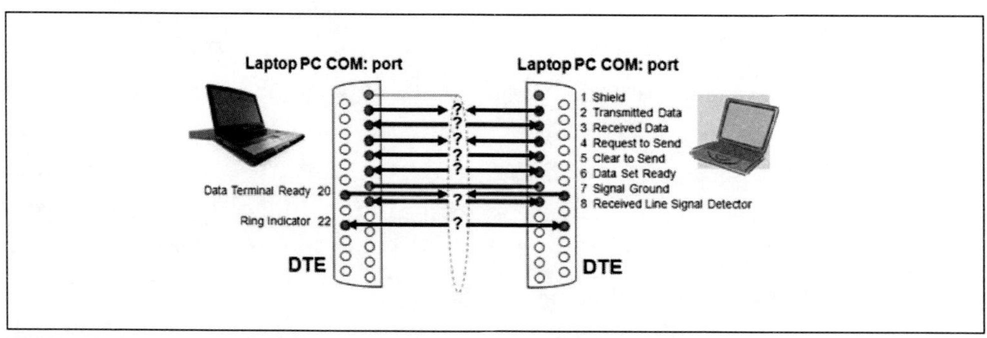

Figure 3-8. EIA-232 DTE Connection Incompatibility

In order to connect between DTEs, it is necessary to have some means to rewire the pin-to-pin ordering, so that, for example, transmit goes to receive. The typical practice is to install a null modem adapter at one end or the other (not both) between the straight-through cable and the DTE port. A null modem adapter re-routes the signal wiring and may also, depending on the design, provide some level of handshake signal "loopback." There are several null modem designs and you must make sure that the one you use is compatible with the communications functions of your software. While the one shown in figure 3-9 will almost always work, a commercial null modem adapter generally will work. Figure 3-9 shows a null modem design that provides a total hardware handshake loopback. Only transmit, receive, and signal ground are actually connected between the two computers. All the hardware handshake signals are looped back to the originating computer. This creates the illusion of a functional DCE to the software in each PC.

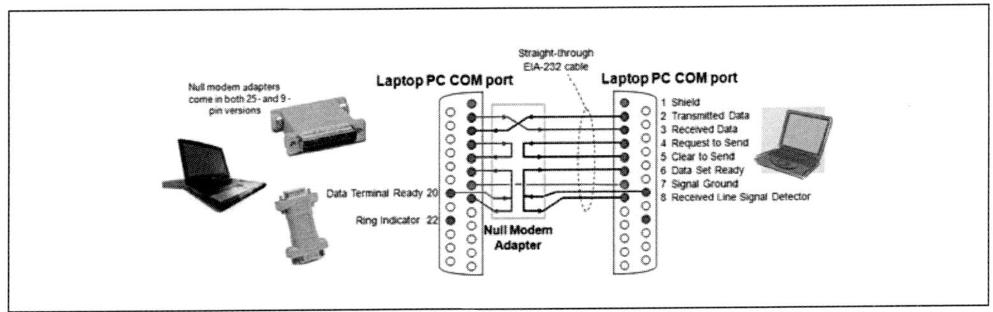

Figure 3-9. Using a Null Modem Adapter to Connect PCs

In 1977, as mentioned earlier, EIA recommended a new standard for media connections that started at 20 Kbps and was designed for even higher speed data. EIA-449 was to replace EIA-232 at all speeds. It used a 37-pin main connector, relegating the secondary channel connections to a separate plug, which, in most communications networks, was never used. Surprisingly, with all its problems and limitations, EIA-232 continues to survive and is still commonly used to make local point-to-point connections between laptop PCs and intelligent devices (e.g., Ethernet switches, RTUs, PLCs, protective relays, etc.) for configuration, calibration, firmware updating, and diagnostic purposes. Many such devices have a "console" port that is an EIA-232 port.

EIA-449: Interface Standard

In contrast to RS-232, EIA-449 used the EIA-422 and EIA-423 standards for media and electrical descriptions. EIA-449 described only the pins, definitions, connector specifications, and functions and referred to EIA-422 and EIA-423 whenever electrical connections to the media interface were described. EIA-449 was originally intended to phase out RS-232 but EIA-449 itself has been replaced by EIA-530 for circuits above 20 Kbps (most implementations do not use EIA-530, rather their own connections, since only two wires are required for EIA-485 and four wires are required for EIA-422).

TIA/EIA-422 and 423

TIA/EIA-232 specifies the pinouts, line characteristics, and input and output impedances; in fact, it specifies the entire interface. As an interface may include more than point-to-point circuits, the newer standards tend to describe just the interface's physical and synchronization characteristics. A number of standards were developed that defined the electrical interface and could then be referenced from the physical standard. The development of these new standards prevented duplication and contention over small changes in different written standards. Two electrical interface standards now in place are TIA/EIA-422 and TIA/EIA-423 (although you are unlikely to ever encounter 423).

TIA/EIA-422: Balanced Interface

TIA/EIA-422 is an electrical media standard, specifying input and output impedances, line lengths, rise and fall times, signaling speeds, voltage levels, and so on. The standard calls for a ±200 mV to ± 6 V signal, with the most positive condition being the logic 0 state.

TIA/EIA-422 is a "balanced-to-ground" specification. This means that both of the transmit terminals and both of the receive terminals have the same resistance to ground (figure 3-10). A balanced-to-ground (also simply referred to as *balanced*) line does not have to contend with the charge/discharge of uneven line capacitances, which it would have to do if it were unbalanced. Noise is normally found in the common mode (between the 0 V DC reference and the signal line), but the balanced line uses differential inputs and outputs, which by nature have a high common mode rejection ratio (see glossary for definition), so noise is reduced as a factor in signal quality. A differential transmitter puts the transmitted signal on one line and a polarity-inverted copy of the transmitted signal on the other. At the receiving end, the difference between these two signals is determined and amplified. This action greatly diminishes any common mode noise (common node noise is equally present on both lines).

These factors (and some others) enable a balanced line to operate at a higher speed than an unbalanced line. TIA/EIA-422 is specified for 20 Kbps to 10

Mbps. All the data provided in the standard is relative to a twisted copper pair (22 AWG) media. Because the signal voltage is measured relative to one pin (usually A), the polarity of the output pins is important. Any interface will typically mark the two pins as + and – (or perhaps A and B).

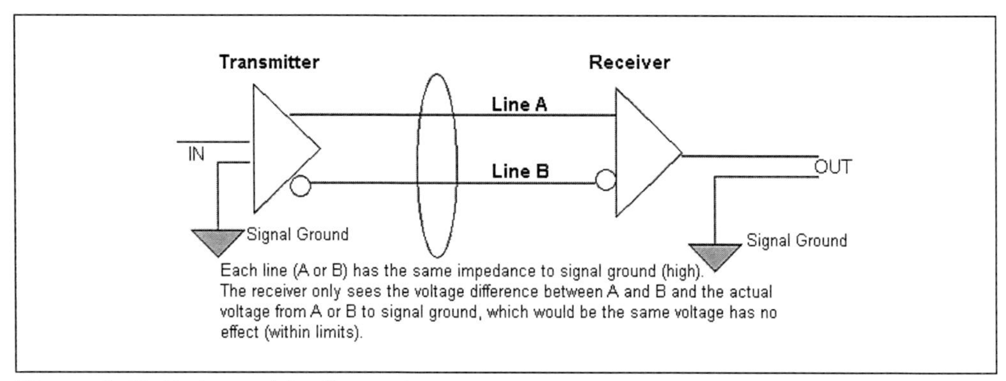

Figure 3-10. Balanced-to-Ground

Using a balanced system requires the terminals of the communications line (whether the send and receive pair for 422 or a shared pair for 485) to have an equal impedance to the signal common. A "1" condition is determined when A is negative with respect to B; a SPACE is determined when A is positive with respect to B.

EIA-422, which is specified for 1 transmitter and up to 16 receivers, is mostly used today for point-to-point transmission involving four wires (two pair) for duplex transmission. One transmitter can drive 16 devices because the standard does not allow for contention, that is, the condition where two (or more) stations try to transmit at the time and they are contending for the medium. If two transmitters become active at the same time, but in the opposite polarity, a short circuit will result unless the circuitry specifically supports this condition: EIA-422 does not. This is the reason for 1 transmitter and 16 receivers—only 1 transmitter sends at a time. As a result, if you want a network that contains more than one transmitter, you must ensure that the network will not allow more than one transmitter to be active at a time.

TIA/EIA-423: Unbalanced Interface

TIA/EIA-423, on the other hand, is specified as unbalanced-to-ground. That is, the return for both transmit and receive is at a common potential: the circuit 0 V DC reference point, which is referred to as *signal ground* or *signal common*. Figure 3-11 illustrates an unbalanced-to-ground (unbalanced) system.

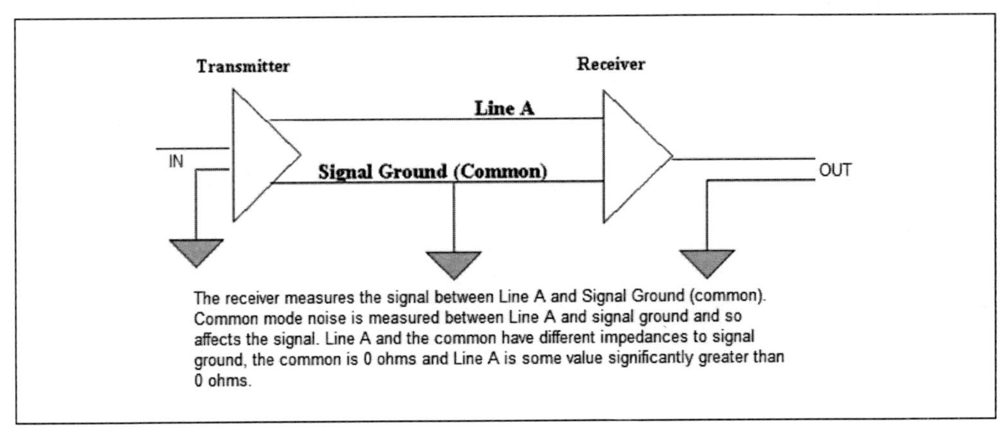

Figure 3-11. Unbalanced-to-Ground

An unbalanced line will run more slowly (for a given medium and distance) than a balanced line. This primarily results from the different charge/discharge times of the cable capacitance and the fact that noise is also referenced to the signal ground, meaning it is in the common mode. EIA-422/423 transmitters and receivers are electrically identical. When a 422 circuit is connected with one lead common to the signal return and a single "hot" lead, there is no discernible difference in operation below 20 Kbps. Whereas the TIA/EIA-422 is specified for 20 Kbps to 10 Mbps, TIA/EIA-423 is rated at 20 Kbps maximum (as is EIA-232, which is also unbalanced-to-ground). EIA-423 is relatively obscure and not being utilized, while deployment of EIA-232 continues. The reader is not likely to come across an EIA-423 circuit.

Note that TIA/EIA-422 refers to a copper twisted pair as its standard media. TIA/EIA-232 is an unbalanced line standard and the TIA/EIA-423 receiver should be able to accept the TIA/EIA-232 signal. It is entirely probable that both balanced and unbalanced lines may be found in the same cable: data and timing signals (because of the high line rate) will be balanced lines, while control lines may very well be unbalanced.

There are, of course, recommended maximum transmission distances before the signal becomes degraded (i.e., where the resistance/reactance effects of the copper pair distorts the signal envelope and detection becomes difficult). Most specified distances are shorter than those found in actual use. Whereas

TIA/EIA-232-C specified a maximum length for all data rates (15 m, 50 ft.), TIA/EIA-232-D had a sliding distance scale based on speed. TIA/EIA-232-E and F have no distance limit, but instead use the line capacitance as the limiting factor. With modern unshielded cable, distances could easily extend to 40 m (132 ft.). TIA/EIA-422 and TIA/EIA-423 specify a maximum distance (4000 ft.) (although, that is not at the maximum data rate).

At 10 Mbps (TIA/EIA-422), the distance between a transmitter and receiver is limited to 3 m (10 ft.). This is unfortunate, since unshielded twisted-pair (UTP) cables used in Ethernet (10BASE-T) operate at 100 m (390 ft.) at 10 Mbps, while 100BASE-T will operate at 100 Mbps over that same 100 m distance. Both use balanced line pairs. TIA/EIA-422 and TIA/EIA-423 do not specify protocol, timing sequence, quality limits, or pin assignments.

TIA/EIA 485-A

The standards described in the preceding sections had, as their basic architecture, point-to-point communications. In this section, we describe multipoint communications and architecture that has more than two stations, as an adjunct to most of the previous standards.

TIA/EIA-485 was written to describe a transmitter-receiver combination that is capable of multipoint operation. This standard will allow any combination of up to 32 transmitters (generators) and receivers on the same two-wire line. It specifies a balanced line for data transmission and reception, much as does EIA-422, however, where EIA-422 allows only one transmitter and specifies that all other devices must be receivers, EIA-485 uses tri-state logic and ties the transmitter and receiver to the same wire pair. This is a bus arrangement, so the addressing must be taken care of within the software, as must the response to all commands. The major differences between EIA-422 and EIA-485 are detailed in table 3-3.

Table 3-3. EIA-422/485 Differences

Function	EIA-422	EIA-485
Minimum output voltage	200 mV into 100 ohms	1.5 V into 60 ohms
Current (short to ground)	150 mA maximum	
Current (short to ± source)		250 mA maximum
Rise Time	<10% bit time	<30% bit time into 54 ohms, 50 pF load.

If the differences between EIA-422 and EIA-485 do not appear to have deep significance, one can at least appreciate the fact that the EIA-485 standard was designed from EIA-422, and that a circuit can be easily designed to accommodate both types of transmitter/receivers. The primary difference between the two is that EIA-485 can support contention.

EIA-485 does not say what software is required to affect a serial multipoint network and does not specify timing requirements, protocol, or pinouts. The typical pinout is the subminiature 9-pin D connector but it may be a "D" block or any other termination that the manufacturer adopts. There must be some means of directing who may speak, who must listen, and in what order; there must also be direction as to what to do if nobody is talking. These are the same issues that arise over and over again, regardless of the type of network used. EIA-485 has been used for multipoint local area networks using both four-wire (full-duplex) and two-wire (half-duplex) variations, as shown in figure 3-12. An EIA-485 network is probably one of the least expensive networks to implement with interface cards for PCs (they generally take the place of one of the COM ports), provided that no more than 32 nodes (communicating devices) will be needed. The industry-standard 56- to 115-Kbps data rate is typically supported in commercial adapters, so it should be no surprise that this serial scheme is used in many commercial applications, including instrumentation systems.

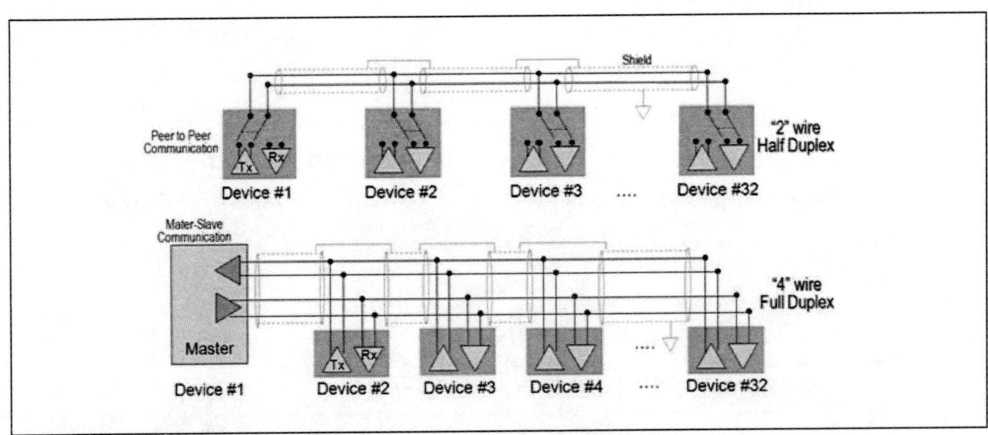

Figure 3-12. EIA-485 Local Area Network Configurations

There were several early efforts to develop local area networks for DCS and PLC systems based on the network design in the EIA-485 standard or on a variation of that standard. Note that the bit rate allowed (56 Kbps to 115 Kbps) was dependent on the overall length of the network wiring.

Example 3-2

Keeping the previous description of EIA-485 in mind and the knowledge that it supports contention, evaluate the following statements, identifying necessary actions a protocol would have to consider.

1. Upon power-up, or initialization, who starts talking?
2. What method will give each station a fair amount of time, or priority?
3. Can one station talk to only one other station, or to all stations?
4. What addressing scheme will be used?
5. If a station fails, will all be affected?
6. How is a new station added?

These actions are provided by the Data Link layer and, in particular, the Media Access Control. EIA-485 is only a Layer 1 standard, therefore, Layer 2 services (the Data Link layer) must be provided externally, usually by vendors or integrators. The actions outlined above cannot be met by a Layer 1-only device.

Due to the susceptibility of EIA-232 to common mode noise and to ground loop problems (in addition to its distance limitations—the speed vs. distance trade-off), it is not uncommon to see plug-in converters used to turn a serial EIA-232 COM port into either an EIA-422 or an EIA-485 port. Figure 3-13 shows how this can be accomplished. Note that none of the EIA-232 hardware handshaking signals can be passed through the conversion, only the transmitted/received data; therefore, a software handshake is required.

TIA/EIA-530

This standard was intended to gradually phase out EIA-449 at the higher (above 20 Kbps) data rates. It specifies the TIA/EIA-422 balanced standard for its Category I (data) circuits, and states that Category II (flow control) circuits may be unbalanced TIA/EIA-423 types. The main function of the TIA/EIA-530 standard is to provide a pinout, similar to that of EIA-232, for signaling

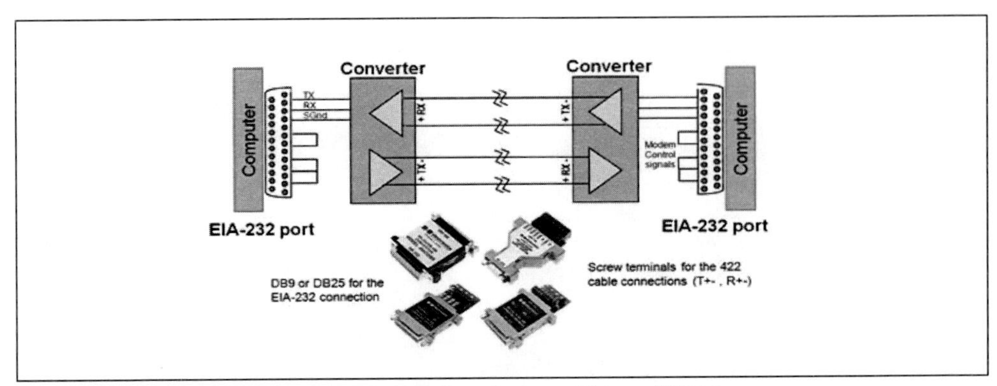

Figure 3-13. Converting an EIA-232 Port into a 422 or 485 Line

above 20 Kbps (the upper limit of the EIA-423 standard receiver). The specified 25-pin connector is similar to the one described in the section on TIA/EIA-232, yet the pinout is quite different. It is doubtful, however, that you will encounter this standard in today's market.

The functions of the named pins are much the same as in RS-232. Table 3-4 illustrates the differences between RS-232-C, EIA-232-E, and EIA-530 pinouts. Note that EIA-530 pins are the same as EIA-232-E pins.

Table 3-4. Differences Between RS-232-C, EIA-232-E, and EIA-530 Pinouts

Signal	RS-232-C-	EIA-232(E)	EIA-530
Frame Ground	1		
Shield Ground		1	1
Transmit Data	2	2	2(–) 14 (+)
Receive Data	3	3	3 (–) 16 (+)
RTS	4	4	4 (–) 19 (+)
CTS	5	5	5 (–) 13 (+)
Data Set Ready	6		
DCE Ready		6	6 (–) 22 (+)
Data Term Ready	20		
DTE Ready		20	20 (–) 23 (+)
Signal Ground	7	7	7
Carrier Detect	8		
Receive Line Signal Detector		8	8 (–) 10 (+)
TX Timing DCE	15	15	15 (–) 12 (+)
RX Timing DCE	17	17	17 (–) 9 (+)

Table 3-4. Differences Between RS-232-C, EIA-232-E, and EIA-530 Pinouts

Local Loopback	18	18	18
Remote Loopback	21	21	21
Ext TX Clock	24		
TX Timing DTE		24	24 (–) 11 (+)
Test Mode	25	25	25

Interface Signal Functions

Table 3-4 illustrates that certain signals are common to all the interface functions. In the next few paragraphs, we will describe these "hardware handshake" signals for typical operations. This knowledge is useful for determining whether a serial interface is operating correctly (or if it will even operate). The order in which the signals are described here is generic. However, when using modern devices, you will encounter different sequences and not all equipment uses all of these signals. It is quite possible to find a serial communications cable that has only three lines, although five or six lines is more typical.

A device should not attempt to transmit information if there is no functional communication channel or if the receiving device is not able/ready to receive the information. There must be some mechanism that allows both the sender and the receiver to coordinate with each other. With older "dumb" devices, this usually entailed some form of hardware-implemented signals that could be used for coordination. Several of the signals in the various EIA standards were used specifically for handshaking (e.g., Request to Send, Clear to Send, Data Terminal Ready). Hardware handshaking enabled devices to signal their state and ability to receive (or desire to transmit) to the corresponding device at the other end of the channel. With today's "smart" devices, it is usually still possible to employ hardware handshaking; although it may also be possible to exchange short, special-purpose messages to establish inter-device coordination (this is called *software handshaking* or "X-On/X-Off" handshaking). In most cases today, the requirement (or option) to use one or the other form of coordination is a software or configuration issue. Nevertheless, in all instances, it is important that all devices are configured to use the same flow control handshaking mechanism.

The most common interface between a computer and, for example, a telephone modem is an EIA-232 connection with all of the hardware handshaking signals. The following paragraphs discuss some of these signals and how they are used.

Ensuring Operability

In most PCs (including laptop PCs), the serial COM ports are connected to a special integrated circuit, called a *UART*, that can send and receive 8-bit data values via an EIA-232 circuit. These UART chips interface to the various data and handshaking signals of the EIA-232 circuit and allow software to examine and set/clear the applicable signals. If a communications program (e.g., Microsoft HyperTerminal or open source PuTTY) is configured to examine and control those 232 signals, then hardware handshaking will be necessary. It may be possible to "fool" the software by jumpering some signals to others, to simulate hardware handshaking (e.g., by using a null modem adapter). The following are hardware handshaking signals:

- **DTE Ready** – Data Terminal Equipment Ready. This must be TRUE, which signifies that the DTE is ready to communicate.
- **DCE Ready** – Data Circuit-Terminating Equipment Ready (formerly known as DSR; Data Set Ready). This must be TRUE. Sent from the DCE to the DTE, this signal determines whether the DCE is in the DATA mode and ready to communicate and (as used in EIA-232) whether the DCE is authorized to set up a link once it has detected ringing.

Set Up Link

If it is desirous to use a dial-up circuit on the public switched network (analog), the handshaking signals should have the following states:

At the originating DCE

- **DCE Ready** – This state should be TRUE. This may be set at power-up or upon RTS (Request to Send).
- **RTS** – This state should be set to TRUE by the DTE when it is ready to transmit the message. At this point, the DCE dials up, establishes a connection, and so on.

At the answering DCE

- **RI (Ring Indicator)** – The Receive Line Signal Indicator changes to TRUE, which indicates that a call request is being made.

  Upon receiving the RI and after synchronization, the DCE Ready TRUE will be sent to the DTE. If non-switched (e.g., leased telephone) lines are used, RI is not used.

- **DCE Synchronize** – Both ends have DCE Ready as TRUE. The originating DTE has RTS TRUE. Then the originating DCE signals the DTE with CTS TRUE to begin transmitting. As long as DTE Ready is TRUE, then data will be transferred to the DTE.

  In practice, hardware flow control depended on only one set of the two handshaking pairs (DTE/DCE Ready or RTS/CTS) to be used for handshaking. To meet European standards (V.24 Circuit 133) in the E version of EIA-232, RTS (if used as Circuit 133) was set to TRUE as a ready to receive signal, rather than having a relationship with CTS. Contemporary transmission uses software, as opposed to hardware handshaking, while DCE/DTE Ready, along with RTS/CTS, are connected in the software to provide UART actuation.

Signals Required with Software Flow Control

- **TD (Transmit Data)** – Referenced to pin 7 (Signal Ground); the DTE outputs data to the DCE.
- **RD (Receive Data)** – Referenced to pin 7 (Signal Ground); the DTE receives data from the DCE.
- **Signal Ground** – The 0 V DC reference (signal return).

Synchronous Communication

If a device uses a UART for serial communications (as do all desktop and laptop PCs with COM ports), then it is performing *asynchronous* serial communications in which there can be gaps (variable-duration time delays) between data being sent. The data is sent as one or more octets (bytes), with a special **start** bit and **stop** bit appended to each octet at the sending end by the transmitting UART. The start and stop bits are removed and discarded at the receiving end by the receiving UART. Thus, all data is sent as a multiple of octets, with each carrying two additional special-purpose (non-data) bits.

Even at its most efficient, asynchronous transmission can only approach an 80% efficiency level (efficiency = data bits sent/total bits sent), thus, an alternate scheme was developed: *synchronous* transmission. With synchronous transmission, there are no start or stop bits and messages can be of any bit-length without needing to be an even octet multiple. The HDLC (high level data link control) protocol is a commonly used synchronous protocol, in which there is a unique 8-bit pattern (01111110) called a *flag character* used to identify the start and end of a message (a frame).

Within a message (frame), there cannot be a bit sequence that looks like the flag character pattern. The transmitting side's Data Link layer keeps track of consecutive 1 bits in the message and any time there are five 1s sent in a row, the transmit Data Link layer inserts a 0 bit into the message (which will be removed and discarded at the receiving end—a process called *bit stuffing*) to prevent the receiver from mistaking the data as a flag character.

Specialized transmitting and receiving hardware is needed for synchronous communications because the UART chip in a typical PC cannot support synchronous transmission. With asynchronous transmission, the start bit on each data octet provides a synchronization event for the receiver's clock. With synchronous transmission this is not possible, so the various interface specifications allow for a clock signal source.

Synchronous data transmission is commonly found today in all high-speed local area networking technologies. A special way of encoding data, called *Manchester Encoding*, provides a way to embed a clock signal within the transmitted data to allow continuous synchronization between the transmitter and receiver, even at extremely high bit rates (e.g., 10 Gbps [gigabits per second]).

Clock Source in a Synchronous System

The timing in a synchronous system is bit-oriented. There are no start and stop bits; either the DTE or the DCE supplies the transmitter clock (which is determined by bit timing).

If the DTE supplies the clock

(E)TC External Transmitter will be used to clock the transmit bit timing.

If the DCE supplies the clock
TC Transmitter Signal Element Timing will be used to clock the transmit bit timing.

In either case
RC Receiver Signal Element Timing will be used to clock out the received data. The receiver clock is generally synchronized to the transmitter bit timing by the signal transitions of data.

PC Serial Communications

Just as EIA-232 use was greatly expanded by the personal computer (PC), other buses, internal and external to the computer, have also appeared. Some are actually networks themselves, rather than just an extension of an internal bus. For a long time, external line speeds required only the EIA-232, then the EIA-422/485 connections or perhaps the SCSI extended bus. However, much higher speeds and the constraints media imposed on these signals required a rethinking of the serial signal organization and the software that drives it. USB was the answer.

Universal Serial Bus

The Universal Serial Bus (USB) was originally developed in 1995 by industry-leading companies. The concept was to define an external expansion bus that made adding peripherals to a PC as simple as plugging in a network jack. The design goals were low cost and true "plug-and-play" operation. Both were achieved by using an external expansion architecture. Figure 3-14 illustrates a laptop computer with a root hub showing connections to a USB hub, a printer, a mouse, a wireless access point, and a ready-to-connect flash drive.

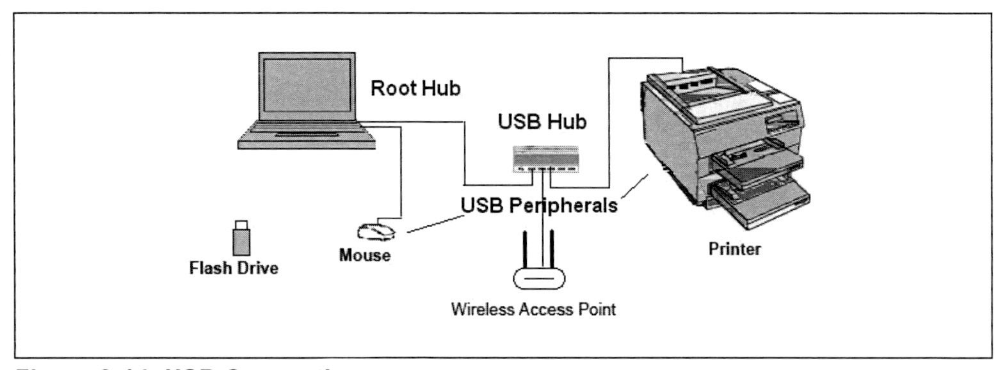

Figure 3-14. USB Connections

USB is currently in version 3.0 (announced in 2008) and runs at speeds approaching 5 Gbps (super speed – SS). It is backward-compatible with the 1.x and 2.0 version devices that run at either 480 Mbps (high speed – HS), 12 Mbps (full speed – FS), or 1.5 Mbps (low speed – LS). External hubs (counting the root hub) can be nested up to five deep and USB allows up to a maximum of 127 devices (including the root hub). Most PCs today come with two, four, or six (and sometimes more) USB ports. (Note that often only one of these ports will support HS and the remainder will support only FS, or even LS speeds.)

One distinct advantage USB has over the older technologies is that it is truly plug-and-play. You may have to load a driver; however, if you use the Windows® XP operating system, any of the Windows Server versions, Windows Vista operating system, Windows 7 operating system, or Windows 8 operating system, you will find software support for all the USB 2.0 specifications. In addition, USB 2.0 was modified to allow peer-to-peer operation. Therefore, a PC is not necessary for some device-to-device transactions. Linux® has had USB support for 2.0 since at least 2005. Additionally, USB is hot pluggable; that is, capable of being removed or inserted with the device power on.

USB version 3 operates at a higher speed and is the newest release. Software and hardware for USB 3.0 are still evolving, although all manufacturers intend to add version 3.0 capabilities. Apple®, Microsoft (as of Windows 8), and Linux (version 2.6.31 and later) support USB 3.0 functionality, although the performance depends upon the underlying PC hardware. There are PCI Express cards available to add USB 3.0 hardware to PCs.

The USB attachment cable may have either a Type-A or a Type-B connector. A Type-A port is usually found on the PC, and Type-B ports are generally on the devices. As smart devices shrank, there was a need to reduce the connector size and so a "Mini," and then a "Micro," version of the A and B connectors were introduced. Figure 3-15 illustrates the various 1.x/2.0 connector types (Micro-A is not shown). For ease of viewing, the connector images in figure 3-15 are not to scale.

There are several physical differences between the USB 3.0 connectors and the USB 1.x and 2.0 connectors. Version 3 introduced a new Type-A connector with a second row of pins for USB 3.0 connectivity, as well as a basic set of pins for USB 1.x and 2.0 compatibility. Note that the Type-B and Micro-B con-

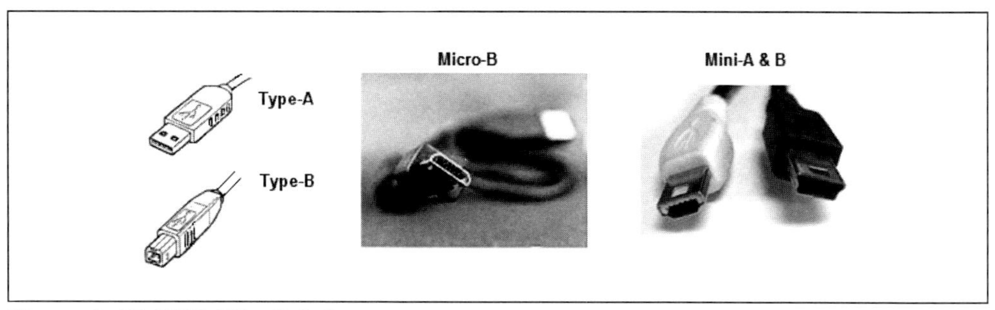

Figure 3-15. USB V1.x/2.0 Connectors

nector and port versions in USB 3.0 are NOT backwards compatible with the USB 1.x and 2.0 versions.

These connectors and ports are physically incompatible with the USB 1.x and 2.0 connectors and cannot be intermixed. Because USB 3.0 has not (yet) found its way into industrial applications, we will focus only on the standard USB 1.x/2.0 Type-A and Type-B connectors, whose pinouts are shown in figure 3-16. Note that as of the Mini and Micro connector versions, there are five pins rather than the original four pins in the USB 1.x standard connector.

USB 1.x/2. Standard Pinouts

Pin	Name	Cable Color	Description
1	V_{BUS}	Red (or Orange)	+5 V
2	D-	White (or Gold)	Data -
3	D+	Green	Data +
4	GND	Black (or Blue)	Ground

Note: Mini and micro connectors have five pins and Ground is on pin 5, whereas pin 4 is named ID and used to differentiate host/device.

Type-A Type-B Mini-A Mini-B Micro-A Micro-B

Figure 3-16. USB Type-A and B Pinouts

One of the most powerful aspects of USB is the ability of a PC to automatically detect the connection of a device to the USB LAN and then to interrogate the device and identify its class code (available functionality) so that the appropriate driver can be loaded to interface with the device. The standard defines a large number of device types, each with a class code number to identify them, as shown in table 3-5.

Table 3-5. USB Device Class Code Numbers

Class	Use	Description	Examples or Exceptions
00h	Device	Unspecified	Device class is unspecified, interface descriptors are used to determine needed drivers
01h	Interface	Audio	Speaker, microphone, sound card, MIDI
02h	Both	Communications and CDC control	Modem, Ethernet adapter, Wi-Fi adapter
03h	Interface	Human interface device (HID)	Keyboard, mouse, joystick
05h	Interface	Physical interface device (PID)	Force feedback joystick
06h	Interface	Image	Webcam, scanner
07h	Interface	Printer	Laser printer, inkjet printer, CNC machine
08h	Interface	Mass storage (MSC or UMS)	USB flash drive, memory card reader, digital audio player, digital camera, external drive
09h	Device	USB hub	Full bandwidth hub
0Ah	Interface	CDC-data	Used together with class 02h: communications and CDC control
0Bh	Interface	Smart card	USB smart card reader
0Dh	Interface	Content security	Fingerprint reader
0Eh	Interface	Video	Webcam
0Fh	Interface	Personal healthcare	Pulse monitor (watch)
10h	Interface	Audio/video (AV)	Webcam, TV
DCh	Both	Diagnostic device	USB compliance testing device
E0h	Interface	Wireless controller	Bluetooth adapter, Microsoft RNDIS
EFh	Both	Miscellaneous	ActiveSync device
FEh	Interface	Application-specific	IrDA Bridge, Test and Measurement Class (USBTMC) protocol, USB DFU (Direct Firmware Update)
FFh	Both	Vendor-specific	Indicates that a device needs vendor-specific drivers

The PC acts as the USB bus "master" and uses a periodic poll to both service devices and to identify changes in the LAN configuration (e.g., device connection or removal). USB implements connections to storage devices using a set of standards called the *USB mass storage device class* (MSC or UMS). This was, at first, intended for traditional magnetic and optical drives and has been extended to support USB flash drives. It has also been extended to support a wide variety of devices that have internal file storage (e.g., digital cameras, MP3 players, printers, etc.).

At the moment, USB, aside from PC workstations and their peripherals, has not achieved a significant presence within industrial settings. That said, however, there is an area where USB has begun to impact (both good and bad) industrial use: the flash drive (pen drive, USB drive, etc.). The flash drive has all but eliminated the use of floppy disks, as well as the CD and the DVR for inter-device data transfer. You may currently find flash drive devices ranging from 8 GB to 128 GB at affordable prices (and downright cheap if you look back only a few years). The upside is that most maintenance technicians can now carry around with them the diagnostics and software for any number of uses (no more floppies—yay!). The downside is that flash drives make software and data theft, and the spread of malware, much easier to accomplish.

IEEE 1394

IEEE 1394 is a fast external bus standard that supports data transfer rates of up to 400 Mbps (in 1394a) and 800 Mbps (in 1394b). The standard allows for rates of 1.6 and 3.2 Gbps. Different manufacturers use the 1394 standard but do so under different names. Apple (the original developer) uses *"FireWire,"* a trademarked name. Sony has trademarked *i.link*.

IEEE 1394c-2006 (FireWire S800T)

IEEE 1394c-2006 was published on June 8, 2007. It provides 800 Mbps over standard Ethernet connectors using Cat 5e cable (copper unshielded twisted pair), along with automatic negotiation to allow connection to either IEEE 1394 or IEEE 802.3 devices. While this appears to be a lofty goal, as of November 2008, no products or chipsets include this capability.

IEEE 1394-2008 (FireWire S1600 and S3200)

This classification of products (scheduled to be available before the end of 2008) uses the S1600 and S3200 modes that compete with USB 3.0. The devices use the same 9-conductor connectors as existing FireWire 800 and are fully compatible with existing S400 and S800 devices. As of 2012, there were few S1600 devices released, with a Sony camera being the only notable user.

IEEE 1394d – Proposed (Future Enhancements)

This was a project to add single mode fiber as an additional transport medium to FireWire.

While future iterations of FireWire are expected to increase speed to 6.4 GBps and use additional connectors, such as the small multimedia interface, it will have to survive in the marketplace against USB 3.0. Industrial applications are generally limited (at the present) to machine vision and digital recording applications.

A single 1394 port can be used to connect up to 63 external devices in a daisy-chain connection, in which each device is connected to the previous device. IEEE 1394 supports *isochronous* data, which means it delivers data at a guaranteed rate. This would appear to make it ideal for industrial control devices that need to transfer data in real time. The IEEE 1394 protocol uses an 8B/10B signaling scheme, where 8-bit combinations are represented by 10-bit patterns. This minimizes the number of consecutive 0s allowed in a data stream. This scheme means, however, that the protocol is only 80 percent efficient, in terms of the line data rate.

Although fast, IEEE 1394 is relatively expensive, particularly when compared to USB. Like USB, 1394 supports plug-and-play and hot plugging. Various cables are keyed so devices cannot be wrongly connected together. The original 1394a used a 6-pin (supplying power) or 4-pin (no power) connector, while 1394b uses a 9-pin connector.

IEEE 1394 appeared to be a promising bus for industrial networks and its adherents insisted it was competitive with Ethernet. However, it appears that the success of USB 2.0 (and now 3.0), along with high-speed Ethernet, has made this technology largely obsolete.

SCSI (Contemporary Use Only)

SCSI (Small Computer System Interface) has been around since 1978. It has a high data transfer rate and can be used as an external bus. It is, however, more expensive than USB, IEEE 1394, or Serial ATA (known as SATA), now in Version 3.2 (a very-high-speed serial interface for hard drives).

SCSI has moved in the past several years from a parallel configuration to a serial connection, although the Ultra 320 parallel still transfers data at a higher rate over a longer distance than the SCSI serial versions. The serial versions allow for more devices by a factor of six or more.

SCSI's primary use is as a multiple disk interface, particularly in fault-tolerant arrays, such as Redundant Array of Independent Disks (RAID) in configurations as RAID 10 or RAID 50. It is not normally included in modern PC design and must be added through an expansion card.

Table 3-6. High-Speed Serial Hard Disk Interfaces

Name	Bus	Throughput	Max. length	Max. Devices
Ultra3	Parallel	160 Mbps	12 m	16
Ultra-320	Parallel	320 Mbps	12 m	16
SSA	Serial	40 Mbps duplex	25 m	96
SSA 40	Serial	80 Mbps duplex	25 m	96
FC-AL 4 Gb Fiber Channel Arbitration Loop	Serial	400 Mbps duplex	3 m	127
SAS 3 Gbit/s	Serial	300 Mbps duplex	6 m	16 K

Not one of the versions of the SCSI standard has ever specified the kind of connector that should be used with a particular interface.

SATA

SATA (Serial ATA, ATA is AT Attachment) replaces the parallel ATA (now known as PATA) for connecting EIDE (Extended Integrated Disk Electronics) drives to PC motherboards. SATA uses a 7-pin serial conductor for signal and a 15-pin connector for power. These are specifically keyed so they cannot be connected incorrectly.

SATA currently has data rates from 0.5 Gbps and 16.0 Gbps, however, these data rates have a maximum length of 1 m (3.3 ft.); therefore, SATA should not be considered for long external peripheral connections. eSATA (external serial ATA) should be used for distances up to 2 m (6.6 ft.).

Summary

This chapter considered some of the more commonly used serial data communications interface standards. Essentially, all data in modern PCs and their peripherals is transmitted serially. A modem transmits data serially; most networks transmit data serially. There are standards and industry associations for most of these serial architectures.

This chapter has shown how a standard, RS-232, now TIA/EIA-232-F and originally rather limited in scope, has kept pace with technology. Designed in the early 1960s, it is still in wide use. RS-232 was extended to the D suffix, then changed to TIA/EIA-232-E and then F to meet international standards. It will be around for a while as an unbalanced (to ground) standard. EIA-442/423 and EIA-485 are standards that describe the electrical characteristics of transmitters (generators) and receivers. EIA-485 has electrical characteristics similar to EIA-422, except that it supports contention—up to 32 transceivers—and is used as a network. EIA-530 is a mechanical standard for a balanced-to-ground signal system that uses a connector identical to RS-232, except in pinouts.

The evolution of the PC from curiosity to mainstay has brought with it a number of connectivity standards. Although Ethernet (explained in detail in the next chapter) is a serial technology, others, such as USB (now in 3.0), SCSI, and SATA, will play an important part in the industrial arena as PC technology moves into this environment.

All of the standards discussed here are in everyday use and, increasingly, in industrial applications where simple, low-cost, reliable communications are required.

Bibliography

Axelson, J. L. *Serial Port Complete.* Madison: Lakeview Research, 1998.

Electronic Industries Association. *EIA RS-232-C.* Arlington: EIA, 1968.

---. *EIA-422-A.* Arlington: EIA, 1972.

---. *EIA-423-.A* Arlington: EIA, 1972.

---. *EIA-499-A.* Arlington: EIA, 1972.

---. *EIA-485.* Arlington: EIA, 1982.

---. *EIA-530.* Arlington: EIA, 1986.

---. *EIA RS-232-D.* Arlington: EIA, 1987.

---. *TIA/EIA-232(-E.* Arlington: EIA, 1990.

---. *TIA/EIA-232.F.* Arlington: EIA, 2000

Thurwachter, C. N. *Data and Telecommunications Systems and Applications.* Upper Saddle River: Prentice Hall, 2000.

Wikipedia. Information on subminiature D connectors. Current as of 2014.

Wikipedia. Discussion of FireWire (IEEE 1394), USB and SATA. Current as of 2014.

4 Local Area Networks

How We Got Here

Thus far, we have looked at some communication models and at some of the prevalent serial communications standards. In the natural progression from stand-alone to networked systems, we now come to a fork in the road.

All data communications were originally point-to-point, much as airplane travel was before the advent of the airport as a hub. There were established networks—telegraph, teletypewriter, telephone, television, railroads, and interstate highways—but most data communications were performed over a point-to-point link. Even if those links were made into a network, they didn't have a shared medium of transmission.

Data communication in the 1970s was a costly venture, as bandwidth, disk storage, and printers (of any kind) were expensive. Due to their cost, high-bandwidth lines were (and are) often shared. When facing similar pressure from resources costs, airlines found that they could achieve greater efficiency and reduce expenses for both the passenger and the airline by feeding into a hub and then transporting passengers from hub-to-hub. Likewise, telephone companies improved their efficiency by using central offices, as opposed to direct subscriber-to-subscriber connections.

Data communications followed a similar pattern by utilizing wide area data networks to increase efficiency and reduce costs. Wide area data networks use trunk lines (analogous to airline flight paths and phone lines) and have "store and forward" switches (analogous to hub airports and switchboards). However, bad things can happen when sharing (air travel, telephone, or bandwidth) networks: weather, mechanical problems, and traffic jams can result in reduced efficiency.

The aforementioned fork in the road comes at this point: do we want to network locally (as in a plant or office area) or do we want to go through the public (wide area) network? Do we want to utilize a local area network (LAN) or a wide area network (WAN)? Do we want to process information in the "cloud" or locally? In this chapter, we discuss the basics of local area networks.

Definitions

For the past 25 years, the process industries have had large numbers of "smart" devices that communicate electrically to monitor or control a process or processes. The combination of such devices and the electrical connectivity that enables them to communicate is, by definition, a network. However, industrial networks are referred to by many different names (some are trade names): data highways, fieldbuses, distributed control systems (DCSs, the network puts the "distributed" in DCS), and so on. Most networks are named after their communications infrastructure or their function. Industrial LANs seem to be everywhere and every instrument manufacturer has a proprietary variant. In addition to the vendor-proprietary LANs, there are now several standards for industrial LANs, including: FOUNDATION Fieldbus (H1 and HSE), PROFIBUS and PROFINET, plus Modbus/TCP, and EtherCat. An administrative or central data collection computer or even a process control computer with smart terminals, smart devices, and programmable controllers connected to it, constitutes a local area network.

Programmable controllers can be used to form a distributed control system (DCS) when they are connected by a local area network (also known as a *data highway*) with a third-party human-machine interface (HMI) to provide integration and a graphics interface. Any number of manufacturers offer DCSs with differing degrees of smart peripherals. These, too, are local area networks. So, what is the definition of a local area network?

There are several. We prefer the following definition: If a company owns the medium of transmission, the infrastructure, and three or more devices that are in communication, it has a LAN. An alternate definition of local area network is this: a system with more than three nodes that uses a protocol with defined rules and performs a specified set of functions. It may be suggested that almost any network of more than two devices qualifies. In many ways, that suggestion is almost correct. What, then, prevents this definition from describing a wide area network (WAN), such as the telephone system? In most cases, in a WAN the medium or infrastructure is leased, rented, or otherwise not owned by the company.

A commonly used method for determining network type (although not necessarily technically correct in some cases) is to differentiate network types based on geographic coverage. Thus, we have networks geographically typed as: PANs (personal area networks), LANs, MANs (metropolitan area networks), and WANs. However, a case can be made that a LAN can include WAN nodes (particularly through remote access services [RAS]). A LAN can be described as one that is not bounded by geographic lines but by functional lines. An example would be a network used in an enterprise resource planning (ERP) system. Regardless of the geographic distances covered by the devices connected into the network, it has one main function: the control of production at that facility.

Originally, there were only two forms of networks: LANs and WANs. The underlying technologies used for each were quite different and made the differences pronounced. LANs were geographically distance-limited but could operate at (relatively) high bit rates. WANs could span large geographic distances but usually offered low bit rates. Almost all WANs were based on telephone company phone system technologies, the analog forms of which offered very limited bandwidth. Today our WAN technologies (e.g., Synchronous Optical NETwork [SONET], Asynchronous Transfer Mode [ATM], and Frame Relay) operate a very high speeds, as well as over great distances, making the operational differences between WAN and LAN less clear.

Today, networks can be both wired and wireless and they tend to fall into four categories, based on geographic coverage, with various given technologies being predominant in each category, as explained in table 4-1.

Table 4-1. Network Types

Network Category	Geographic Coverage	Predominant Technologies
PAN – Personal Area Network	0 to 2 m	Bluetooth, USB, Ethernet
LAN – Local Area Network	2 to 300 m	Ethernet, Wireless Ethernet (Wi-Fi), Mesh wireless networks
MAN – Metropolitan Area Network	1 to 20 km	ATM, SONET, WiMAX, FDDI
WAN – Wide Area Network	50 to 4000 km	SONET, ATM, Frame relay

IEEE 802 LAN Model

The IEEE 802 model is a Layer 1 and 2 model concerning itself, from inception to this day, with variable size packet (as opposed to fixed cell, i.e., ATM) networks.

One should first look at the Open Systems Interconnect (OSI) ISO 7498 model and the IEEE (Institute of Electrical and Electronic Engineers) 802 model, to discover how these models complement one another. The 802 standard describes network functions (see figure 4-1).

(Note that the 802 model is believed by many to be named for February 1980, when the committee was first established—others state it was the next available IEEE number.)

You will note from figure 4-1 that there are only two OSI layers of function; Layer 2 is subdivided into two sublayers. These are the Logical Link Control (LLC; described in IEEE 802.2) and the Media Access Control (MAC; described in most of the other standards, such as 802.3 and 802.5).

IEEE 802.2 (LLC sublayer) describes how to interface externally to an 802 system, how to bridge (a Layer 2 function dividing segments of a larger network), and so on. The MAC standards are concerned with who talks and when, how you address each network device on the media, and how you frame the data for transmission. The current 802.2 standard also provides for the sharing of a LAN by multiple protocols and services, each having different upper-layer protocol "stacks"—ISO model Layers 3 through 7 (e.g., Transmission Control Protocol/Internet Protocol [TCP/IP] or Internet Protocol version 4 [IPv4] and Internet Protocol version 6 [IPv6]), with the ability to route messages to the right protocol stack. It also provides for messages that are link-

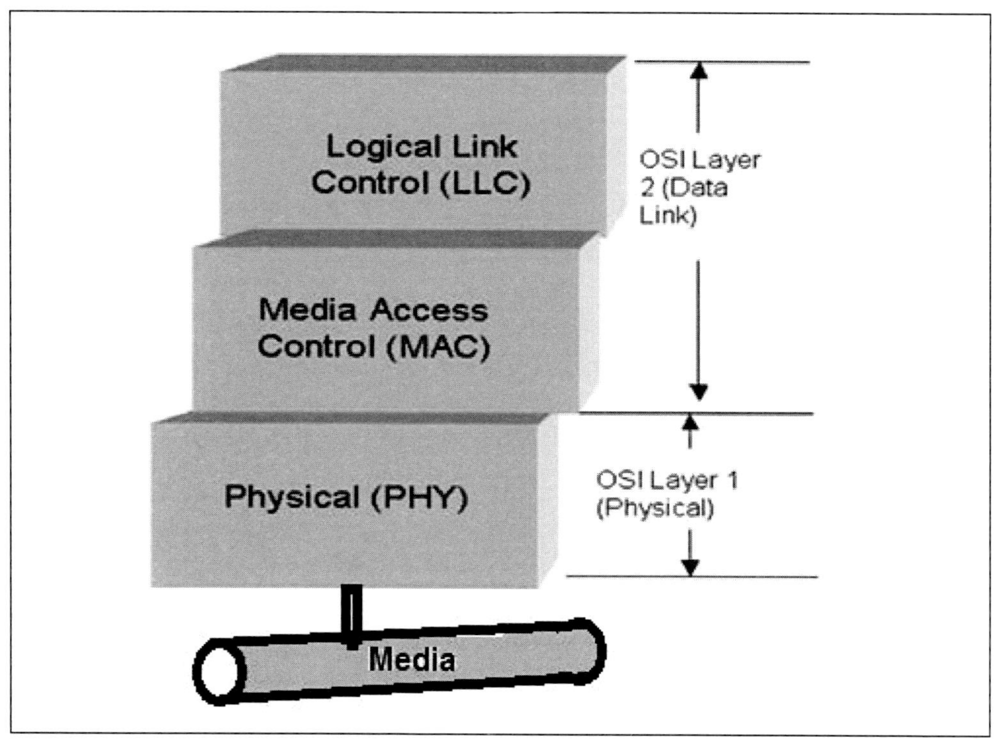

Figure 4-1. The IEEE 802 LAN Model

layer specific to LAN components, such as switches and routers, and do not get delivered to an upper-layer protocol stack (e.g., Link Layer Discovery Protocol [LLDP], Rapid Spanning Tree Protocol [RSTP], etc.)

The Physical layer (Layer 1) is responsible for placing data on the medium, extracting data from the medium, and synchronizing the network devices.

An underlying function of any network (WAN included) is that it is shared by multiple intercommunicating parties. This means that message traffic flowing over the LAN medium will be from any of a number of sources and directed to any of a number of destinations. This brings us to the basic concept of how to share medium. There are two basic methods for sharing a common medium (there are others, but these two are the most basic): (1) creating separate subchannels on the medium and assigning them to different intercommunicating parties; (2) or taking turns using the full bandwidth of the medium. These two approaches are technically called *broadband signaling* and *baseband signaling*; they are described next.

Layer 1, The Physical Layer

How data is placed onto and taken off of the medium determines the type (broadband, carrier band, or baseband) and method for sharing the network. In general, if modulation-demodulation is used—that is, if it uses a carrier signal and data modulation/demodulation—it is a broadband or carrier-band system (carrier band is a type of broadband system). If there are multiple modulated carrier signals placed concurrently on the channel and spaced apart at different frequencies, or if separate time slots are observed for different communicating parties, this is a broadband system. If data is placed digitally (even though it is conditioned) at a bit rate that approaches the bandwidth capacity of the channel, it is a baseband system.

Broadband

Broadband refers to separating signals by frequency (a process called *frequency division multiplexing* [FDM] or *analog broadband*) or time (*time division multiplexing* [TDM] or *digital broadband*). The main bus (trunk) connects the devices in the network and it will have a large number of different signals on it. Broadband services can include network technology for multiple LANs on one trunk or they can consist of carrier-supplied services, such as Integrated Services Digital Network (ISDN) and variations of Digital Subscriber Line (xDSL), which are supplied to networks from a WAN provider. The bulk of the technology found in an analog frequency division, multiplexing (broadband) LAN evolved directly from cable TV systems, right down to the connectors and the 75-ohm cable. Analog or digital broadband trunks offer a multitude of services—data, voice, and television—all on the one cable.

Using FDM networks on a broadband trunk requires two channels for each network: a transmit channel and a receive channel. All addresses on the network transmit on the transmit frequency and all addresses on the network receive on the receive frequency. Some means of converting the transmit frequency signal is required for the receivers to detect it. This is the function of the head-end remodulator (a device which takes in information on the transmit channel and rebroadcasts it on the receive channel). Having receive and transmit channels means the bus is directional; that is, all transmit frequencies are inbound (to the head end), and all receive frequencies are outbound (away from the head end). The bandwidth available to broadband is approximately 300 to 400 MHz. This allows transmission of quite a few network channels, as well as other services (such as alarm systems, security cameras, etc.). Just as you may have 200 analog television channels on cable TV (and significantly more on a digital cable system), you may also have a large number of networks on one cable. Since much of a network's capital outlay is tied up in the cabling and in its life-cycle cost, having a number of LANs on one cable may be a cost-effective solution.

On the other hand, the more modern cable and satellite TV systems use digital techniques. The video signal is converted to digital and is then compressed (far easier to do with a digital signal than an analog signal). A variety of modulation schemes are applied (same ones used in data modems to increase the data throughput of a voice channel). As it is, six (or more in some cases) standard definition (SD) TV channels (or two high-definition [HD] channels) can fit into one of the older 6 MHz analog TV channels. Indeed, channels now have a channel number (same as the old analog channel) and a sub-channel designation that identifies which of the six (or two) sub channels they are transmitting on.

Baseband

In baseband transmission, there is only one digital signal on the bus or shared medium. Separation for more than one network is achieved by physical spacing (i.e., using different cables.) Most baseband systems are now run on unshielded twisted-pair (UTP) cable, however, a few still run on coaxial cable. In the industrial environment, they may be run on fiber optic links. Switched media (which reads network addresses and assigns a destination port, as opposed to shared media, in which information is available to all receivers) has gained acceptance as a lower cost, high-performance alternative to statistical-based media access schemes (dependent upon the timing of stations contending for access). Switching is accomplished by using a fast-switching matrix, which is a set of electronic switches arranged so there can be a path from any port to another port depending upon a signal-contained address. The matrix provides a connection to individual nodes, akin to point-to-point technology. Users think they have a private line because the switching is done for only a few packets at each switch connection.

There are a number of commonly used methods for sharing a common channel or medium, called *media access*. These are almost always protocols but they might be ruled by convention. Earlier protocols had an appointed master and the nodes could only speak when spoken to. Some protocols allow each node to speak in a specified order; of course, this can end up wasting available bandwidth if these message origination times go unassigned. Think of this process as being like crossing a street using the traffic light. If you wait for the green light before crossing, even if there is no cross traffic, you are guaranteed an opportunity to cross within a minute or so, regardless of how much traffic exists. Another method of sharing is a contention method, wherein everyone is allowed to grab the channel for a short period of time if it is currently idle. This approach can lead to "collisions," when two (or more) devices attempt to initiate communications concurrently. This kind of sharing method is like getting to an intersection and going right across if there is no cross traffic or dashing across as soon as there is a break in traffic. If there is a lot of cross traffic, you may have to wait a while (possibly quite a while) to cross. And there is the possibility of a collision! The first method mentioned (specified order) is how polling works. Contention is how the original Ethernet worked.

Baseband technology has advantages:

1. No modem is required (although there must be some form of transmitter/receiver circuitry).
2. It is easier to install than broadband.
3. *The overpowering advantage* is that it is the least expensive per node (100M/1G/10G/40G/100G/400G bps) by a significant amount.

Compared to broadband, baseband has some disadvantages:

1. Limited capacity
2. Limited distance per segment

Both disadvantages are outweighed by advantage #3.

Carrier Band

Used in industrial environments, carrier band (a related technology) has only a single carrier on the bus, which is used to translate (in frequency) the digital signal away from a DC reference, i.e., to modulate the digital signal to the frequencies available with the medium. Carrier band is a broadband technology but it has some of the restrictions of baseband; notably, it only has a single channel. Carrier band eliminates all of the costs associated with multiple channel devices but also all of the capacity. Carrier band uses coaxial cable and for the most part has been replaced by other twisted pair schemes; it does not, at this time (2014), have a significant market share in instrumentation and control.

Topologies

Whether a system is baseband or broadband has an effect on the system's topology, or how the system is laid out. In practice, most broadband systems use the bus topology (described in the following paragraphs), while most baseband systems now use star topology. That is not to say that they are always that way, particularly if the medium is fiber optic, where the topology may be some form of point-to-point (daisy chain or ring) or star.

Topology Definitions

Network topology is the arrangement of the various links and nodes of a network, as well as the schematic of the various components within a network. A geographical topographic map shows elevations (the lay of the land) and the topology map for a network illustrates the "lay of the LAN" (as it were). While *topology* refers to the physical layout, a network systems' *architecture* refers to the *logical connections* between intermediary and end devices. Logical connections may differ significantly from the physical connections (the topology). The term *logical connection* usually refers to the way in which the network operating software views the dataflow of a device. The three major topologies used in LANs are illustrated in figure 4-2 (star), figures 4-3 and 4-4 (ring), and figure 4-5 (bus). Another topology often seen in modern LANs is the "tree," which is essentially a multi-level star (refer to figure 4-6). With the use of Ethernet switches arranged in a multi-tier configuration, the multi-level star topology is in common use today.

Star Topology

The star topology has been a fundamental computer network topology since the computer joined networking. As seen in figure 4-2, there is a central point (or star, analogous to the sun) and all devices (analogous to planets) connect only to that central point. In one of the first instances of the star topology (circa 1950), a central computer, usually a mainframe, time-shared all inputs and outputs. Time-sharing is another way of saying time division multiplexing (TDM), which was described previously. The computer's operating time is measured in instruction cycles. In a multi-tasking or multi-user system, each task (such as the terminal interface program) is assigned a time slot and a protected location in memory. Each time slot has its own program code, data, and the like. The computer uses an algorithm that schedules its resources; tasks are temporarily assigned a time slot to support each particular (in our case) terminal.

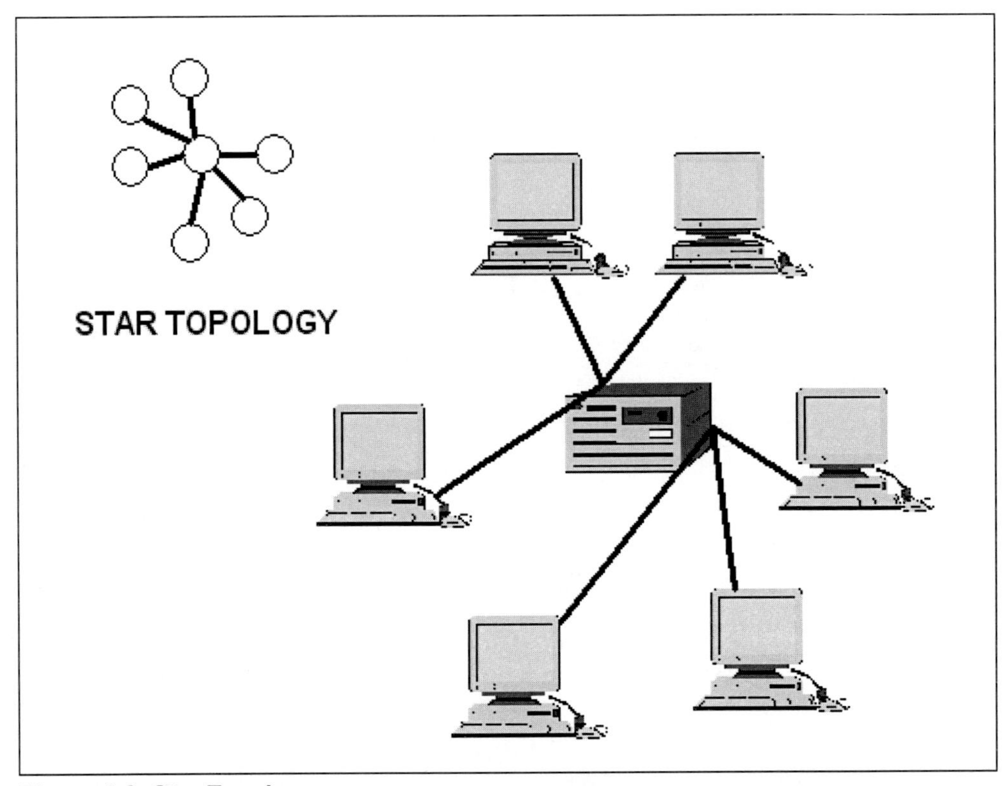

Figure 4-2. Star Topology

In many instances of star topology, all linked devices input to the computer and the computer outputs to all linked devices. The access method generally employed in this topology was called *polling*. At predetermined intervals, the central point, the device in control located at the center of the star (in this case, the time-shared computer), polls each node (terminal) to transmit as necessary during the scheduled time slot interval.

Note that with the star topology, any node at the end of an arm can only communicate with any other node through the central point. This draws attention to one critical problem with a star topology: a failure of the central device causes the loss of all communications. For this reason, many star configurations are built with some level of redundancy at the central device (e.g., Supervisory Control and Data Acquisition [SCADA] systems usually have a redundant central host).

The star topology is the way computer networks were arranged until the decreasing cost and size of electronics allowed processing power to be spread out among users. The star topology is identical to the topology of direct digital control (DDC) systems. These systems were originally hampered by the lower reliability of earlier computer systems and, except at the lowest level (closest to the terminal or end device), are not utilized in any significant way by industrial systems today. Presently, it is preferable not to be dependent upon a single entity in any system. Under present practice, loop controllers, regardless of whether there are one or eight loops controlled by the controller or even a programmable logic controller (PLC), end up being in a star topology: the controller or PLC connected to the field instruments. A modern Ethernet network system using switch technology is an example of a physical star network topology with the switch as the star; however, the logical topology will normally be that of a bus.

Ring Topology

A physical ring is just what it sounds like: it is comprised of point-to-point connections and each device in the network is in the path of data travel. Each device must be active on the ring or there must be some provision for bypassing the device in the event of failure or inactivity. Token rings (that is, rings in which there is a circulating pattern of data packets around the ring) simplify protocol and the maintenance of synchrony throughout the ring. Due to the fact that fiber optic cable is not open to easy line taps (no bus), terminations are difficult (due to the difficulty of ensuring a good optical connection), and most fiber connections are considered a point-to-point connection, the ring topology is ideal for fiber optic-based networks.

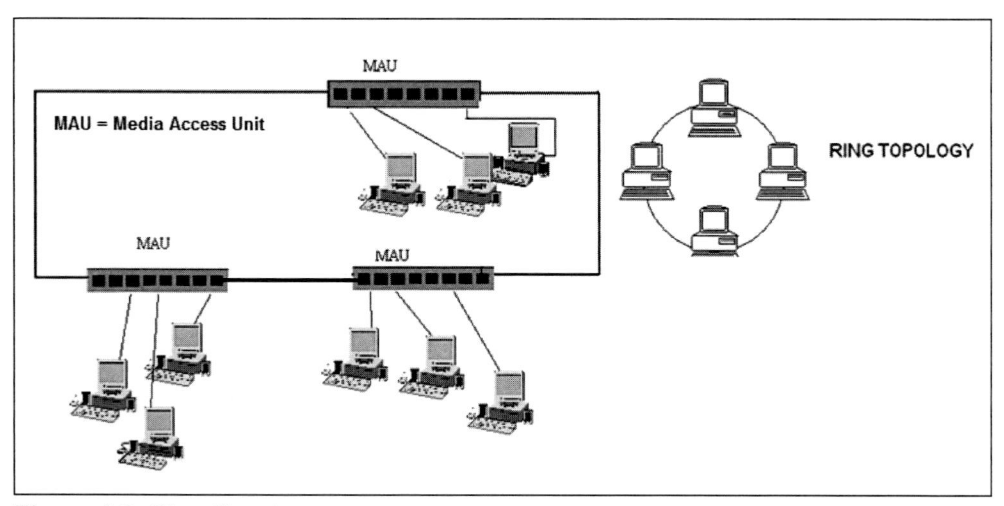

Figure 4-3. Ring Topology

The Media Access Unit (MAU) is the device connected in a ring. Each end of the MAU has a Ring In (RI) and a Ring Out (RO) connector. This is the physical ring, as opposed to the logical connections ring. The computers are connected (logical ring) through electronics that ensure each computer is on and ready for network traffic before being connected to the ring. To clarify, the MAUs are actually switches; that is, the MAU is the physical ring object and the individual devices are connected via the switch electronics. If a workstation computer is turned off or is otherwise removed, the ring remains unbroken due to the MAU being the physical device.

In most ring topologies there is always a message being transmitted, even if it is just the special "token" message being passed to the next node in the ring. A node's failure to "hear" a token (or other) message within a defined time window will cause some form of token regeneration to be activated. Nodes often have a time window based on their unique address, so one node will always reach its time window limit before the other nodes and it will start a new token message circulating. When a node receives the token, it can elect to pass it on (if it has no messages to send) or hold the token and send one or more messages.

When it is done sending messages, the active node sends a token message, which allows the next node in the ring to have a chance to send messages. This process continues until all nodes have had an opportunity to send messages, and then repeats endlessly. Messages passing around the ring are generally delayed one bit per node, so that when the current "transmitting" node starts sending, it will start to receive its own message after N bits have been transmitted (where N is the number of nodes in the ring.) This allows the transmitting node to check its own message and to send the message to all nodes almost simultaneously.

As the bits of a message pass through each node, they are copied into a local receive buffer, so that each node eventually has a copy of the full message to process. Ring systems do not send a full message from node to node to node, as the time required to send/receive the full message by all the nodes in the ring would introduce long time delays in getting the message around the ring to all other nodes.

There is a special type of ring topology that provides a degree of fault tolerance. A ring normally only needs a one-way pathway around all the nodes.

Each node has a simplex connection to the next node in the ring. In order to provide the ability to continue operating if a node dies or the ring medium is damaged or cut, there are various types of network hardware that support dual, counter-rotating rings (e.g., FDDI [Fiber Distributed Data Interface], ATM, and SONET). With counter-rotating rings, a loss of connectivity between successive nodes (e.g., a contractor with a backhoe cuts a cable, both the primary and the backup lines) will cause those nodes to go into "loop back" mode, where messages arriving on the primary ring will be sent back out onto the backup ring, effectively creating a new ring of twice the length. Figure 4-4 shows the difference between a simple ring and a dual counter-rotating ring.

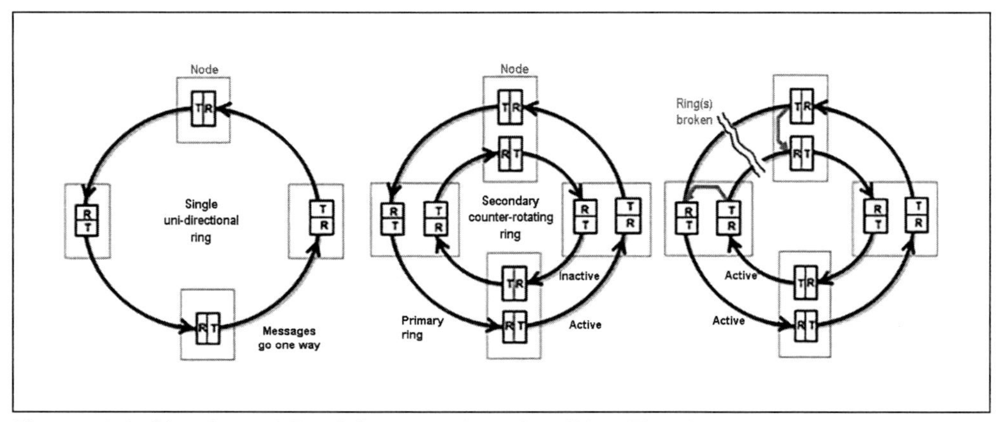

Figure 4-4. Simple and Dual-Counter Rotating Ring Topology

Bus Topology

A bus may be a single pair of conductors or it may look a great deal like the serial extension of a high-speed computer bus. In either case, a bus is a common physical connection to all nodes. The bus topology is used extensively in industry because of the installation flexibility it affords and because the failure of a node to bypass when it's inactive is not necessarily devastating to network operations. Most industrial bus systems used to be token passers and were used in a logical ring; that is, they were physically connected as a bus. However, the system software had them using circular token passing to control access, as if in a physical ring.

It should be noted that Ethernet (in the twisted-pair configuration) is a logical bus, yet the physical topology where the switch is connected to the nodes is a star. Original Ethernet was a physical bus based on coaxial cable with taps and drops to each node. In a bus, all nodes have the ability to directly communicate with all other nodes, with no need for message passing through some other node. A node on the bus can be turned off without shutting down the bus and preventing others from continuing to communicate. A bus usually allows for only half-duplex communications, as only one node can be transmitting and all the rest must be receiving. Collisions can occur if multiple nodes attempt to transmit concurrently. To prevent collisions, some bus systems use logical token passing as a means for controlling bus access (the right to transmit) by the various nodes.

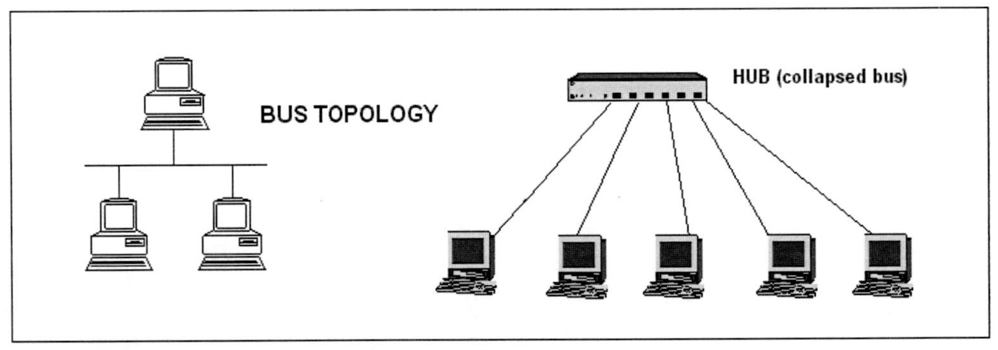

Figure 4-5. Bus Topology

This protocol application allowed a network to operate like a logical token ring. This scheme eliminated the bus contention/collision problems but does require additional software functionality to handle the token-passing logic.

Originally, all Ethernet was coaxial cable-based (1985). Coaxial cable-based Ethernet was eventually replaced by wiring hubs, where the drop to each node was a UTP cable with an 8p8c plug and connector (mistakenly called RJ-45 due to visual similarity—it is not electrically the same) at each end. A hub is a dumb device that electrically works much like a bus (half-duplex, collisions); the main difference is that a hub failure causes total communication failure because the hub is an electronic device. Today, hubs have been replaced by intelligent devices called *switches*. Ethernet switches are essentially powerful computers with large numbers of Ethernet network interface cards (NICs), whose function is to take incoming Ethernet frames from connected computers and pass them back out to their intended destinations based on the MAC addresses in the frames. In most cases, each connected computer gets a separate, full-duplex connection (via a UTP, Category6a cable) to a port on the switch. Ethernet switches can be interconnected to create a multi-layer (multi-tier) tree, in order to accommodate a large number of physically distributed nodes. Figure 4-6 shows a typical three-tier switch topology.

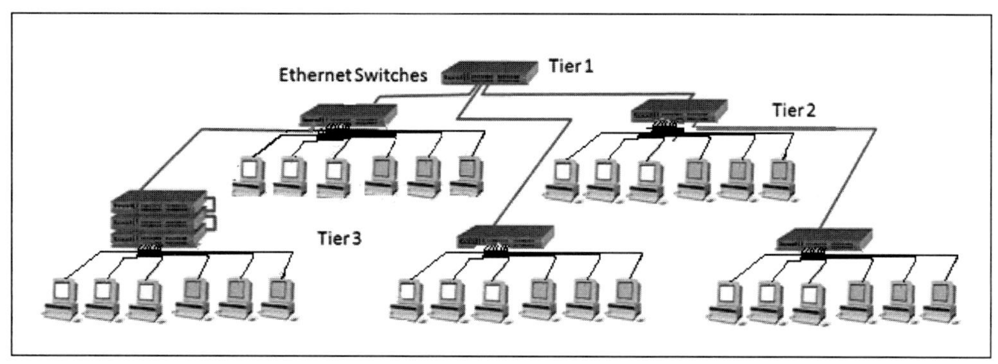

Figure 4-6. Tree (Multi-level Star) Topology

Transmission Media

There are basically three types of transmission media available: copper, fiber optic, and wireless. In the next section we will discuss Ethernet media, as all three types of media are currently recommended for Ethernet systems.

Ethernet Media

The traditional twisted-pair telephone wiring was investigated carefully by early network designers because of the tremendous savings that could be realized if the already installed media could be used instead of special cabling. However, even the first networks ran at speeds (3600 bps duplex, 1961, and later Ethernet [DIX] at 10 MBps, 1980) that required either specific adjustments to use with telephone media or required the use of different media altogether. At current network speed and interface requirements, the medium will almost always be a dedicated twisted-pair cable (Category 6a [Cat-6a] or higher), wireless, or fiber optic, rather than the installed telephone wiring (which is typically Category 1, 2, and 3). Categories are, for our purposes, defined in IEC 11801 and assign the maximum data rate a cable is capable of carrying over a stated distance. As Ethernet assumes a larger share of the industrial interconnect market, its Layer 1 specification, listed in table 4-2 for types 10, 100, 1,000, and 10,000 (10 Gb [gigabit]), will likely dominate. Table 4-2 lists the date accepted, the 802.3 committee responsible for the specification, and the data rate and use for selected Ethernet speeds.

Table 4-2. Selected 802.3 Amendments

Date	802.3	Description
1990	802.3i	10BASE-T, 10 Megabit over UTP
1993	802.3j	10BASE-F, 10 Megabit over fiber optic cable
1995	802.3u	100BASE-T, Fast Ethernet at 100 Megabit
1998	802.3z	1000BASE-X, Gigabit Ethernet over fiber optic cable
1999	802.3ab	1000BASE-T, Gigabit Ethernet over twisted pair
2003	802.3ae	10 Gigabit Ethernet, over fiber optic cable
2005	802.3-2005	Reissue to include previous approved committee amendments
2006	802.3an	10GBASE-T, 10 Gigabit Ethernet over UTP

Aside from IEEE 802.3 Ethernet, shown in table 4-2, other Layer 1 standards used in industry include:

- ISA/ANSI 50.02/IEC 61158-Types 1 and 3/FOUNDATION Fieldbus and PROFIBUS-PA
- IEC 61158 Type 2, ControlNet

802 and Industrial LANs

An observation about Layers 2 and above: Until recently, most industrial LANs were three-layer LANs, in which the layers were 1, 2, and 7 (the Application layer). They did not need Layer 3 (Network) because they did not intend to go anywhere but on their own network. Each LAN was an island of automation. Layer 4 (Transport) was not needed because most networks were connection-oriented and most error handling and packet sequencing was taken care of by Layer 2. All other error handling was taken care of by proprietary actions in Layer 7. Layer 5 (Session) was handled in the proprietary Layer 7 or perhaps by a real-time dispatcher (through Layer 7). Layer 6 (Presentation) was not needed, since all machines on these networks used a common syntax and data representation.

Having successfully conquered the office and business environment, Ethernet has continued on to assume a large share in the industrial arena. The IEEE 802 standard now encompasses more than local area networks, which are used generally as Layers 1 and 2 in many networking applications. Table 4-3 lists some of the IEEE 802 standards and standard subcommittees (as of 2014). This is by no means a complete list, just a selection of the main committees. Further information can be found on the web pages listed in the bibliography at the end of this chapter. Table 4-3 shows that what started out as a Layer 1-and-Layer 2 set of network definitions has been extended, yet 802 is still all about the Layer 1 and Layer 2 implementation and leaves the routing, reliability, and applications interfacing to the higher layers, which are defined elsewhere.

IEEE 802.3 is the main concern in this section. We discuss industrial systems, including FOUNDATION Fieldbus, extensively in chapter 6, "Industrial Networks and Fieldbuses." A discussion of IEEE 802.3 is essential here because of the increasing Ethernet applications in the industrial area. IEEE 802.3 specifically describes a baseband signaling system using legacy 50-ohm cabling and has been revised to include 100-ohm unshielded twisted pair (10BASE-T, 100 Mbps Ethernet–100BASE-T, 1000 Mbps Ethernet–1000BASE-T, and 10 Gb Ethernet [over fiber]).

IEEE 802.4 is not covered in this text (although it was in the first edition) since it is obsolete for industrial control (it was the basis of the Manufacturing Automation Protocol [MAP]). It specifically described three types of medium signaling—two broadband and one carrier band—and specified 75-ohm cabling for the trunk cable.

Although IEEE 802.5 has limited applicability in the industrial control area (only older, legacy systems), we have covered it here to illustrate the advantages and disadvantages of using a ring topology. Most industrial networks today are based on high-speed (100 Mbps or faster), switched Ethernet. Most (but not all) also use TCP/IP layered onto Ethernet to provide the network and transport functions. Some vendors place node limits on their systems to ensure deterministic performance. Others disallow third-party devices on their systems to prevent performance problems. As of today, all the PLC manufacturers and DCS vendors support Ethernet-based systems and products.

Table 4-3. Overview of IEEE 802.x Committees and Their Focus

Standard	Sub- Committee	Focus
802.1		Overview
	802.1d	Specifics for the spanning tree algorithm used in transparent bridging
	802.1p	Traffic class expediting and multicast filtering—allows Layer 2 switches to prioritize traffic at the MAC layer
	802.1q	VLAN
	802.1at	Stream Reservation Protocol (SRP) – QOS
802.2		Defines the Logical Link Control
802.3	-2005	CSMA/CD networks (2005 contains all previous amendments)
	802.3x	Full duplex and flow control
	802.3af	Ethernet-based telephony
802.4		Token-passing bus (legacy systems)
802.5		Token-passing ring
802.6		Metropolitan area networks (MAN)—a dual isosyn-chronous high-speed ring (inactive)
802.7		Broadband implementations (inactive)
802.8		Fiber optic network technologies (inactive)
802.9		Standards for integrated voice and data (inactive)
802.10		Interoperable LAN and WAN security (inactive)
802.11		Wireless LAN (Wi-Fi certification)
802.12		Demand priority access (the other 100 MB Ether-net)
802.13		Unused
802.14		Cable broadband implementations (inactive)
802.15		Wireless personal area network (PAN)
	802.15.1	Bluetooth
	802.15.4	IEEE 802.15.4-2003 Low-rate wireless personal area network (WPAN)
802.16		WiMAX (Wireless MAN)
	802.16e	(Mobile) broadband wireless access (Mobile WiMAX)
802.17		Resilient packet ring
802.18		Radio Regulatory Technical Advisory Group (TAG)
802.19		Coexistence Technical Advisory Group (TAG)
802.20		Mobile broadband wireless access (under study)
802.21		Media-independent handoff (changing seamlessly between different networks, e.g., Bluetooth to GSM to 802.11, etc.)
802.22		Wireless regional area network

Wireless LANS

An increasing number of IEEE committees are devoted to wireless transmission, as it has become a preferred method of installation where installing copper or fiber optic would be expensive, inconvenient, or impossible. The following are some of the wireless committees and their specifications.

802.11

The original 802.11 standard is now a legacy; there is no 802.11 standard anymore, just the amendments, due to the marketplace availability of much higher data speeds. The original 802.11 standard, approved in 1997, provided for 1 or 2 Mbps transmission in the 2.4 GHz band using either frequency hopping spread spectrum (FHSS) or direct sequence spread spectrum (DSSS). Although the standard allowed for either DSSS or FHSS, commercial implementations have primarily used DSSS, partly due to the efforts of Apple to promote that technology. FHSS is more commonly seen in other wireless applications, such as mesh wireless networks. A mesh network is comprised of multiple switching nodes (or routers) with multiple paths between each set of nodes (in other words, there is more than one route to arrive at the same destination). In Ethernet networks, the nodes will probably be routers and in the wireless networks described, the multiple nodes are routers.

802.11a

Standard 802.11a, approved in 1999, used the 5 GHz band, which provided up to 54 Mbps and used an orthogonal frequency division multiplexing (OFDM) encoding scheme rather than FHSS or DSSS. Though the 5 GHz band is better from an interference standpoint than 2.4 GHz (which is used for portable telephones, Bluetooth, and segments of the amateur band), the higher frequency does not provide the same range as 802.11b. Range has been greatly improved since that time and current adapters have nearly the same range as the 11b devices, although the inability to penetrate structures remains. Although 802.11a was approved at the same time as 802.11b, technological problems and the limited 5 GHz range prevented it from gaining acceptance in the marketplace as quickly.

802.11b

The first of the 802.11 amendments for which commercial product became available was 802.11b (sometimes called Wi-Fi). It extended the original 802.11 to provide 11 Mbps transmission (with a fallback to 5.5, 2, and 1 Mbps) in the

2.4 GHz band using DSSS. The revised standard, 802.11b, was accepted in 1999.

802.11g

In June 2003, 802.11g was ratified. It works in the 2.4 GHz band (as does 802.11b) but it has a maximum data rate of 54 Mbps (with a net throughput of about 27 Mbps) and is backward compatible with 802.11b. In mixed networks (both 802.11b and g components together), the data rate is significantly reduced (11 and 5.5 Mbps); alone in a network, 802.11g achieves high data rates of 6, 9, 12, 18, 24, 36, 48, and 54 Mbps by using OFDM. Unfortunately, like 802.11b, 802.11g is subject to the same interference in the 2.4 GHz range. The 802.11g standard has been highly successful in the commercial marketplace.

802.11n

At the beginning of 2004 the IEEE formed a new 802.11n task group to develop a new wireless modulation method capable of achieving a data rate of up to 540 Mbps. Proposed standard 802.11n was built on previous 802.11 standards by adding multiple antennas (multiple-input multiple-output [MIMO] technology) to make possible spatial multiplexing (using individual encoded data to separate transceiver antennas) and other coding schemes. Due to competing schemes, 802.11n took a while to develop but became a published standard on October 29, 2009. However, many suppliers produced "pre-n" products for the marketplace using draft standard technology and the promise to update firmware once the standard was finally approved. Due to the increasing popularity of "n" equipment, this was most likely accomplished successfully.

802.11ac

IEEE 802.11ac-2013, published in December 2013, is an amendment to IEEE 802.11 that builds on 802.11n. Changes, compared to 802.11n, include wider channels (80 or 160 MHz vs. 40 MHz) in the 5 GHz band, more spatial streams (up to 8 vs. 4), higher order modulation (up to 256-QAM [quadrature amplitude modulation—a combination of phase and amplitude modulation] the number describing the available states in one quadrant of a cycle vs. 64-QAM), and the addition of multi-user, multiple-input multiple-output technology (MU-MIMO). As of October 2013, high-end implementations support 80 MHz channels, three spatial streams (a technique in wireless communication to transmit independent and separate data signals—streams—from each of multiple transmit antennas) and 256-QAM, yielding a data rate of up to 433.3 Mbps per spatial stream, 1300 Mbps total, in 80 MHz channels in the 5 GHz band. Vendors have announced plans to release so-called *Wave 2* devices with support for 160 MHz channels, four spatial streams, and MU-MIMO in 2014 and 2015.

802.11ad

IEEE 802.11ad is a published standard (December 2012) that defines a new physical layer for 802.11 networks to operate in the 60 GHz millimeter wave spectrum. This frequency band has significantly different propagation characteristics than the 2.4 GHz and 5 GHz bands where Wi-Fi networks operate. Products implementing the 802.11ad standard are being brought to market under the WiGig brand name. The certification program is now being developed by the Wi-Fi Alliance, instead of the now defunct WiGig Alliance. The peak transmission rate of 802.11ad is 7 Gbps (gigabits per second).

802.11af

IEEE 802.11af, also referred to as "White-Fi" and "Super Wi-Fi," is a standard, approved in February 2014, which allows WLAN operation in the TV white space (unused) spectrum in the VHF and UHF bands between 54 and 790 MHz. It uses cognitive radio technology to transmit on unused TV channels, with the standard taking measures to limit interference for primary users, such as analog TV, digital TV, and wireless microphones. Access points (APs) and stations determine their position using a satellite positioning system, such as GPS, and use the Internet to query a geo-location database (GDB), provided by a regional regulatory agency, to discover what frequency channels are available for use at a given time and location. The physical layer uses OFDM and is based on 802.11ac. The propagation path loss, as well as the attenuation by materials such as brick and concrete, is lower in the UHF and VHF bands than in the 2.4 and 5 GHz bands, which increases the possible range. The frequency channels are 6 to 8 MHz wide, depending on the regulatory domain. Up to four channels may be bonded in either one or two contiguous blocks. MIMO operation is possible with up to four streams used for either space-time block code (STBC) or multi-user (MU) operation. The achievable data rate per spatial stream is 26.7 Mbps for 6 and 7 MHz channels and 35.6 Mbps for 8 MHz channels. With four spatial streams and four bonded channels, the maximum data rate is 426.7 Mbps for 6 and 7 MHz channels and 568.9 Mbps for 8 MHz channels.

802.11ah

IEEE 802.11ah is (at the time of this writing) an emerging standard that defines a WLAN system operating at sub-GHz license-exempt bands (e.g., the 900 MHz bands), with final approval slated for March 2016. Due to the favorable propagation characteristics of the low frequency spectra, 802.11ah will provide improved transmission range, compared with the conventional 802.11 WLANs operating in the 2.4 GHz and 5 GHz bands. 802.11ah can be used for various purposes, including large scale sensor networks, extended range hotspots, and outdoor Wi-Fi for cellular traffic offloading, although the available bandwidth is relatively narrow due to the lower frequency carrier used.

All the preceding WLAN technologies are essentially used to bridge one or more wireless clients to a wired LAN, or possibly to each other, via a *wireless access point* (WAP; also called a *wireless router*, *wireless bridge*, or *hot spot*, the commercial name for a WAP). An access point allows a user to connect to the wireless network, as does a "hot spot." While a router may have access point capabilities, it is concerned with routing packets to various other routers (access points) and is a Layer 3 device (modern routers also contain an Ethernet switch, generally 4 ports or less). A wireless bridge connects two (or more) wireless systems together as one larger network and does not route—consider it a network extender—a Layer 2 device. They are designed to operate over moderate distances (5 to 200+ m) and provide a bit rate comparable to wired LANs (although the available bandwidth is shared by all the clients.) Figure 4-7 shows how APs can be integrated into a wired LAN.

Because communications are radio based, in order to minimize errors due to poor signal strength, all these technologies will drop their bit rate to a given client as the measured signal strength of the client's transceiver decreases. As a result, at the maximum range from the access point, the actual bit rate will be much lower than the rates specified in the paragraphs above. Table 4-4 shows the basic specifications (including bit/data rate drop back speeds) for the current standards. The 802.11af and 802.11ah standards are still too new to be included in this table.

Since communications are radio based, there was a concern on the part of various IEEE committees about communication confidentiality and integrity. The IEEE initially introduced a set of optional security measures called *WEP (Wired Equivalent Privacy)*, which turned out to provide weak and easily

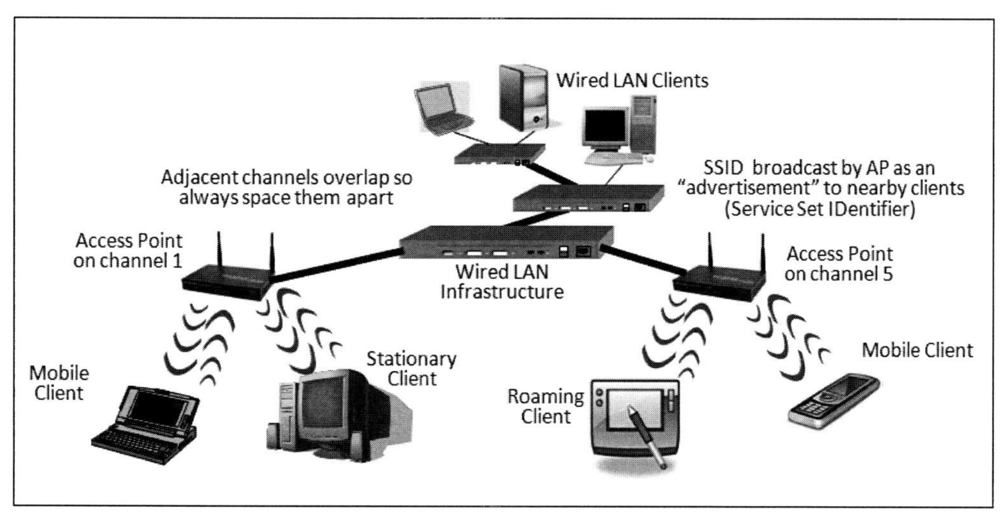

Figure 4-7. LAN with Integrated Wireless Access Points

Table 4-4. Summary of IEEE 802.11xx Standards

802.11 version	Release	Freq. (GHz)	Bandwidth (MHz)	Data rate per stream (Mbit/s)	Allowable MIMO streams	Modulation	Approximate indoor range		Approximate outdoor range	
							(m)	(ft)	(m)	(ft)
a	Sep 1999	5	20	6, 9, 12, 18, 24, 36, 48, 54	1	OFDM	35	115	120	390
		3.7[1]					—	—	5,000	16,000
B	Sep 1999	2.4	20	1, 2, 5.5, 11	1	DSSS	35	115	140	460
g	Jun 2003	2.4	20	6, 9, 12, 18, 24, 36, 48, 54	1	OFDM, DSSS	38	125	140	460
n	Oct 2009	2.4/5	20	7.2, 14.4, 21.7, 28.9, 43.3, 57.8, 65, 72.2	4		70	230	250	820
			40	15, 30, 45, 60, 90, 120, 135, 150		OFDM	70	230	250	820
ac	Dec 2012	5	20	up to 87.6	8					
			40	up to 200			Highly dependent on data rate,			
			80	up to 433.3			power and streams being used but			
			160	up to 866.7			not designed for long distances			
ad	~Feb 2014	2.4/5/60		up to 7,000						

1. IEEE 802.11y-2008 permitted operation of 802.11a in the licensed 3.7 GHz band. It allows a range to 5,000 μm. It has only been licensed in the United States by the FCC as of 2009.

compromised security. Since then WPA (Wi-Fi Protected Access) and WPA2, as well as IEEE 802.i, have been introduced to significantly upgrade wireless LAN security. In a later chapter, we will discuss wireless security in detail.

Aside from being used to bridge/route wireless Ethernet traffic to and from a conventional wired LAN infrastructure, there are also other modes in which wireless access points (WAPs) can be configured to operate. One is as a wireless repeater or range extender. In this mode (as shown in figure 4-8), one or more APs can be "slaved" to a central AP, to provide the ability for their local wireless clients to access the central AP by relaying messages back and forth to the central AP.

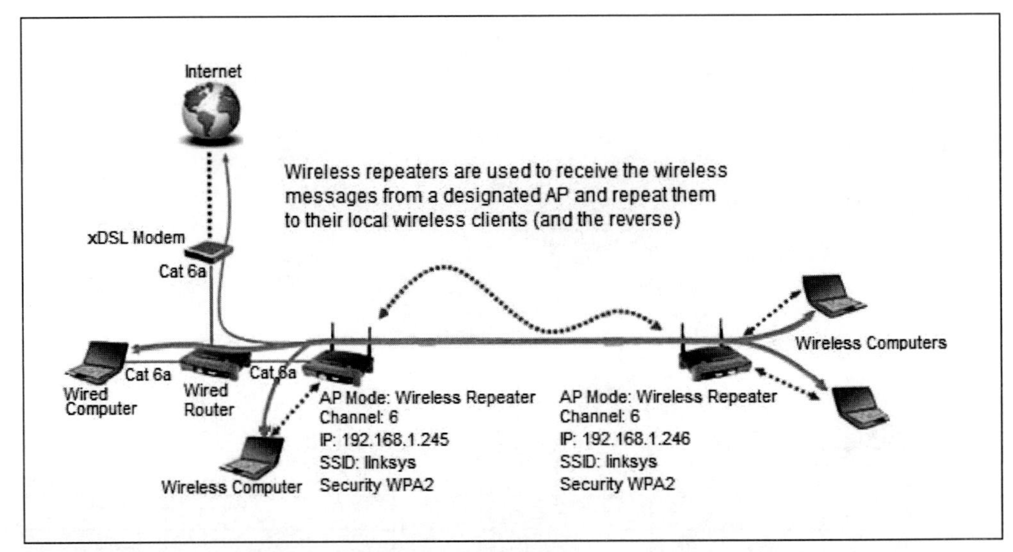

Figure 4-8. Wireless AP as Wireless Repeater

A similar, but slightly different, mode is as a wireless bridge that links a remote switch to the central AP. In this mode, the remote AP cannot have local wireless clients, only a wired connection to a switch into which remote clients can be connected. The remote AP creates a wireless point-to-point connection to the designated central AP but does not act as a local access point to its geographic locale.

802.16 WiMAX

WiMAX is a metropolitan area network (MAN) standard meant to cover much larger areas than the 802.11 standards (it has been called "Wi-Fi on steroids"). Standard IEEE 802.16 was approved in 2002 for operation over a broad frequency range of 10 to 66 GHz. A mobile version, IEEE 802.16e, has also been approved. WiMAX is intended for two distinctly different markets: rural distribution of broadband services and as a contender for broadband delivery of data to and from cell phones. WiMAX has also been used on large oil production platforms and at large industrial facilities to provide site-wide wireless communications capability. Use of WiMAX requires a different wireless NIC than does 802.11, due to technical incompatibilities.

Wireless Mesh Networks

A wireless mesh network (WMN) is a communications network made up of radio nodes organized into a mesh topology. Wireless mesh networks often consist of mesh clients, mesh routers, and gateways. The mesh clients are often laptops, cell phones, and other wireless devices; the mesh routers forward traffic to and from the gateways which may, but need not, connect to the Internet. The coverage area of the radio nodes working as a single network is sometimes called a *mesh cloud*. Access to this mesh cloud is dependent on the radio nodes working in harmony with each other to create a radio network.

A mesh network is reliable and offers redundancy. If a node fails, the rest of the nodes can still communicate with each other, directly or through one or more intermediate nodes. When there are communication obstacles such as buildings, storage tanks, or process units, a mesh can be designed to route around them. Wireless mesh networks can be implemented with various wireless technologies including 802.11, 802.15, 802.16, cellular technologies, or combinations of more than one type. This type of infrastructure can be decentralized (with no central server) or centrally managed (with a central server). It is becoming more common to see industrial facilities using wireless mesh networks to provide a Wi-Fi cloud that provides coverage over the entire facility without having to run cabling all around the facility and particularly into hazardous areas. Figure 4-9 shows an example of an extensive wireless mesh network (excluding wireless clients for simplicity's sake) with several connectivity options including point-to-point, microwave, and satellite.

LAN Infrastructure

A LAN consists of devices other than the servers and clients. Aside from the physical media, there are also active components that form a LAN and these components generally share a common definition, although sometimes it gets stretched by vendors' or users' usage. Some of these components are merely dumb electronic devices, while others contain impressive computing power and associated software, requiring extensive configuration. We discuss two of these devices—the repeater and the hub—in this section.

Repeater

A repeater is a Layer 1 device that regenerates (repeats) the input signal, thus restoring its amplitude and clock sequence. It is used to extend a particular segment of a network. Care must be taken that the overall delay caused by the

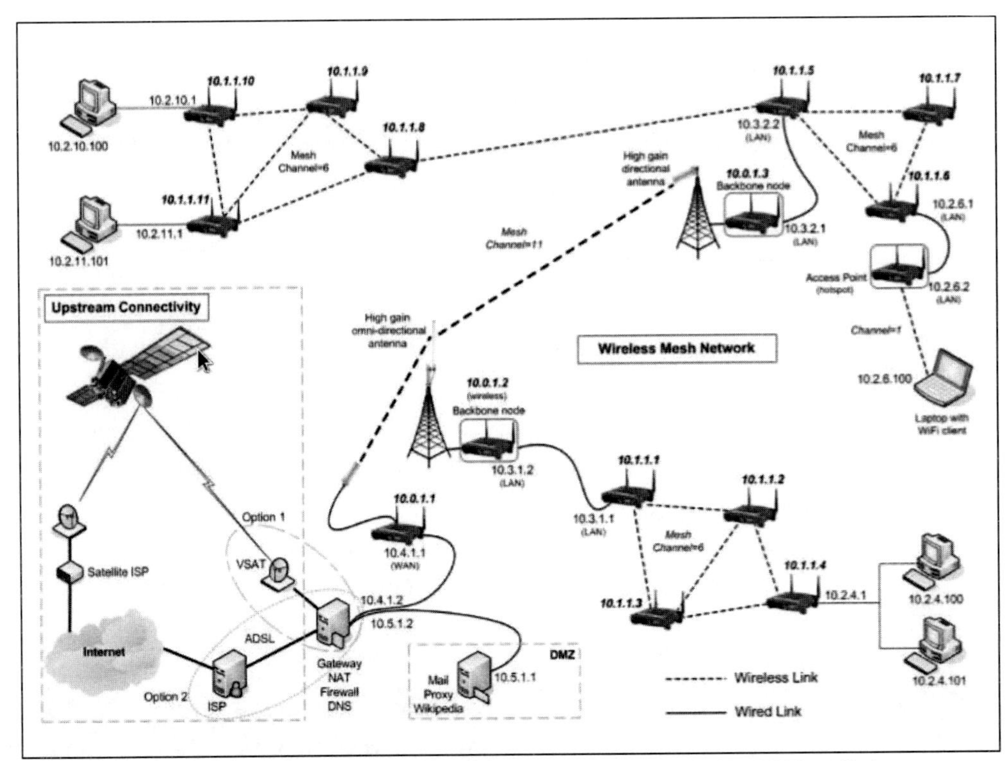

Figure 4-9. Wireless Mesh Network Example (Courtesy of the Wikipedia)

repeater propagation and the length of the additional segment do not exceed the network's round-trip times; otherwise the message delivery time-out interval may be exceeded. A block diagram of a repeater is illustrated in figure 4-10. Note that a "squarer" circuit is one that squares up the leading and trailing edge of a pulse; generally, it is an overdriven (with positive feedback) amplifier, which is also called a *regenerator*.

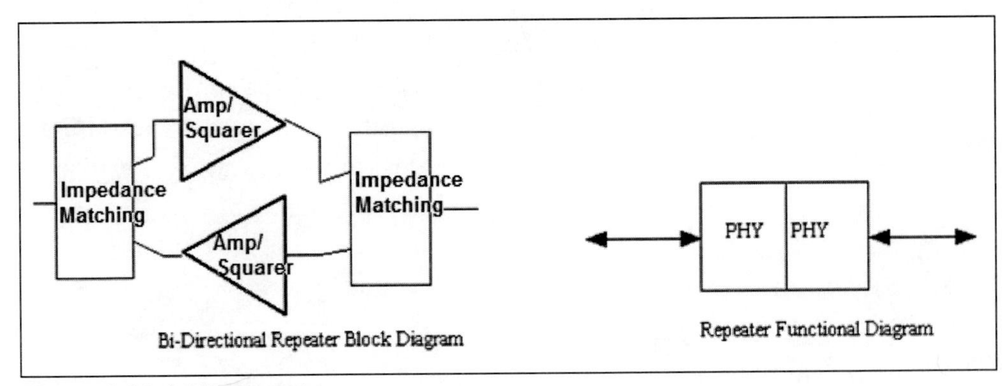

Figure 4-10. LAN Repeater

There are two types of repeaters: a simple amplifier that is designed to boost (and possibly clean up) a degraded signal (as shown in figure 4-10), and a receiver/transmitter device that sits between LAN segments listening for signals/messages on either segment and then totally regenerating a new, duplicate signal/message on the other segment. This type of device does introduce some message latency.

Hub

Basically obsolete now, hubs first became popular with the advent of 10BASE-T Ethernet, which featured a twisted-pair hub consisting of 4 to 24 twisted-pair connectors (8p8c jacks used with the 8p8c plug) in commercial use. Hubs can also take the form of optical fiber hubs but, in either case, hubs may be called *wiring concentrators*. In the twisted-pair version, what you find behind every RJ-45 connector is a multiport repeater. In simple terms, the hub looks like figure 4-11.

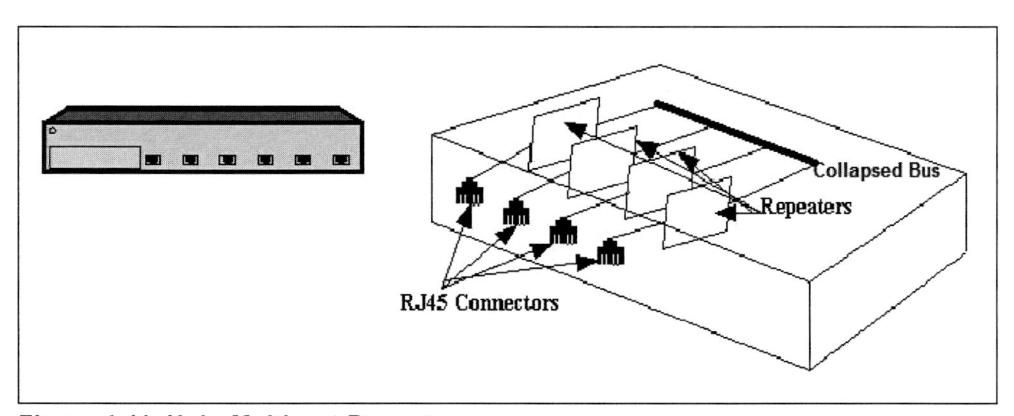

Figure 4-11. Hub: Multiport Repeater

As figure 4-11 shows, each of the 8p8c connectors is connected to a repeater. It is here that the impedances are matched. Each repeater then connects to the shared media. This is why most Ethernet hubs have power: they are repeating hubs. In fact, in most cases they are called (correctly) *multiport repeaters*. Again, remember that hubs act much like a bus, in that communications are half duplex, all connected devices hear every message transmitted and collisions can occur. If the hub fails (e.g., loses power), all communications stop.

Layer 1 and Layer 2 Devices

The limitations of a repeater (multiport or otherwise) arise from the fact that it only knows 1s and 0s and attaches no meaning to either state, it just regenerates the signal condition; that is, it outputs a signal with the correct rise and fall time and voltage state achieved. One way to extract better performance is to have a bridge between the two network nodes. A bridge consists of a repeater with intelligence that reads the Layer 2 information: the destination and source adapter addresses; the "type or length" field; and the frame check sequence (FCS), which is a cyclic redundancy check (CRC). The destination address indicates where the frame is intended to go; the source address indicates which adapter sent the frame; the type or length of the frame indicates what follows; and the CRCC enables the bridge to determine if there were any bit errors and discard such frames rather than passing them.

Bridge

Containing two sets of Layers 1 and 2, a bridge is a device that connects two network segments of the same type, perhaps to connect different segments or to split a single segment. A bridge provides physical connections, namely, MAC addressing and signaling. Modern bridges are "learning bridges," that is, they look at all message traffic on both segments and their Layer 2 (Ethernet MAC) addresses. After less than 30 seconds, they can identify on which side of the bridge a Layer 2 address resides. After that, *they only pass traffic across the bridge that requires a destination on the other side* (which is why a bridge is not just a repeater). This significantly increases performance by reducing the segment loading and reducing collisions. We discuss bridges at some length in chapter 8, "Internetworking." Figure 4-12 is a block diagram of a bridge.

Ethernet Switch

A switch has been described quite briefly previously. A switch is essentially a bridge with a large number of Ethernet NICs. Like a bridge, the switch learns the MAC addresses of the devices on each of its ports and, just like a bridge, a switch examines the Layer 2 addresses to decide where the destination node is located (the port on the switch to which it is connected). Because logic, memory, and arbitration procedures are built into the switch, each node thinks it has total use of the network. Switches have been described (reasonably accurately) as being multi-port bridges. Since each device gets a private connection and arriving messages can be queued in memory until they can be retransmit-

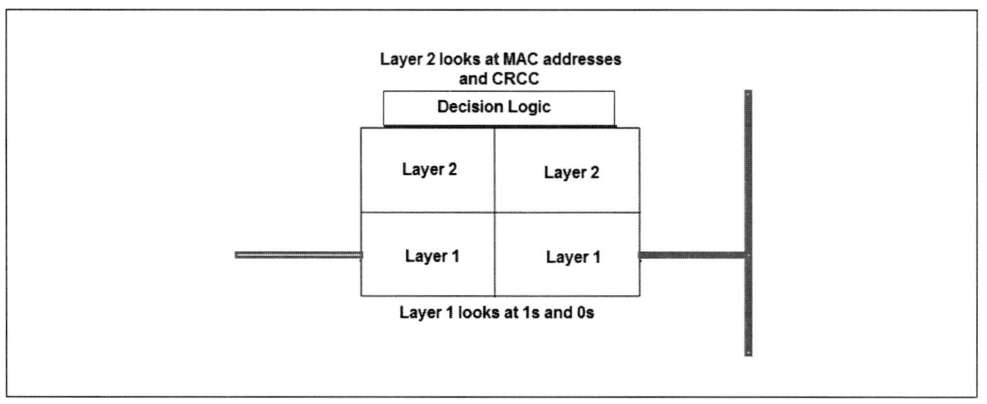

Figure 4-12. Bridge Block Diagram

ted out to the appropriate port, there can't be any collisions; however, this can introduce message delivery latencies (which is why tagged frames and the IEEE 802.1q standard were added). Figure 4-13 illustrates a simple block diagram of an Ethernet switch.

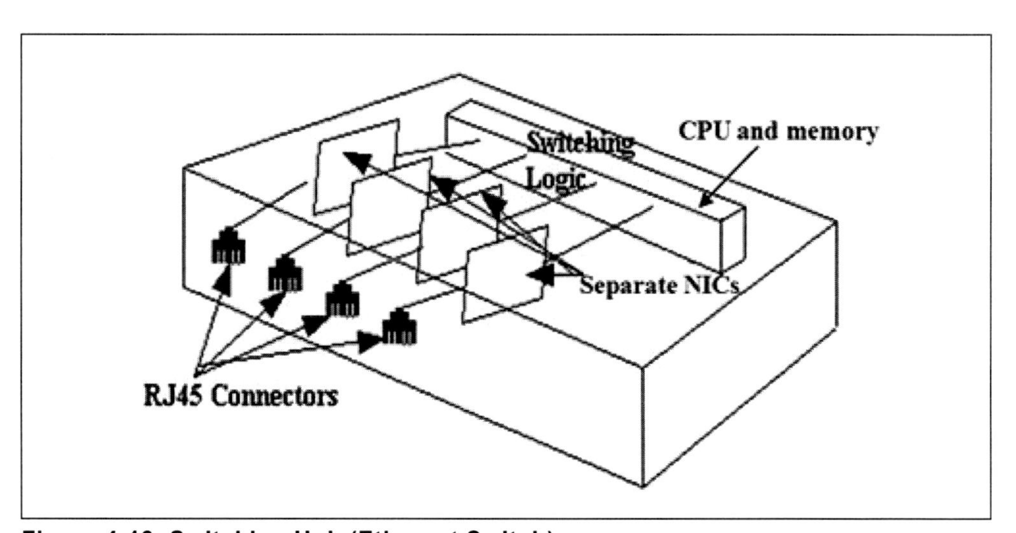

Figure 4-13. Switching Hub (Ethernet Switch)

Although Ethernet switches and bridges have different functions, a switch (using two ports) can be used to perform a bridging function to link two LAN segments. Today, only legacy systems still use bridges. Switch technology has come down in price enough to make bridges and hubs obsolete.

Layers 1, 2, and 3 Devices

The addresses used at Layer 2 are unique to any device on its network. At Layer 2, we can be transported to any addressable device on our network, a network which is identified as either a single segment or segments bounded by addresses. If we should wish to go to a different network, there is no provision to do so in Layer 2. However, the EtherType code value, in the Ethernet frame or the Control in HDLC, does dictate what follows in the frame data area. The Ethernet frame aids in locating where the network addresses (Layer 3) may be found.

Router

A router contains logic that examines Layers 1, 2, and 3 information and will connect two or more networks, providing network addressing, error correction, physical signal conversion, and conversion to compensate for differences in signal frame size (the format, organization, and content of the frames used by the various networks connected to the router).

Routers often perform a gateway function in converting messages from the format used on one network to the format used on another, in addition to dealing with signaling and media differences. (Gateways will be discussed shortly.) If two of the same types of networks are connected with a router, then the gateway function isn't needed.

A router may be a separate computer-based device designed specifically for routing functions or a set of software and network interfaces added to a general-purpose computer or server. In industrial systems, the router is the device or computer on the network that connects to the outside world, be it the communications processor, operator station, gateway, or network monitor point. It will provide routing, if the network is capable of routing. We discuss routers at length in chapter 8, "Internetworking." Figure 4-14 illustrates a block diagram. In general terms, a router accepts messages destined for a device on a another network and sends them to the router that connects that other network to the outside; then, the router that received the messages delivers them to the final destination.

Brouter

If a router is combined with a Layer 2 switch, the result is a departmental router or a Layer 3 switch. This is also called (perhaps for marketing reasons)

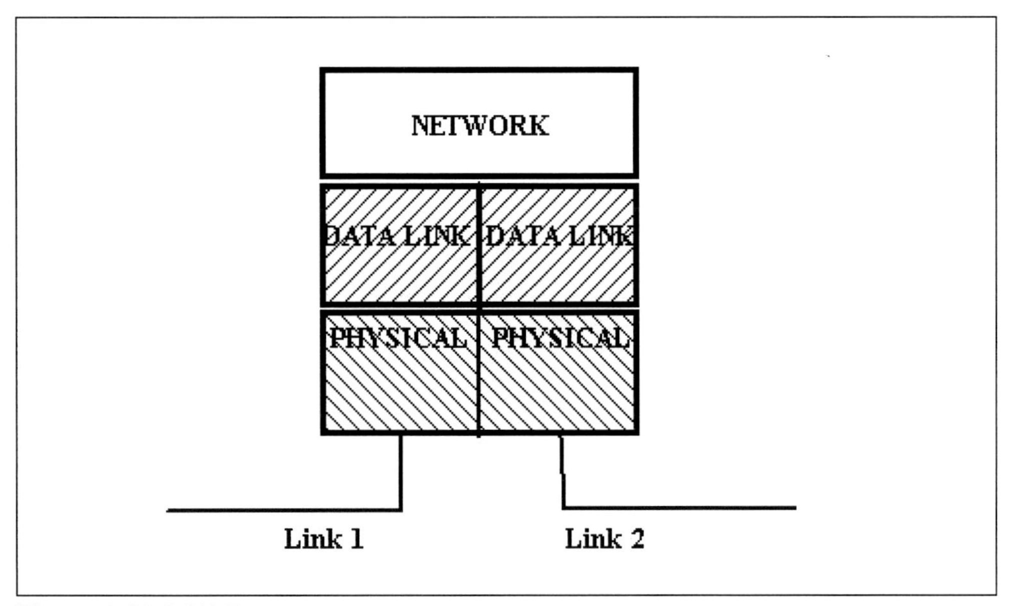

Figure 4-14. LAN Router

a *brouter* (a bridging router), a network device that works as a bridge and as a router. A Layer 3 switch is a device capable of reading Layer 2 and Layer 3 addresses, so it knows whether the message is to stay on the same physical segment of a network, to be bridged to another physical segment (called *bridging*), or to leave for another network (*routing*). The brouter routes packets for known protocols and simply forwards all other packets as a bridge would. Brouters operate at both the network layer (Layer 3) for routable protocols and at the data link layer (Layer 2) for non-routable protocols. These too are discussed at some length in chapter 8.

Gateways

In the real world, systems vary considerably. The organization of 1s and 0s on System A may vary considerably with that of System B. The protocols and addressing may be quite different. In order to accommodate the differences, it takes a dual transceiver, that is, it takes all seven layers for each network to produce the native data, which can then be sent back down the other system. In this context, dual seven-layer systems are called *gateways*.

NOTE On the Internet, the term "gateway" has an entirely different meaning. In Internet terminology, a gateway is the computer/device that makes the connection between the local autonomous network and the Internet. It is frequently a router but can easily be a computer with routing functions enabled.

Gateways consist of two complete seven-layer sets, except for cases where the network architecture doesn't use all seven layers. When networks differ widely, as when a proprietary network connects to an open network, a gateway is used. Technically, a gateway is a transceiver from each network, with the output of the transceivers under computer control so the gateway may perform protocol conversion, timing, different physical signaling, and the like. Figure 4-15 is a block diagram of a gateway.

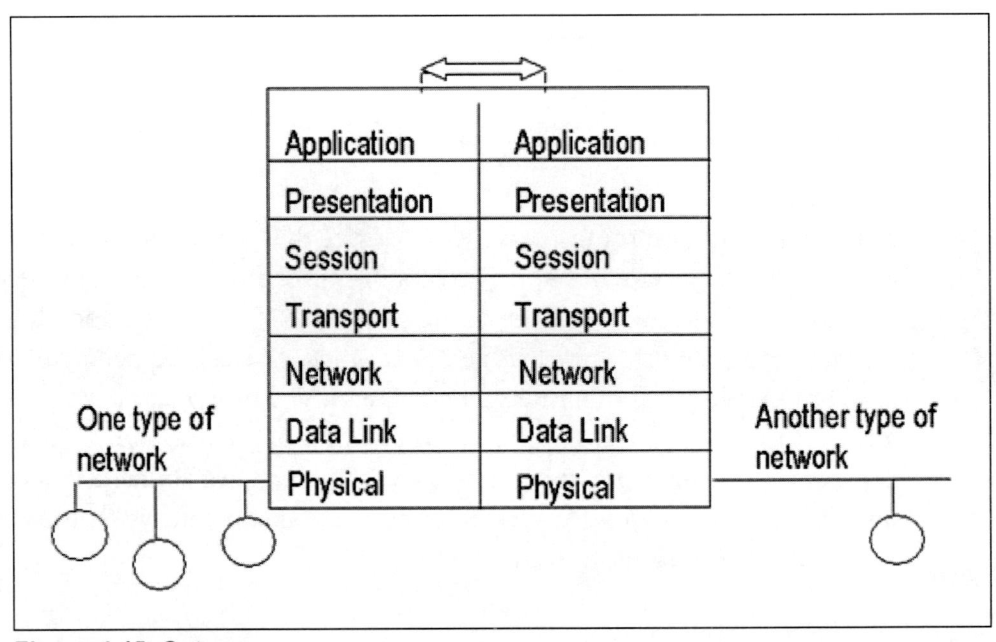

Figure 4-15. Gateway

The reader should understand that few gateways are actually two seven-layer devices, particularly in the industrial field where many of our networks were not routable using only a Layer 1, Layer 2, and Layer 7 model. The word gateway has taken on a functional, rather than a technical, definition. The following are some examples of gateway usage.

- An important and common use of gateways in industrial settings is to connect incompatible networks to allow for data exchange (e.g., a FOUNDATION Fieldbus H1 LAN connected to a ModBus/TCP LAN) using only three layers of the seven-layer OSI Model.

- Gateways are also used to connect incompatible old and new versions of LANs (e.g., FOUNDATION Fieldbus H1 and HSE LAN segments, although this is normally done with a bridging device, not a full seven-layer gateway).
- In the SCADA world, gateway devices can be used to perform protocol conversion so that, for example, a SCADA system that uses asynchronous, serial DNP3.0 protocol to poll its remote terminal units (RTUs) can communicate with an RTU that only supports asynchronous, serial Modbus protocol. A gateway (protocol converter) placed between the SCADA system and that RTU would provide bi-directional protocol/data/command translation. In this particular example, asynchronous, serial Modbus, and DNP3.0 (like many such SCADA protocols) are only three-layer (1, 2, and 7) protocols.
- And lastly, although it is called a *default gateway* (to the Internet) on your TCP/IP equipped computer, it is only a router.

Layer 2 Functions

We briefly mentioned some of the Layer 2 functions when discussing the bridge. At this point, we will take a more detailed look at Layer 2 (Data Link layer) operation.

Media Access

Media access means how a device on a network, such as a LAN, gains control of the network media in order to transmit its information—in other words, who talks and when. Many different methods are available for gaining access and most depend on network philosophy (whether there is a requirement for peer-to-peer communications, scheduling of transmission, etc.), not necessarily on network topology. Before the rise of distributed processing, multi-drop systems (a system with a number of devices connected together, a primitive network) and what there was of networks used the star topology. Peer-to-peer communications were not allowed, even if they could have been accomplished technically. It was an era of centralized control. To be fair, the only computing power available lay in a mainframe or in a minicomputer. All the terminals were dumb; as a result, only one method of access was in general use: polling.

Polling

Although polling may be one of the oldest access methods, it is still used and, in certain situations, quite effectively (e.g., most SCADA systems still use polling to periodically interrogate their field-based RTUs). To use a classroom analogy, if the instructor allows no one to speak unless he or she calls on them, this is polling. In polling, a scheduler, whether a software or hardware entity (it doesn't matter which), determines who should speak, who should listen, and under what conditions. In instrumentation, the *master/slave* access method is an example of polling. Although polling may be used on any topology, it is particularly well suited to the star topology. Polling consists of a primary node (the master station) asking (polling) one or more secondary nodes (slave stations), each in turn, if they have any traffic to send. It does this according to an algorithm that is determined by the application (e.g., SCADA and RTU Poll). The master can be rotated among stations using token passing but the only station that can initiate communications is the (current) master. On most polled networks, it is relatively easy to add, delete, or upgrade applications, since only the centralized control or master needs to be updated.

Event-Driven Polling

Polling methods can be modified. An example of a modified method is interrupt polling, also known as event-driven polling or "hubbing." In this scheme, the master station does not query stations but waits until one of the stations announces its intent to transmit by raising an interrupt or by otherwise signaling its intent. It is much like the classroom, where the instructor may or may not initiate any communication, but in order for you, the student, to initiate communication, you must raise your hand. The instructor then calls upon you. If two students raise their hand at the same time, the instructor uses an arbitration method (usually known only to the instructor) to select one of them first. In event-driven polling, the central station (or master) is in control and can prioritize simultaneous interrupts. This is generally an asynchronous method (i.e., it can occur at any time) and, in general, event-driven polling is how the CPU in a PC is accessed by its peripherals.

Token Passing

In the token-passing access method, a short message with a unique digital pattern (the token) is transferred from peer-to-peer among all the participating peers. Token passing was briefly discussed in an earlier description of the token-passing ring and that material is still applicable to token passing in general. Token passing is deterministic; every station knows (within a maximum-sized window of time) when the token will arrive, so it knows within a specified period of time when it will receive a transmission initiated by another station. Think of an input to a PID algorithm: it must arrive within a specified time or the algorithm will produce incorrect results. Some industrial networks use token passing because of the deterministic results. (Most of the proprietary PLC LANs, such as DataHighway+, also used token passing.) In a typical scheme, only a subset of the nodes on the LAN may pass/share the token and periodically become the master (e.g., all operator workstations and some of the PLCs). The remaining nodes will remain slaves and can only respond to messages initiated by one of the masters.

Carrier Sense Multiple Access with Collision Detection (CSMA/CD)
Think of the classroom again. Better yet, think of the coffee-break room. When a group is gathered together, what are the rules regarding who talks and when? If you have something to say, you wait until the person currently speaking is finished (indicated by a pause of five syllable lengths—100 milliseconds [ms] or more—for English) and then you speak. If two (or more) of you try to speak at the same time, you have a "collision" and certain rules of order operate (e.g., if it is your boss, you generally let him or her speak before you). Eventually, the opportunity to speak will come (or you may think better of it, and keep silent).

In this system, the node on the network that wishes to communicate listens. When there is an idle pattern on the bus, the node transmits its message. If it receives its own message ungarbled, then it knows (and so does everyone else) that it has the bus. The node then transmits packets for an allotted period of time and then is required to go to idle for a minimum period of time, ensuring that others get a chance to transmit. Note the simplicity of the scheme: if you have nothing to say, then you are not using bandwidth and you are not in the loop (just receiving). Only those needing to transmit are contending for use of the bus.

Even at the speeds of electrical signal propagation, a station (A) may start to transmit and another station (B), unaware of the transmission due to its electrical distance from station A, may also start to transmit causing a collision. If two (or more) stations start to transmit concurrently, the stations that are in receive mode will hear garbled messages (refer to figure 4-16). In order to ensure that no receiving station treats such messages as valid, and so that the transmitting stations know that their transmissions have failed, there is a need to detect such collisions (the "CD" part of CSMA/CD). This is typically accomplished by measuring the voltage on the medium. The combination of signals when multiple stations transmit concurrently will produce an invalid voltage level that indicates a collision is occurring. All stations, especially those transmitting, monitor the voltage level on the medium.

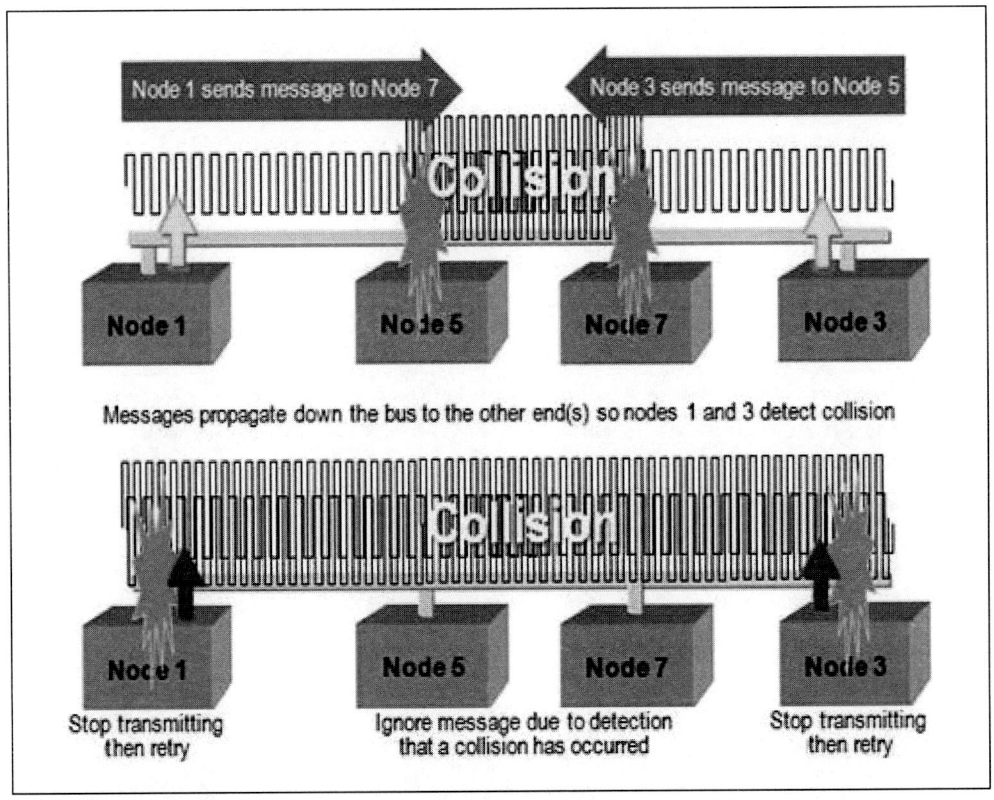

Figure 4-16. Collision Detection Using CSMA/CD

As a result of the time required for a signal to travel the length of a maximum LAN segment and possibly through bridges and additional LAN segments, it is possible for two stations, each at the opposite remote endpoints of a large, multi-segment LAN, to send messages concurrently and to complete their transmissions before a message signaling a collision arrived. To prevent this from happening, the Ethernet frame has a minimum length. At 10 Mbps, this minimum length requirement is accounted for in the minimum payload size of 46 octets. With 100 Mbps Ethernet, the NIC will append the necessary "frame extension symbols" (unless connected to a switch and in full-duplex mode, where collisions can't occur) to achieve a minimum of 512 octets for the frame.

IEEE 802.3/Ethernet: A Layer 1 and 2 Standard

Although IEEE 802.3 is not specifically an industrial protocol as such, it is used throughout industry today. Due to its proven performance, low cost, and high transmission speeds, it has become the de-facto Layer 1 and 2 standard in many industrial networks. With its adoption as the underpinnings of various instrumentation LANs, such as FOUNDATION Fieldbus HSE and PROFINET, it reaches all the way down to the end devices on the plant floor.

Use of IEEE 802.3 as an industrial network protocol was initially highly debated, with arguments presented about its ability to provide deterministic performance due to collisions. (Many such arguments were attempts to justify vendor-proprietary LAN designs—and most of those vendors now use Ethernet in their systems.) The advent of affordable Ethernet switches and 100 Mbps speeds settled that debate and, as the presumptive heir to the industrial networking throne, it bears some detailed examination. It should be pointed out that IEEE 802.3 is a Layer 1 and Layer 2 standard only; Layers 3 through 7 are determined by the particular implementation. Just because two devices have the same Physical and Data Link layers, it does not mean that they can communicate (but it is a start). Most contemporary industrial protocols started life as vendor-specific proprietary designs, including Layers 1 and 2, and have since been migrated to an Ethernet-based design.

As Layer 1 and Layer 2 standards, IEEE 802.3 and Ethernet v2.0 have a Physical layer and a Data Link layer. The Data Link layer is broken into two sublayers: the MAC and the LLC (see figure 4-1). The Physical layer is the part of the LAN model that is responsible for making a connection and signaling to the medium (e.g., trunk cable and fiber optic cable). In a broadband or carrierband system, the Physical layer is the LAN modem (DCE device). In a baseband system, the Physical layer consists of line/signal conditioners found in the network interface. The Physical layer encodes and physically transfers messages between the Physical layers of other units. It provides the procedures and the electrical means for initiating, continuing, and disconnecting physical connections. Wireless devices use electromagnetic transmission but the connections are still physical equipment. Virtual connections are considered equivalent to physical connections for the purpose of discussing functions of the Physical layer.

The Physical layer is responsible for making electrical, wireless, or optical connections to the media and for sending and receiving 1s and 0s. In a switched network, the Physical layer is responsible for ensuring the connections. The Data Link layer (and the MAC sublayer in the IEEE model) is responsible for the media access. Although industrial LANs differ greatly in their adherence to any specific model or protocol because of differences between vendors' products, we will concentrate here on the two major LAN media access methods that were introduced earlier: Carrier Sense Multiple Access with Collision Detection (CSMA/CD) and token passing. Although this discussion will be on the generic (or standard) configurations, we will cover some industrial specifics (like HART and FOUNDATION Fieldbus) in chapter 6, "Industrial Networks and Fieldbuses." Note that 802.3 implements the LAN model: the physical connection, the MAC, and the LLC. The following paragraphs include the physical description, a brief explanation of the Ethernet modulation scheme or line signaling used, and the electrical or optical characteristics.

IEEE 802.3/Ethernet has two older standards at the Physical layer for coaxial cable (referred to as "thick" and "thin," based on the respective cable diameters). With the widespread use of twisted-pair cable, they are no longer specified for new installation, nor are they recommended in EIA 568, the wiring recommendation for commercial buildings.

Since its inception, Ethernet could run on baseband or broadband networks. The majority of Ethernet networks (particularly at 100/1000 Mbps) are baseband. In the shorthand used to describe the implementation of Ethernet (e.g. 10BASE-T), a specific arrangement is used: first the speed (data rates), then the type ("base" for baseband and "broad" for broadband), and finally the distance or media type. Several examples of Ethernet network specifications follow.

10BASE5

This is interpreted as meaning "10 Mbps, Baseband, 500 m end-to-end (per segment)," using Belden 89880 coaxial cable, also known as ThickNet or, if you will, "the frozen orange garden hose"—a reference to the difficulty in handling that this particular cable presents. (Belden 89880 coax meets the impedance requirements of the 802.3 standard and is similar in size to RG-8/U.)

ThickNet/10BASE5 requires line taps, an active transceiver near the tap, and an auxiliary connection to the network interface card (NIC). (The line taps are usually called *vampire taps* because two sharp pins pierce the cable insulation to make contact with the inner and outer conductors and leave a pair of fang-like punctures on the outer insulation layer of the cable.) The cable impedance is 50 ohms and it must be terminated at each end in a 50-ohm non-inductive precision resistor. Figure 4-17 illustrates 10BASE5. It should be noted that this Ethernet standard is obsolete and would only be found in legacy systems and applications.

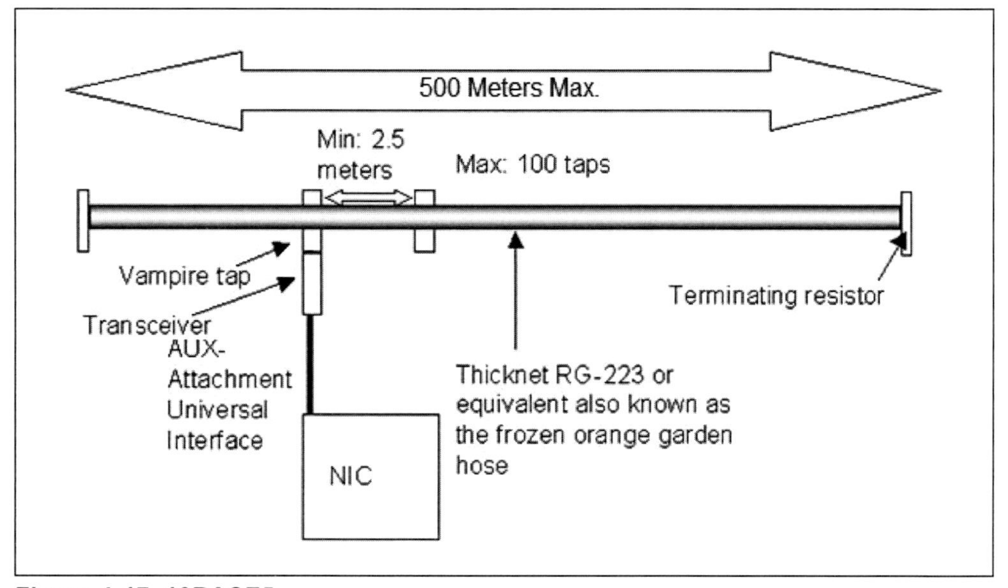

Figure 4-17. 10BASE5

10BASE2

This is interpreted as "10 Mbps, Baseband, 185 m end-to-end (per segment)," using RG58A/U coaxial cable, also called ThinNet. The coaxial cable is terminated in 50-ohm resistors at each end node, the node being connected by a BNC (Bayonet Neill–Concelman) "T" connector, which is attached to the NIC. ThinNet/10BASE2 is usually much easier to work with than ThickNet; however, it does have two cables (rather than the one used in ThickNet from the vampire tap to the NIC) coming into the back of each node, aside from the end ones. Figure 4-18 illustrates 10BASE2. As with 10BASE5, this standard is also considered obsolete and would only be found in legacy systems and applications.

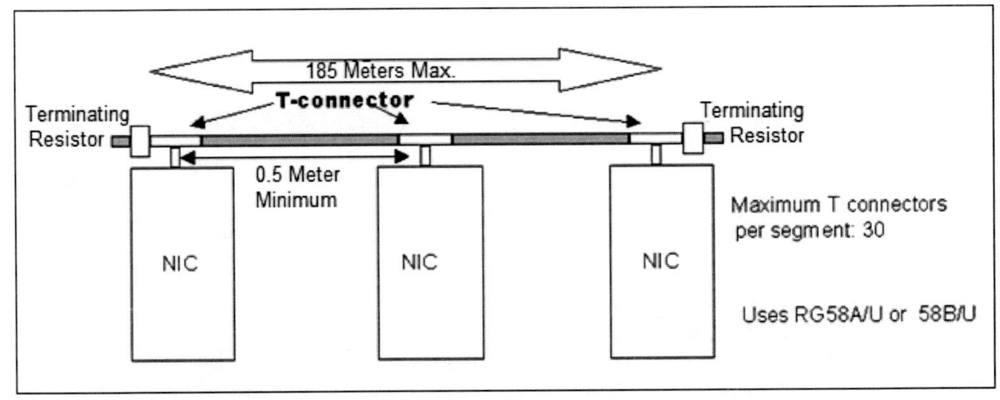

Figure 4-18. 10BASE2

10BASE-T

This is read as "10 Mbps, Baseband, Twisted Pair," the medium being an unshielded twisted pair for interconnection. This system uses a star-wired system whose central point is a hub or switch with RJ-45 connectors. The node NIC is connected to the hub or switch with a 100-ohm, UTP Category 3 cable, using only two pair (Tx pair and Rx pair) of the four-pair (eight conductor) cable. (The extra conductors may be used to provide electric power to end devices—a capability called *Power-over-Ethernet* [PoE].) The connecting cable may be up to 100 m (330 ft.) in length. Termination is taken care of in the hub or switch and the NIC, with some vendors calling the connection a segment. Some hubs/switches have 8 ports; some have 16 ports (or more). Active hubs perform management functions but even passive hubs still require power, as they all contain repeaters, as stated earlier. A switch with management functions is called a *managed switch*. Figure 4-19 illustrates 10BASE-T. Due to the prevalence and low cost of Ethernet NICs capable of "fast" (100 Mbps) Ethernet, this standard is also considered obsolete but most NICs support 10/100BASE-T and can drop down to 10 Mbps if needed.

When a hub is used, there are still the problems of contention and collisions and half-duplex operation. If a switch is used, collisions are eliminated and each connection can operate in full-duplex mode. Note that when it is connected to a switch or hub, the Ethernet NIC of a device will auto-negotiate the necessary settings, including bit rate and full-duplex mode of operation.

10BASE-FL

This is "10 Mbps Ethernet over fiber." It requires one transmit and one receive fiber (fiber pair) and uses an optical switching hub. Depending on the fiber

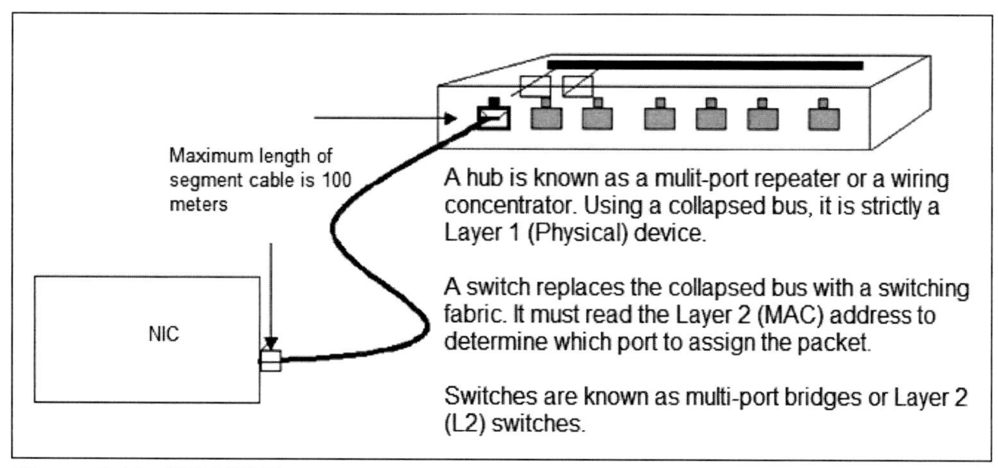

Figure 4-19. 10BASE-T

cable used, 10BASE-FL may have a maximum distance of nearly 2 km. One frequent application is to connect distant switches together.

100BASE-T

This is "100 Mbps, Baseband, Twisted Pair." The 100BASE-T specifications are generally identical to 10BASE-T, with the exception that the connecting cable must be Category 5 (or better), as opposed to Category 3 for 10 Mbps. At present, this is the most popular Ethernet installation. Figure 4-19 represents 100BASE-T as well.

100BASE-FX

This is "100 Mbps Ethernet over fiber." When run in full duplex mode, 100BASE-FX can achieve a distance of 2 km. Half-duplex is limited to 400 m (1320 ft.).

1000BASE-T

This means "1 gigabit per second Ethernet." Originally specified for Category 5 cable, 1000BASE-T requires a tighter specification cable, such as Category 5e or 6a, to operate properly over copper at 100 m (330 ft.), its maximum distance. Note that all four pairs of the cable are used.

1000BASE-CX

This is a shielded twisted-pair (STP) cable with a 25 m (82 ft.) maximum distance. It is no longer considered an active specification.

1000BASE-SX

This is a short-wavelength fiber cable with a maximum distance of up to 550 m (1800 ft.).

1000BASE-LX

This is a long-wavelength fiber cable with a maximum distance of up to 5 km (16,400 ft.).

10GbE – 10 Gigabit Ethernet Over Fiber

This is a standard meant for trunks (high-bandwidth inter-network connections), operated originally only over fiber optic. This form of Ethernet can only be configured for full-duplex operation using switches. CSMA/CD and hubs are not supported.

This standard includes several variations and designations: **10GbBASE-SR**, **10GbBASE-LR**, **10GbBASE-LRM**, **10GbBASE-ER**, **10GbBASE-LX4** (released in 2005), and **10GbBASE-PR**. The following are the fiber versions of 10GbE media:

- **SR** (for short reach or short range) was developed to run on multimode fiber (MMF, in which there are two or more paths for light to reach the terminal end—see appendix C) for a distance of 26 m (85 ft.). A newly developed 50 µm (micrometers) cable at 850 nm (nanometers), which is the wave length of the transmitting laser, has a 300 m (984 ft.) transmission distance.
- **LR** (long range or long reach) runs over single-mode fiber (SMF) at 1310 nm, achieves a minimum distance of 10 km (32,800 ft.), and typically extends to 25 km (82,020 ft.).
- **LRM** (long reach multimode) uses FDDI 62.5 µm cable with a maximum transmission distance of 220 m (722 ft.).
- **ER** (extended range) is a single-mode fiber at 1550 nm that achieves a distance of 40 km (131,233 ft.).
- **LX4** (wavelength division multiplex) achieves a distance of 240 m to 300 m (787 ft. to 984 ft.) over multimode fiber and a distance of 10 km (32,800 ft.) over single-mode fiber at 1510 nm.

- **PR** (released in 2009) is for passive optical networks and uses 1577 nm lasers in the downstream direction and 1270 nm lasers in the upstream direction. Downstream, it delivers serialized data at a line rate of 10.3125 Gb in a point–to–multi-point configuration.

10 GbE – 10 Gigabit Ethernet Over Copper

Due to both monetary and certain installation requirements, a copper version of the 10 GbE was developed for unshielded twisted pair. The standard was known as 10GBASE-T and was published in September 2006.

- **10GBASE-T** was originally intended to run over Category 6 (250 MHz) cable but can actually only achieve a distance of about 55 m (180 ft.). For 10 Gigabit Ethernet (GbE), Category 6e (500 MHz) cable can be used; however, a new Category 6a cable has been developed (625 MHz – ISO Class E) that will achieve the 100 m (328 ft.) distance of 1000BASE-T. The 6a cable was designed to have more twists and the twist rate was varied between pairs to control coupling. This made it a physically bigger cable in part because of the physical separators between the cable pairs. A Category 7 (700 MHz – ISO Class F) cable is recommended for 10 GbE. One version (which uses four pair), the CX4, trades distance for speed and runs 10 GbE for 15 m (49 ft.).

Because 10 GbE is *only supported using switches* and because the different variations require a range of fiber optic media types, most switches allow per-port configuration of the media and the fiber optic connector type using plug-in transceiver modules, as shown in figure 4-20. (The white plug on the left end is protecting/blocking the fiber optic connectors to keep out contamination and prevent inadvertent exposure to dangerous, potentially blinding light.)

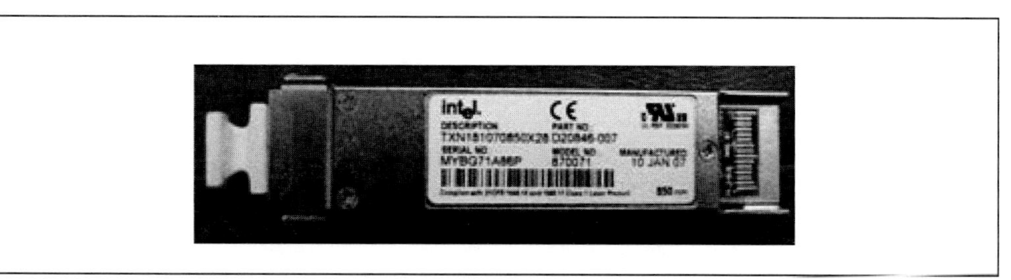

Figure 4-20. Example of an XPF F.O. Transceiver Module

40GbE and 100GbE (40 Gigabit and 100 Gigabit Ethernet)

In a seemingly never ending quest for even greater bandwidth and transmission speed, the IEEE has also specified a set of standards for both 40 Gigabit and 100 Gigabit Ethernet variations. As with 10GbE, these versions can only be implemented using suitable switches and trunk media. The defined standard, as with 10GbE, has numerous variations, mostly having to do with media, light-source power, and wavelength. To date, these extremely high bandwidth technologies have been reserved for data center and telecommunications use and have not made their way into industrial automation systems or products, if only due to their cost.

- 100GBASE-CR10 (copper) is a port type for twin-axial copper cable. It uses 10 lanes of twin-axial cable delivering serialized data at a rate of 10.3125 Gbps per lane.
- **100GBASE-CR4** (copper) is a port type for twin-axial copper cable. It uses four lanes of twin-axial cable delivering serialized data at a rate of 25.78125 Gbps per lane.
- **100GBASE-SR10** (short range) is a port type for multi-mode fiber and uses 850 nm lasers. It uses 10 lanes of multi-mode fiber delivering serialized data at a rate of 10.3125 Gbps per lane.
- 100GBASE-SR4 (short range) is a port type for multi-mode fiber and uses 850 nm lasers. It uses four lanes of multi-mode fiber delivering serialized Reed Solomon Forward Error Checking (RS-FEC) encoded data at a rate of 25.78125 Gbps per lane.
- 100GBASE-LR4 (long range) is a port type for single-mode fiber and uses four lasers using four wavelengths around 1300 nm. Each wavelength carries data at a rate of 25.78125 Gbps.
- **100GBASE-ER4** (extended range) is a port type for single-mode fiber and uses four lasers using four wavelengths around 1300 nm. Each wavelength carries data at a rate of 25.78125 Gbps.
- **100G PSM4 MSA** is a port type which uses eight strands of SMF optical fiber (four in each direction) to give a reach of 500 m (1650 ft.). The 150 m reach of 100GBASE-SR10 is too short for some data center applications, while the 10 km reach of 100GBASE-LR4 is too long. This variation addresses the need for a low-cost port type with a reach of more than 150 m.

- **100G CWDM4 MSA** is a port type which uses two strands of SMF optical fiber (two in each direction) to give a reach of 2 km (6300 ft.). The 150 m reach of 100GBASE-SR10 is too short for some data center applications, while the 10 km reach of 100GBASE-LR4 is too long. This variation also addresses the need for a low-cost port type with a reach of more than 150 m.

- **40GBASE-CR4** (copper) is a port type for twin-axial copper cable. It uses four lanes of twin-axial cable delivering serialized data at a rate of 10.3125 Gbps per lane.

- **40GBASE-SR4** (short range) is a port type for multi-mode fiber and uses 850 nm lasers. It uses four lanes of multi-mode fiber delivering serialized data at a rate of 10.3125 Gbps per lane. 40GBASE-SR4 has a reach of 100 m (330 ft.) on OM3 and 150 m on OM4. There is a longer range variant, 40GBASE-eSR4, with a reach of 300µm on OM3 and 400µm on OM4. This extended reach is equivalent to the reach of 10GBASE-SR.

- **40GBASE-LR4** (long range) is a port type for single-mode fiber and uses 1300 nm lasers. It uses four separate light wavelengths (chromatic encoding), delivering serialized data at a rate of 10.3125 Gbps per wavelength.

- **40GBASE-ER4** (extended range) is a port type for single-mode fiber and uses 1300 nm lasers. It uses four wavelengths delivering serialized data at a rate of 10.3125 Gbps per wavelength.

- **40GBASE-FR** is a port type for single-mode fiber. It uses 1550 nm lasers, has a reach of 2 km, and is capable of receiving 1550 nm and 1310 nm wavelengths of light.

- **25 Gigabit Ethernet** is a standard (proposed) for Ethernet connectivity in a large data processing environment. An industry consortium has been formed to promote the technology and an IEEE 802 study group has formed to develop the standard. This proposal is based on the implementation of 100 Gb Ethernet (CR4) transmission using four 25-Gb lanes.

- **400 Gigabit Ethernet** is presently in a task force group of the IEEE (IEEE P802.3bs 400 Gbps Ethernet Task Force) and is scheduled to be released as a standard in 2017.

With so many different speeds and media choices (let alone 802 versions and full/half duplex), it is important to ensure that the cables you select are compatible with the switch/hub ports and with the NICs in your end devices. We will discuss fiber optic cables and connector standards later in the book, but be aware that there are several types of each and that you need to make sure that everything matches.

Modern NICs and Ethernet switches (and some hubs) use special short messages to tell each other about their capabilities and then both automatically adopt the highest mutual capability settings. This is called *auto-negotiation* and it occurs constantly whenever the link between a device and switch is idle (neither is transmitting). A 16-bit binary number (the "link codeword") is transmitted by both the NIC and the switch using Fast Link Pulses (FLP) (with multiple such numbers, or "pages" for gigabit switches) and this number lists the bit rates, duplex modes, and versions of 802 supported by the devices. Figure 4-21 shows a typical FLP message from a 100 Mbps switch. The link codeword is retransmitted roughly every 16 milliseconds (ms).

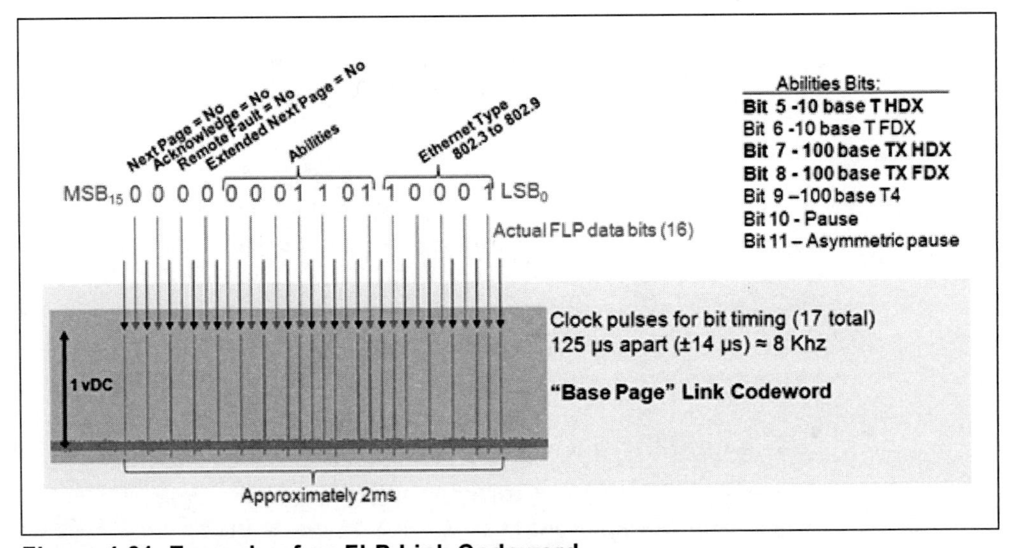

Figure 4-21. Example of an FLP Link Codeword

Note that auto-negotiation automatically occurs and is a firmware function at the Layer 1 level. Media type and connector compatibility must be addressed manually and are not handled by auto-negotiation.

IEEE 802 Media Access Control

The media access control (MAC) sublayer interfaces the Logical Link Control (LLC) sublayer with the Physical layer. The Physical layer is responsible for the actual sending of 1s and 0s. The MAC logic provides the service of sending and receiving 1s and 0s by arranging them into frames. In transmit mode, the MAC:

1. Initiates transmission (802.3),
2. Assembles the frame,
3. Calculates the Frame Check Sequence,
4. Sends the frame, and
5. Ceases transmitting (802.3).

In receive mode, the MAC:

1. Receives the frame,
2. Checks the address,
3. Discards other station frames,
4. Calculates the Frame Check Sequence, and
5. Discards bad frames.

In both transmit and receive modes, the Media Access Control logic also monitors for collisions and executes the applicable recovery logic based on the mode. It is the MAC layer that is primarily responsible for the media access method—token bus or CSMA/CD—hence its name: Media Access Control.

Ethernet CSMA/CD

The Carrier Sense Multiple Access with Collision Detection (CSMA/CD) method is used in many commercial and some industrial LANs, but it defines Ethernet. It has been described in basic operation previously as a contention system. Historically, there were two "Ethernet" variations: Ethernet II (so called *DIX* Ethernet after Digital Equipment Corporation, Intel, and Xerox, the collaborators who co-developed the technology) and IEEE 802.3. However, the minor differences between the two frame assemblies were eliminated in 1997, so there is only one Ethernet now, as defined in 802.3 (2005). Although we will describe Ethernet here on a baseband bus (per IEEE 802.3), it will work as well on broadband buses and wireless systems (from which CSMA/CD was derived).

As we have stated throughout this book, two of the main problems encountered on LANs are: (1) determining who shall talk and when; and (2) how to keep everyone in synchrony. Early efforts at creating LANs used a server; that is, a computerized LAN traffic cop. The server software had algorithms to determine priority, level of access, and so on. Regardless of the way the nodes were connected, it was essentially a star topology because—as in a star system—when the central computer was inoperable (in this case the server), all LAN operations would stop. This could be rectified by using multiple servers or by reducing the server software so the station that was acting as the server could also be used as a station and, therefore, the responsibility for being server could be moved around.

The IEEE 802.3 and 802.5 LANs can be server-based or serverless. Either may use a net control station (the station in charge of the overall operation usually called a *floating bus master*, if it is easily portable between nodes). The reason for using a net control station is that a serverless LAN (peer-to-peer) while efficient, requires management attention at each node for additions, deletions, security, and updates to applications. In practice, a single station or set of stations must have the management responsibility for LAN additions, deletions, security and so forth. This is not a necessity from a network operating point of view but from a strictly human and organizational standpoint. In the IEEE 802.3 and .5 versions, this management responsibility function is provided by software for any station in the station management partition.

In the CSMA/CD method of control, each station listens to the bus when it is idle and discards all packets not addressed to it (except for special "broadcast" messages.) When a station is not receiving or detecting active message traffic, it is eligible to transmit. If there is data to transmit at this station, it first verifies that there is no traffic in process by listening to the network. Data on the network is sent in relatively short packets, so if the traffic is low, there will be a minimal delay (in microseconds) before an idle condition is detected. If the line is idle, the station will first transmit a preamble (eight octets of repeating 1s and 0s) to allow all stations to synchronize. Then it will transmit a frame. As long as there is no contention (two or more stations trying to access the shared medium at approximately the same time), the station will continue with its transmission. There is a chance another station may have elected to use the bus at about the same time and that it may also begin to transmit. This will result in a collision between the signals that will significantly alter each message's content and create a detectable change in voltage level on the bus. The first networked station that detects a collision (CD) transmits another set of frames (a JAM sequence), to ensure that all stations, regardless of distance on the bus, know that a collision occurred. All stations then drop off the line for a period of time (in microseconds) that will randomly differ with each station. It is highly unlikely that the same two stations will be in contention when returning from their individual time-out intervals.

Each time a station makes a consecutive attempt to transmit and detects a collision, the drop-off time is extended up to 16 times, beyond which there is no further extension. At this point, the station determines that either a node has failed and is transmitting all the time (resulting in "jabber") or that there is some other malfunction on the network.

CSMA/CD is perhaps one of the most efficient methods for lightly loaded networks; that is, networks in which traffic occurs in small bursts, such as office environments. In such an environment, most traffic consists of downloading programs, sending and receiving email, browsing web sites, accessing shared printers and servers, plus retrieving or storing documents, all which do not impose a continuous demand except, of course, at the beginning and ending of each business day (and at lunch hour, when web browsing activity tends to increase). Switched Ethernet LANs eliminate the issue of contention and collisions but can potentially introduce latency problems (delays) under certain conditions.

With CSMA/CD, adding or deleting stations (as far as the medium is concerned) is simplicity itself. Each adapter has a unique address, so the adapter just needs to be attached to the network and allowed to communicate. There is no assigned order of polling. Whenever the media access device is told to transmit and the network is idle, it will transmit.

Using the CSMA/CD method in environments that continually transfer data and have a large number of nodes presents some problems. Although the CSMA/CD method is used in industrial settings, there is debate about its use in control applications particularly with field instruments such as a transmitter, controller, or valve. Systems in such environments tend to be heavily loaded; data has a continuous presence, unlike office areas, and a large number of collisions could result, particularly if there were conditions that led up to an alarm. In such circumstances, controllers, diagnostics, alarms, and the like would need to transmit data. A large number of collisions means a significant amount of time would be spent timing out. Under the CSMA/CD scheme, all message packets would probably get through but the time it would take for a PID controller value to be transmitted and received is not predictable. It is possible, although statistically improbable, that the packet might not get through at all. Thus, a CSMA/CD network is called a *nondeterministic* or *probabilistic* network; however, it is important to remember that all DCS systems and PLC systems used proprietary LANs with data transmission rates in the 128 Kbps (kilobits per second) to 1 Mbps range prior to the introduction of Ethernet. The bandwidth provided by switched 100 Mbps Ethernet is an order of magnitude greater and this is one reason why Ethernet has become the de facto LAN standard for industrial networks; the other reason being the use of full duplex switches, which removes all possibility of collisions and lays the "nondeterministic" issue to rest.

The CSMA/CD network's throughput deteriorates under heavily loaded conditions, as do all other forms of network access when loaded. The time-outs in Ethernet are linear and heavy; loading can only slow down, not bring down, the network unless there are too many collisions as a result of a jabbering node or an open bus. For these reasons and others, other methods of access have usually been employed within industrial settings. The current conventional answer is not to use shared media, but rather to use a switch or a "tree" of interconnected switches. Switched Ethernet is deterministic. A collision would only occur is if the node and switch were connected using half-duplex and both wanted to transmit at the same time. That ceases to be a problem if a full-duplex switch connection is used (which is the norm) and the available bandwidth is essentially doubled. Then, the only variable is the latency of the switch itself, which is normally a negligible period of time (and modern switches allow critical traffic to be given priority). The loading problem is addressable, solvable, and manageable. This is why you now see Ethernet predominating on the plant/factory floor.

Industrial Ethernet is here, ready, and able, and we discuss it in more detail in chapter 6, "Industrial Networks and Fieldbuses."

Industrial Token Passing

One of the other methods of access used in industrial networking is token passing. It is deterministic and operates efficiently even at heavy loads. However, it is a much slower protocol, even though it is deterministic, because all nodes are offered a chance to transmit, even if they have no need, and that wastes bandwidth.

Token-Passing Bus

The token-passing bus actually does what its name implies. As an analogy, consider a real bus. To get on, you present a token—no token, no ride. The token will be passed to the next rider when the present token holder gets off, so that next person may ride the bus. That there will be just one rider on the bus at any one time is easy to see; two tokens aren't allowed. If a token is lost or if the bus waits a little too long at any one stop, a new token is issued. Whoever lost the token may not get to ride the bus without petitioning for another turn to ride. The ride is specified in terms of a number of blocks and that is all one gets. Trying to ride more than that will result in removal from the bus. This is about as far as we can use the "ride-the-bus" analogy.

Although most of the legacy IEEE 802.4 standard described a "medium" as a broadband coaxial cable with repeaters, it should be understood that token-passing buses use other media. Industrial-use LANs operate baseband or carrier band on such media as coaxial cable, fiber optic cable, twisted-pair cable, and wireless. The majority of vendors use their proprietary token-passing schemes as baseband, regardless of the medium. The scheme described next is a generalized concept that a token-passing bus might use (it is the one used in IEEE 802.4).

The token-passing bus is a bus topology. However, most token-passing buses are connected logically (their architecture) as a ring (see figure 4-22). One of the important parameters of a token-passing system is the token rotation time—the time it takes the token to make a round trip of the bus. Another important parameter is called *slot time*, which is the delay during which station A sends a message to station B, station B processes it, and station B returns a response. This is different from token rotation time. By dividing this slot time by an octet transmission time, the slot time can be determined in octets.

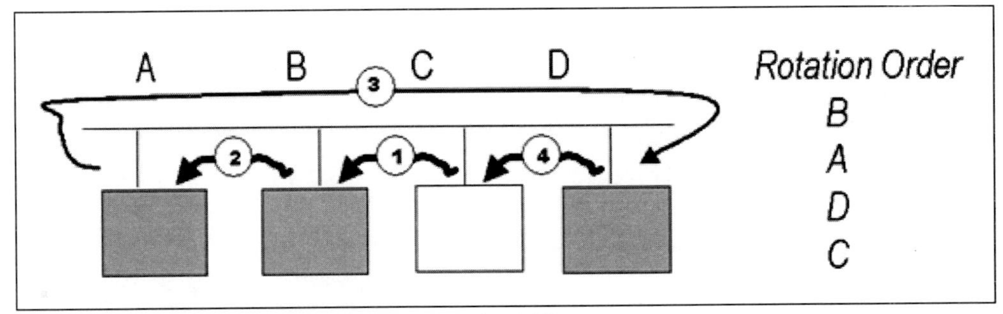

Figure 4-22. Token-Passing Bus as Logical Ring

Before any station can transmit, the following values must have been determined and loaded into the station: (1) station address, (2) slot time, and (3) high-priority token hold time (each station has a token hold time and if it expires, transmission ceases at the end of the transmitted frame).

Some networks have a four-level priority scheme in effect; that is, starting at the highest and going to the lowest, a station transmits its high-priority packets first. If there is time left on the token-hold timer that corresponds to that priority, then the station can transmit its lower-priority packets. Without the priority scheme, all messages have the same high priority and all use the high-priority token timer.

New stations can be added using the "response windows" method. At intervals, a "solicit successor" frame is sent and allows one response frame. In the solicit successor frame will be the range of addresses for stations between the issuing station and the next lower station, its successor. Stations whose address falls into this range can then respond with a set successor frame. If the issuing station receives a valid set successor frame, it places that address as its successor and passes the token to the new successor. If more than one station responds, contention will result. It will be settled through a procedure that ensures that only the appropriate successor is chosen, using the resolve contention frame. The order of token passing goes from the highest address to the lowest. Therefore, to find a successor, the station that issued the solicit successor has to look only at addresses lower than its own. In the end, the lowest of the lower addresses will be selected and the process will repeat itself. However, this process adversely affects token rotation time because the token has been retained by the station that issued the solicit successor command. A solicit successor second frame is used when the recent lowest address station seeks a new successor.

There is also a "ring maintenance" rotation time. If the token rotation time is slower than this time, no additional solicit successor messages will be transmitted (no new nodes will be allowed to enter the logical ring) until the token rotation time is less than the ring maintenance time.

Token passing, then, is a relatively easy-to-implement protocol. It is at its best when it is heavily loaded because each station is assured of a chance at the token. Token passing is deterministic (round-trip time is predictable) under heavy loads, which is precisely the reason why it is used in industrial environments. If token passing is lightly loaded, it is inefficient because it gives every station a chance, even though they may have no data to transmit and will merely pass the token along. It would be more efficient to service only those stations that have traffic (as in CSMA/CD). Note that token ring (time slot allocation) logic can be layered on top of Ethernet to make Ethernet deterministic, and in wireless Ethernet we will see something like that called CSMA/CA where the "CA" stands for collision avoidance (rather than CD, collision detection.)

802.5 Token-Passing Ring

The token-passing ring uses token-passing protocols that are similar to the token-passing bus and it provides quicker response and better priority handling, though at a higher per-node cost (very much higher compared to 802.3). A ring topology suitable for industrial use should meet the following criteria:

1. It should have a 16 Mbps data rate or higher.
2. No single failure can bring down the whole network.
3. Failure of a cable or transmitter should not disconnect the station (i.e., a redundant ring).
4. It should offer support for more than 32 stations.
5. It should provide reasonable per-node costs.

The IEEE 802.5 token-passing ring (sometimes confused with IBM's token ring architecture due to IBM's strong support of the standard) was originally capable of only 4 Mbps and was upgraded to 16 Mbps in 1989. An increase to 100 Mbps was standardized in 2000 and marketed as competition to 100 Mbps Ethernet during the waning years of token ring's existence; although a 1 Gbps speed was approved in 2001, no products were ever brought to market. IEEE 802.5 is now considered to be a legacy technology. Due to media expense and per-node cost, this topology never did find its way into many industrial settings.

The interesting thing about a ring is that it is receiving the same frame as it is transmitting (even at half the speed of light and with a 30-mile radius, not much time elapses in the round trip). This allows the transmitter (when it has the token) to set the "I have data" bit and attach data. As it is processing the transmit frame, the transmitter may read the location where the destination station accepted the frame and change the bit on the received frame to a token, remove the data, and pass the token on to the next station.

Logical Link Control

Together, the MAC and the Logical Link Control (LLC) are responsible for placing and retrieving information without errors on the Physical layer. The specifications in IEEE 802.2 outline the LLC's responsibilities, regardless of which media access/topology is employed, for any 802.X-compliant network. Most proprietary networks have their own version of Layer 2. However, for the purposes of discussion, the 802.2 LLC will be examined here.

The functional definition of the Data Link layer includes framing data blocks or packets (measured in 8-bit octets), determining the check character, and determining network addresses. The two sublayers, MAC and LLC, must ensure error-free (at the bit level) data transmission and reception. The MAC receives instructions from the LLC and performs protocol functions along with frame (packet) error checking. The LLC section interfaces with the user program and provides the MAC's services to the user program. It interprets the data frame for Link Service Access Points (LSAPs), these (usually) being the source and destination LLC; it also identifies the type of service.

Introduction to Types of Service

"Connection-oriented" service means service that is point-to-point oriented. More precisely, a connection-oriented service starts out by establishing a persistent data channel. Once connection is established, data transfer is effected and can continue until the connection is terminated. This transfer may be node-to-node, end-to-end, or one-layer-to-another-layer. On the other hand, "connectionless" service places addressed packets on the media (datagrams) without establishing a persistent data channel (this is called *send and forget*). Both 802.3 and the Internet (IP protocol) are essentially connectionless. There can also be a hybrid between these two just-mentioned services. The rest of this section discusses these types of service.

- **Type 1**: **Connectionless service** (unacknowledged, connectionless) allows two LLCs to exchange data without establishing a persistent channel connection. There is no message sequencing or error recovery, both of which are taken care of by the higher layers. This type of service essentially hands off end-to-end reliability to another layer, typically Layer 4 for TCP/IP or Layer 7 for many industrial applications (this is not referring to the CRCC, which is a packet bit error correction). This is known as *best effort* transmission.

- **Type 2: Connection-oriented service** establishes a data link and provides message sequencing, flow control, and error recovery. It does not require the services of a higher layer, all these services are performed in the Data Link layer. In other words, there will be *no* datagrams with Type 2. Since they need acknowledgment and have less overhead, the early (1970) industrial systems typically used Type 2, even on token-passing buses. Simply put, any application or service that establishes a data path prior to transmission is connection oriented. Note, that does not mean every case that is connection oriented is also Type 2. All Type 2 services are connection oriented, but not all connection oriented services are Type 2.

- **Type 3: Connectionless, acknowledged service** is limited to frame acknowledgment, limited flow control and recovery, and (typically being a Layer 2 service) limited to a single (not routed) segment. This form is the one adopted for most modern industrial systems. The single-frame basis means that the transmit station will keep its last transmitted frame (or frames depending on the system) until the called station acknowledges this frame. If the frame is not acknowledged the transmit station will retransmit the frame a set number of times and then assume that something is awry. In such an event, automatic or manual recovery methods (such as raising an alarm to alert a human) are needed to determine the reason for the lack of acknowledgment. The software could also determine that—due to the lack of response—the called station was out of service and apply maintenance procedures.

802.2 Information (When Running 802.3 Networks)

Whether you are using 802.3 or any other 802.X frame, the 802.2 information (frame content and structure) is identical at the LLC interface. In this discussion, an 802.3 (Ethernet) frame is used and the LLC header comes after the Layer 2 addressing and control octets (in this case 802.3 length/type field). (This format applies to any 802 number network—802.3, 802.5, 802.11.) Figure 4-23 breaks out an 802.3 frame. The 802.2 information is located right after the type/length octets in the block labeled "Data." If the decimal value of the length field is 1,500 or less (to 64), it indicates that this is the length of the data portion of the frame and that 802.2 information (an LLC header) follows at the beginning of the data area (which slightly reduces the available data area size).

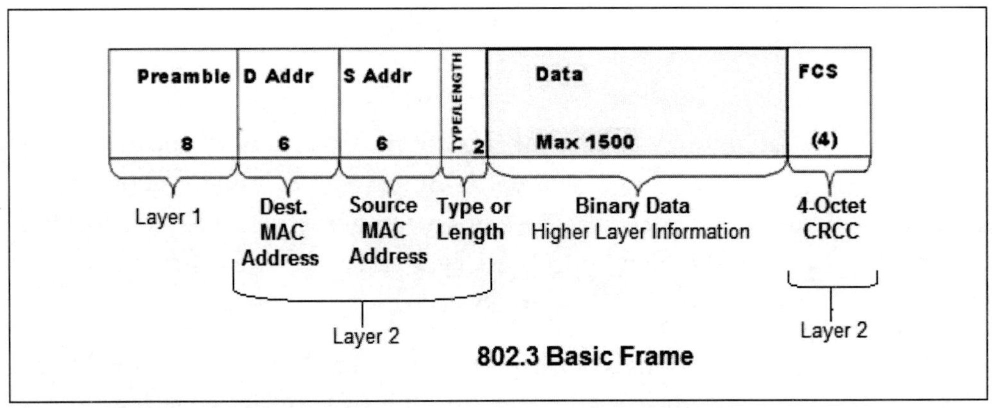

Figure 4-23. 802.3 Ethernet Frame

At its original introduction, Ethernet (IEEE 802.3) did not provide any means of knowing what was being transported in the data area of a frame or any means of sharing the medium with other protocols. It was presumed that the sending program and receiving program would be using a common data/message scheme and could understand each other's messages. However, since the original introduction of Ethernet (1980), it has become important to be able to examine the contents of an Ethernet frame as it traverses a network in order to perform firewall functions, "sniff" (sample) and analyze message traffic, optimize message routing, and manage link-layer message traffic. For this reason, if the length field contains an invalid length value (anything over 1,500 decimal) then this was to be interpreted as an "EtherType" code number. Table 4-5 shows a short list of example EtherType codes. The full list can be found on line.

Table 4-5. EtherType Codes for Various Protocols

EtherType	Protocol
0x0800	Internet Protocol version 4 (IPv4)
0x0806	Address Resolution Protocol (ARP)
0x8035	Reverse Address Resolution Protocol (RARP)
0x809B	AppleTalk (Ethertalk)
0x80F3	AppleTalk Address Resolution Protocol (AARP)
0x8100	VLAN-tagged frame (IEEE 802.1Q) & Shortest Path Bridging IEEE 802.1aq
0x8137	IPX (Novell Net)
0x8138	IPX
0x86DD	Internet Protocol Version 6 (IPv6)
0x8870	Jumbo Frames
0x888E	EAP (Extensible Authentication Protocol) over LAN (IEEE 802.1x)
0x8892	PROFINET Protocol
0x88A4	EtherCAT Protocol
0x88AB	Ethernet Powerlink
0x88CC	Link Layer Discovery Protocol (LLDP)
0x88E5	MAC security (IEEE 802.1AE)
0x88F7	Precision Time Protocol (IEEE 1588)

By examining the EtherType code in the frame, a software application can identify what is being held in the data/payload area of the frame. For example, if the code is 0x800 (IPv4), then it will be an IPv4 header which can be examined for source and destination network (IP) addresses, the length of the entire packet, options or not, and to determine which protocol follows the IP header. If the TCP or UDP packet header follows, then both the source and the destination ports can be determined. All which is generally needed to evaluate typical firewall rules.

The original commercial Ethernet DIX had the length/type octets with values above 0600 hex. The IEEE 802.3 length/type octets had the LLC information following the actual packet length (equal to or less than 1,500 decimal), which caused a number of problems due to a short Service Address Point (SAP). To resolve the issues, the IEEE devised a workaround to enable an EtherType code to be carried in the data area of the frame. This work around is explained in the next few paragraphs.

In old-style IEEE frames, where the length/type octet is actually the frame length in octets, the IEEE followed with octets containing an IEEE 802.2 LLC header. This length/type set of control octets serves two purposes: (1) it is a way to differentiate messages in various protocols that are sharing the LAN (the DIX Ethernet way); (2) it provides the length of the datagram. The data area of the frame is reduced by the length of the LLC header (which is placed at the beginning of the frame's data area after the two length/type octets).

Table 4-6 shows what that LLC header looks like. Part of the header is the optional SubNetwork Access Protocol (SNAP). On networks using IEEE 802.2 LLC, SNAP is a mechanism for multiplexing (time sharing) more protocols than can be distinguished by the 8-bit 802.2 Destination Service Access Point (DSAP) and Source Service Access Point (SSAP) fields. SNAP supports identifying protocols by EtherType field values; it also supports vendor-private protocol identifier spaces. It is used with IEEE 802.3, IEEE 802.4, IEEE 802.5, and IEEE 802.11.

Table 4-6. 802.2 Logical Link Control Header

Data Octet	Field Usage	Comments
Octet 1	Destination Service Access Point (DSAP)	If these two octets contain 0xAAAA or 0xABAB then this is a SNAP message. Otherwise read Octets 7/8 as the EtherType.
Octet 2	Source Service Access Point (SSAP)	
Octet 3	Control field	Indicates the type of LLC message
Octets 4-6	Originators Unique ID (OUI)	0x0 or MAC address of sender
Octets 7 & 8	SNAP/EtherType	Interpret based on Octets 1 and 2 value
Octets 9+	Data for SNAP message	Number of octets is message specific

Gigabit Ethernet Jumbo Frames

As the speed of Ethernet reached gigabit rates, many of the equipment manufacturers of switches wanted to see a change to allow an Ethernet frame to carry a much larger data payload. Although several variations have been proposed (and some vendors have devised their own proprietary designs), there is general support for a "jumbo" frame that supports a roughly 9,000 byte (octet) data area (six times the size of a standard frame). There is an EtherType code to designate a jumbo frame (0x8870) and all such frames are usually of fixed size (configured on the NIC), so no length value is actually required. Other than the data area being roughly 9,000 octets in length, there is no difference in the basic frame structure. A special type of 32-bit frame check is used.

802.1q Tagged Ethernet Frames

Another variation in the Ethernet frame was generated to add more functionality to Ethernet switches. Vendors wanted the ability to provide different classes of service/priority to some messages (e.g., streaming video/audio and VoIP) in order to minimize latency and delay through the Ethernet switch. Vendors also wanted to be able to provide a means for logically isolating message traffic to create a virtual local area network (VLAN) across interconnected switches. These changes required the addition of two data fields to the Ethernet frame. This data, a four-octet long IEEE 802.1q header, is inserted into the frame between the source MAC address and the Length/EtherType code. The first two octets are an EtherType code of 0x8100 that tells the receiving switch that this is a "tagged" frame. The second two octets contain a VLAN number and a priority value. Figure 4-24 shows the revised frame structure. The EtherType code (or payload data size) that defines the payload contents (or size) is not overwritten; it follows the inserted 802.1q header octets.

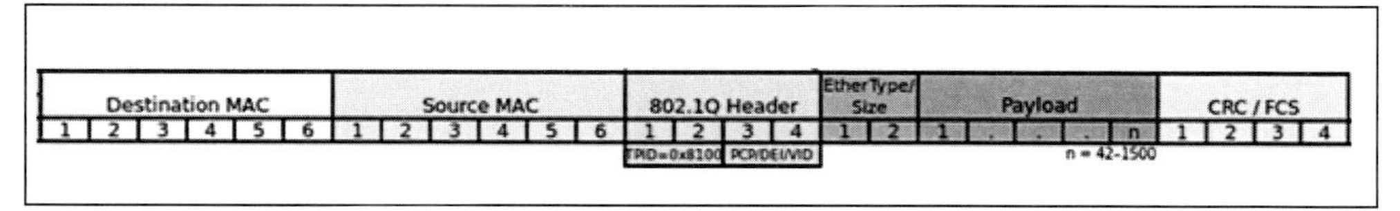

Figure 4-24. 802.3 IEEE 802.1q "Tagged" Ethernet Frame

IEEE 802.1q tagging will be discussed later in the book. The important thing to understand is that the tag field is typically added by the switch itself when an end device sends a frame into the switch. The tag is used within the switch and will be passed with the frame if the frame has to travel to another switch to reach the intended destination device. When a frame is finally delivered to an end device, the tag field is removed by the final switch.

In concept, IEEE 802.5 (token-passing ring) has a similar frame, although the number of octets and the use of a poll octet instead of a type/length octet are both different, and by the time the information reaches the LLC, it is the same interface. That the 802 LLC should present the same interface, regardless of the MAC-Physical layer, is the rationale for 802.2. In the 16-Mbps version of

802.5, the frame length can be up to 17.1 K octets (or about 12 times the 802.3 frame size) although, as mentioned, there are "jumbo" frames for 802.3 of 9 K octets for use with 1/10 Gb switches. At the LAN level, the 1,500 (decimal) frame size (payload in octets) is not necessarily constraining, given modern processor speeds and the off-loading of network functions to the network adapter. However, it does use up processing time on overhead. Many Gigabit and 10 Gigabit switches (at the high end) will support frames in excess of 9,000 octets (some as high as 14.2 K octets and there are even discussions of a 64 K octet jumbo frame that can hold the maximum size IPv4 datagram). Since the frame size has the most deleterious impact on WAN performance (as big messages have to be broken into many smaller pieces and transmitted separately for higher-speed Ethernet) jumbo frame sizes are popular.

It is the type/length field that is the most interesting part of the Ethernet frame. If the value of these two octets is less than 1,500 decimal, this *IS* the length value and it indicates that an 802.2 LLC header (usually three or four octets of LLC header information) will follow at the beginning of the data area. If the decimal value of the two octets giving the data area length is greater than 1,500, the value can't be the length, so it must be an EtherType code.

An Ethernet frame may be 1,518 octets (1,522 if 802.3q-VLAN tagging is used). Some exceptions to this rule were made by 802.3 subcommittees (1997) for compatibility with then-existing systems. For example, one bit pattern in the LLC (0xAAAA or 0xABAB) is for SNAP and another is for Novell (two octets of all ones—the Netware CRCC). Neither will be discussed here, but a great deal of reference material can be found on the Internet by pointing your search engine to "SNAP" or "IEEE 802 LLC."

LAN Layer 3 and 4 Software: TCP/IP

First, as mentioned earlier, it should be acknowledged that until recently, most industrial systems didn't need a Layer 3 (Networking). They were never intended to leave their "island of automation." With the advent of modern, customer-driven, partnered, just-in-time manufacturing, the need for plant-wide (the word used now is *enterprise*) communications became evident. Some industrial systems make it possible to move between several industrial networks, but that is more akin to bridging than to TCP/IP's intent. Although we discuss routing in detail in chapter 8, it should be stated here that a routed system is much more reliable than a large network consisting of bridges and repeaters, and it is much easier to converge (recover after a link or device fails).

Transport Control Protocol (TCP; fits in the OSI Model Layer 4) and Internet Protocol (IP; fits in the OSI Model Layer 3) are two of a suite of protocols developed for the Department of Defense's Advanced Research Projects Agency (ARPA) project. They were intended to be the "lowest common denominator" for connecting diverse systems together in a private wide area network of considerable speed (for the 1970s). These protocols were designed with the then-existing systems in mind and certainly not for what the Internet has evolved into.

It is a myth that the system was designed from the start to withstand a nuclear attack. At the time, it was introduced as a routed system. A paper not connected with ARPA was the one that touted the fact that routed systems were best able to withstand a nuclear attack and its infrastructure damage. A routed system is segmented and routed, so alternate paths can be established if the primary path should become disabled. ARPAnet is now the Internet, but remember, it has design roots in equipment from several technology generations ago. It was never intended to host as many nodes as it now does, and it certainly was never originally intended to run on a local area network.

There are many advantages to TCP/IP. It is an open, free standard. It is user driven. Almost all major operating system vendors had, and have, the protocol stack included in the system at no extra cost. It is the language of the Internet. It provides for a robust, reliable method of data transmission and reception. However, it has its disadvantages as well. When running on a LAN, it requires considerable overhead and has known security holes. Nonetheless, when layered on top of high-speed, switched Ethernet, it has become well accepted for use in industrial automation. As was already mentioned, most vendor-proprietary LANs, and also older standard LAN designs, have been migrated to Ethernet-TCP/IP (OSI Layers 1 through 4) with varying Application layer (OSI Layer 7) protocols. All the commercial operating systems support both Ethernet and TCP/IP. TCP/IP has also become the standard for every corporate WAN and LAN, essentially eliminating all the competition.

How did TCP/IP become the standard? By default: it was here, it worked, and it was and is well understood; but most of all, it is inexpensive per node. The only competitors (1980) other than proprietary systems (whose vendors either charged for their protocols or designed them so they could only communicate with that vendor's machines) were the OSI-compliant systems. However, between the time the standard specifications for the OSI-compliant systems were completed and the time when cost-effective hardware became available (1995), the window of opportunity slammed shut and only one viable technology existed in hardware, software, and specification: TCP/IP.

Keep in mind that Ethernet (IEEE 802.3) is a Layer 1 and Layer 2 set of protocols. TCP and IP are Layer 3 and 4 protocols. And note that IP is a Type 3 protocol (connectionless with acknowledgment) and does not guarantee datagram delivery.

Up through the end of the 1990s (until connection to the Internet was important for business applications), most industrial networks did not use Layer 3 or 4 but went straight to Layer 7 (Application). IP is currently used widely as IPv4 (version 4). IPv6 (version 6), although now running the Internet, is only slowly being adopted into industrial systems and products.

To understand why IP version 6 is needed, we must first look at IP version 4. IP—the inter-network protocol—was designed to be the means used to exchange message traffic between different autonomous networks. It includes the concept of a global address (an IP address) that identifies entities across all interconnected networks, rather than their local address (e.g., an Ethernet MAC address on a LAN).

When TCP/IP was being devised, the engineers involved assigned a 32-bit binary number as a unique global address that could be assigned to entities, in addition to their local hardware address. This permitted the creation of nearly 4.3 billion unique addresses. At that time, the total number of computers in the world was only a few thousand, due to their size, complexity, and cost, so 4.3 billion seemed like a safe figure. Those numbers not used up until 2013 only because drastic workarounds had been devised to keep IPv4 going.

In summary, IP is one of the protocols spoken by the routers that form the Internet and it is used to deliver datagrams from computer to computer. IP can also be used between and among computers on the same local network, which is why nearly all IT departments use it for their networking requirements on corporate WANs.

Figure 4-25 provides a super-simplified conceptual model of the early Internet. All nodes, those that form the Internet itself and those on every interconnected autonomous network (including their gateway routers) would have been assigned one of those 4.3 billion unique IP addresses. One autonomous Ethernet LAN *could* have nodes with the same MAC addresses as some nodes on different autonomous Ethernet LANs (although that is not supposed to happen) and IP would not care, since it deals in the global (IP) addresses. (But no two nodes on the same Ethernet LAN can have the same MAC address!) A node on the left network (figure 4-25) can send messages to a node on the right network because the IP address of that destination node would allow the routing of the message across the Internet to that other network's gateway and then to the destination node. How exactly that occurs will be a later discussion, but the messages will be in the form of IP datagrams.

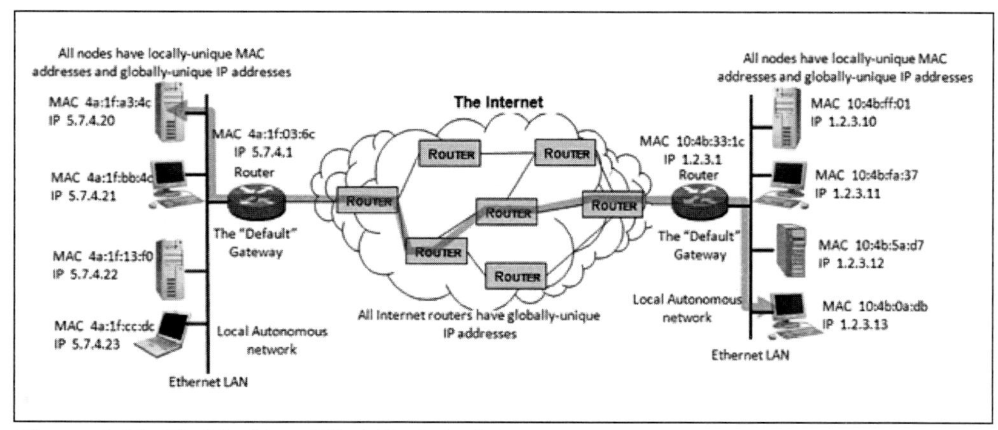

Figure 4-25. Highly Simplified Model of the Internet

An IPv4 datagram can be up to 64 kb in length and the first portion of the datagram is a header that contains the information needed to get the datagram delivered to the desired destination computer. Note that some of the fields in the header (see figure 4-26) were reinterpreted (e.g., "Time to Live" is now a hop count where it originally was in a time value) as the Internet truly became global (in the time frame of 1995 onward, the TCP/IP structure was designed in 1974) and others fell out of use due to communication technology advances (e.g., type of service.) The protocol has several options that can be appended to the basic header and can be used for a range of purposes. Some useful ones are: option 130 (security), option 7 (record route), option 82 (traceroute), and of course option 0 (end of option list). A complete list of IPv4 options can be found on line at: https://www.iana.org/assignments/ip-parameters/ip-parameters.xhtml.

Figure 4-26 is a diagram of the IPv4 header. Each field is used for a particular purpose. The Version field holds the value 4 (this is IPv4). The Fragment ID is used to uniquely identify this datagram, especially when it has to be broken into smaller pieces for delivery. This, plus the fragment offset (up to a total of 8,096 eight-byte blocks; specifies the offset of a particular fragment relative to the beginning of the original unfragmented IP datagram), enables the receiver to find all the datagram's pieces and put them back together again. The protocol number tells us what is in the data area (like the EtherType did with Ethernet frames). This could be a TCP or UDP (User Datagram Protocol) packet (and headers) or an ICMP (Internet Control Message Protocol) message. Table 4-7 lists some of the protocols that can be carried in (that is, appended after the header of) an IP datagram. A complete list can be found on line: http://en.wikipedia.org/wiki/List_of_IP_protocol_numbers.

Table 4-7. Example IPv4 Header Protocol Numbers

Protocol Number	Protocol Name	Abbreviation
0	IPv6 Hop by HopOption	HOPOPT
1	Internet Control Message Protocol	ICMP
2	Internet Group Management Protocol	IGMP
4	IP within IP (Tunneling)	IPinIP
6	Transmission Control Protocol	TCP
17	User Datagram Protocol	UDP
41	IPv6 encapsulation	ENCAP
89	Open Shortest Path First	OSPF
132	Stream Control Transmission Protocol	SCTP

The Time to Live (TTL – now measured in hops) is used to determine if the IP datagram (or any fragments thereof) is lost. Each router counts that value down as it moves across the Internet and if it reaches zero prior to delivery, the datagram (or fragment thereof) is discarded and a note to that effect is returned to the sender. But at this point we are primarily concerned with the network (source/destination) addresses, the 13th through 20th octets. Notice that they are 32 bits in length.

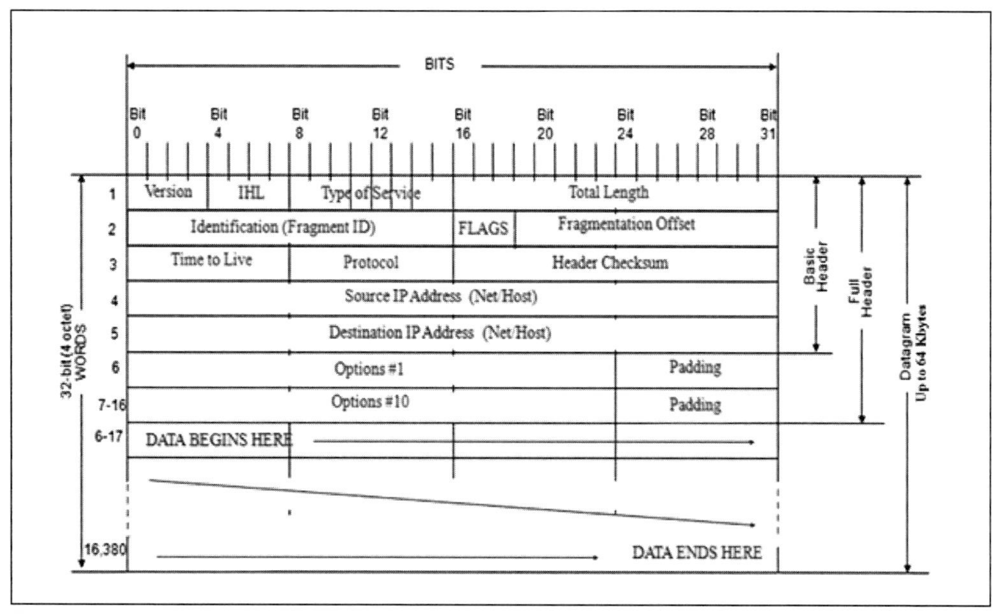

Figure 4-26. IP v4 Header

The 32-bit IP address (IPv4) assigned to a node/router/gateway was originally (1974) designed for the router (rather than the user) and was composed of two parts: the binary number was split in two with the high-order bits treated as a network number and the low-order bits treated as a unique (within that network) node ID number. The split could originally occur at one of three bit positions: (starting at the most significant [left-most] bit) after the first 8 bits, the first 16 bits, or the first 24 bits. The remaining bits to the right could be used to assign unique node numbers on the network. A good way to think about this is to think of the network portion like a telephone number area code and the node number as the 7-digit phone number. Thus a person in Chicago and a person in Los Angeles could both have 555-1212 as their phone number but you get the right person if you add their unique area codes in front of their identical phone numbers.

The IPv4 addressing scheme was originally intended to help routers perform. Routers originally had tables, describing how to get messages to any autonomous network's gateway router, that were entered manually (not determined by software in the router) and the router processing speed was not as fast as it is today. If some way could be devised so that a router could determine the processing needed in the first of these four octets, it would benefit the performance of the router significantly. Figure 4-27 illustrates the general addressing scheme originally used in IPv4 to determine router processing. A Class A

IP address used the high-order eight bits for the network address. A Class B IP address used the high-order 16 bits for the network address. The Class C IP address used the high-order 24 bits for the network address. We won't discuss class D or E addresses, as they were reserved for special purposes.

NOTE Class-based addressing has been replaced by Classless Inter-Domain Routing (CIDR), in an effort to postpone running out of addresses.

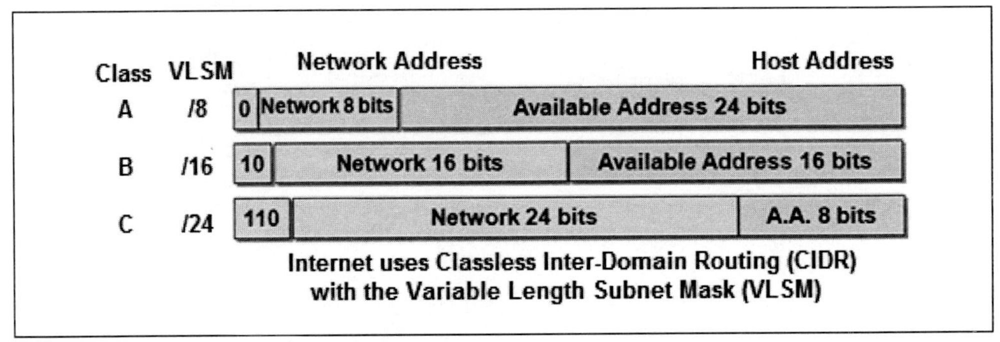

Figure 4-27. Class-Based Address Scheme

Since a 32-bit binary number is nearly impossible for all but an Über-geek to remember, the decision was made to use a different means for writing the IP address: dotted decimal notation. In this scheme, the 32-bit binary number is broken into four octets and each is written as its decimal equivalent, with a period between each number. Figure 4-28 shows how the class address ranges and host ID number look using dotted decimal notation. But don't be confused; an IPv4 address is a 32-bit binary number, with two parts, regardless of how you write it down. Since, in class-based addressing, the split between network and host portions always fell on an octet boundary, the dotted decimal notation worked quite well.

Note that the first three (four if we include class D and E addresses) high-order bits of an IP address would tell you immediately which class it was, so the router knew with the first address octet where the split was between the network and host ID parts of the address. This scheme also prevented any overlap of (duplicated) network numbers across the three classes. There were some other rules about IP addresses:

- Address 127.0.0.1 (the first valid class A address) was reserved for local loopback purposes, so two applications could use TCP/IP to

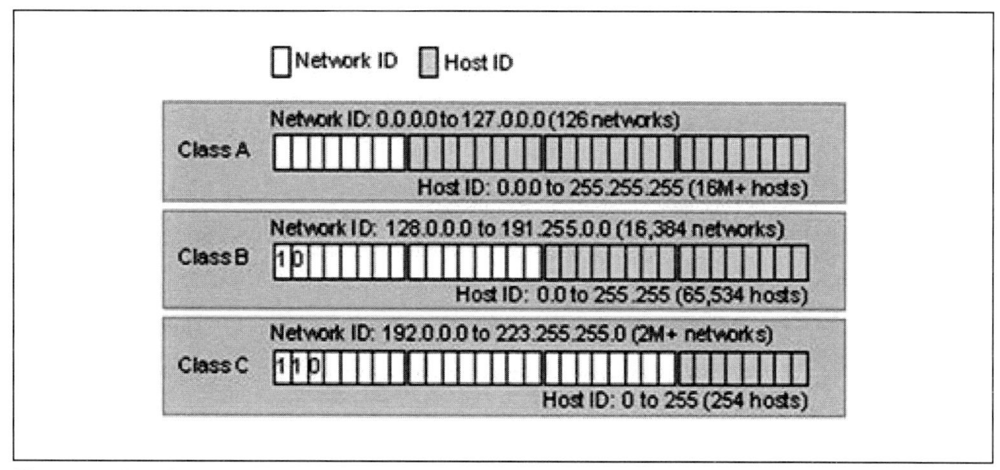

Figure 4-28. Class-Based Addresses in Dotted Decimal Notation

communicate within the same computer without a need to go out onto a LAN.

- Any address with all 1s (or all 1s for the node ID) is a local network broadcast message.
- Any address with all 0s for the node ID is invalid.

It was (and is) not a formal rule, but the general practice is to assign the lowest valid node ID address (node ID = 1) to the default gateway (router) that connected the local network to the Internet.

A problem with this class-based scheme was that corporations, universities, and other organizations wanting to connect to the Internet were generally given a Class B address range (which gave them 65,534 unique addresses to assign to their computers.) Most never needed anywhere near that many (maybe only a few hundred to a few thousand), but with 4.3 billion addresses to play with, who cared if some (most) were wasted? It turns out that this was a mistake because we did run out of IPv4 addresses.

Although the IP v4 addressing scheme seems adequate for any industrial plant—after all, a couple of dozen Class B addresses would be all that is needed—the designers of the Internet did not foresee the day when every computer, smart device, telephone, and probably microwave and refrigerator would require an IP address. The class A addresses were and a classless method of addressing was put into effect; yet, even then, it was clear that there would not be enough addresses for the anticipated demand (the whole world). Enter IPv6, the current version of the addressing scheme which enhances and updates IP v4. As of this date, the Internet itself is running IPv6 and all the popular commercial operating systems (i.e., Windows, Linux, and OSX) support IPv6, but, due to the huge installed base of systems that are based on IPv4, the total transition to IPv6 will take some time. IPv4 and IPv6 are not compatible and require separate v6 versions of protocols, such as ICMP. There are work-arounds that allow IPv4 and IPv6 to coexist during the transition, such as tunneling (carrying IPv4 datagrams as payload in an IPv6 datagram—which is what happens on the Internet, and also the reverse) and running both concurrently on the same computer ("dual stacks").

In order to keep IPv4 working until IPv6 was ready, the Internet Engineering Task Force (IETF – the folks that control the specifications for the Internet) introduced several fixes including:

- CIDR classless IPv4 addresses
- Special local-use-only (non-routing) IPv4 addresses
- Network address translation (NAT)

Classless Inter-domain Routing and Subnet Masks

Classless inter-domain routing (CIDR) eliminated IP address classes, or, more specifically, the need to split the network and host parts of the IP address on an octet boundary. By allowing an IP address to be split at any point, the range of addresses for a given network number could be kept small. Figure 4-29 shows an IP address where the split provides only the lowest six bits for assigning local node numbers (64 potential host ID addresses). The problem was that since there were now no classes, how did the router know where to make the split between the network and host ID portions?

The solution was to add another 32-bit binary number, called a *subnet mask*, and use it to indicate which bits in the IP address form the network portion and which bits form the host/node ID portion. A subnet mask uses 1s to indicate the bits of the network portion and 0s to indicate bits of the host/node portion. By using Boolean AND operations with the subnet mask, it is possible to isolate the two parts. The subnet mask (as part of Layer 3 message forwarding decision-making) may be used to determine whether a destination IP address is on the local network (has the same network number) or not and, if not, if it needs to be sent to the default gateway for routing.

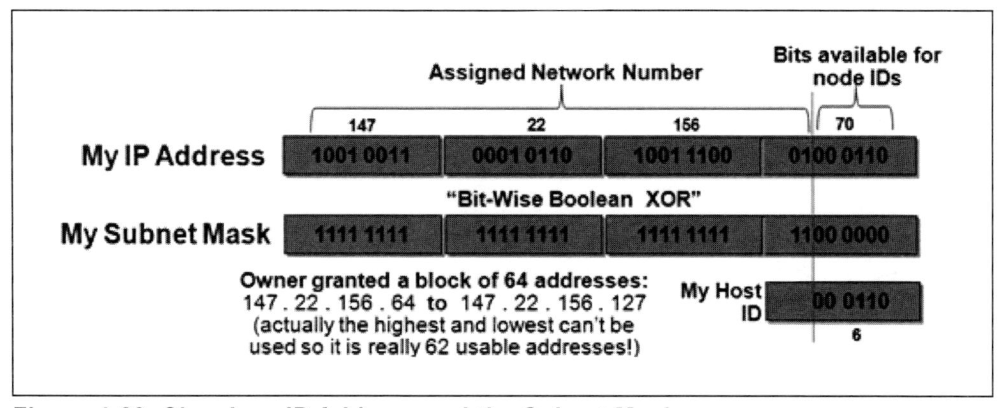

Figure 4-29. Classless IP Address and the Subnet Mask

Non-Routing (Private) IP Addresses

By the mid-1990s, most corporations/organizations had implemented TCP/IP within their own networks, as their Layer 3 and 4 architecture. The IETF had to provide a means for allowing those networks to have IP addresses without using up the rapidly-dwindling supply of available addresses. Finding the solution began by recognizing that nearly all such private networks only had one connection point to the Internet: the default gateway device/router. Instead of giving all the computers on the internal network "public" IP addresses (the way it used to be done), why not just give one "public" address to the gateway device/router and give all the computers on the internal network "private" IP addresses? The IETF designated several IPv4 address ranges as being non-routable and only for use on internal networks. Table 4-8 lists the three address variants allocated for this purpose.

Table 4-8. Non-Routable IPv4 Addresses

RFC1918 name	IP address range	number of addresses	*classful* description	largest CIDR block (subnet mask)	host id size
24-bit block	10.0.0.0 – 10.255.255.255	16,777,216	single class A	10.0.0.0/8 (255.0.0.0)	24 bits
20-bit block	172.16.0.0 – 172.31.255.255	1,048,576	16 contiguous class Bs	172.16.0.0/12 (255.240.0.0)	20 bits
16-bit block	192.168.0.0 – 192.168.255.255	65,536	256 contiguous class Cs	192.168.0.0/16 (255.255.0.0)	16 bits

Most organizations prefer to use the 10.X.X.X version of non-routable IPv4 addresses for their corporate networks. Home users will find that their PCs have been given an IPv4 address in the 192.168.X.X format. Again, these are not public IP addresses and any datagram handed to an Internet router with a destination that is one of these addresses would be discarded with no attempt to make delivery. However, within an internal TCP/IP-based WAN or LAN, these addresses are perfectly usable. The question then arises: If the Internet will not recognize or deliver messages to or from computers with these IP addresses, how come you can send and receive email from people all around the Internet? The answer is Network Address Translation.

Network Address Translation

As was mentioned earlier, most private (autonomous) networks have one point (or multiple points) of connectivity to the Internet at the default gateway device that routes internal traffic out to the Internet (and the reverse). The IETF proposed the idea of allowing internal users with non-routable IP addresses to "borrow" the valid IP address of the gateway device for sending and receiving messages over the Internet. Figure 4-30 shows one example of how network address translation (NAT) can be performed. In this example, the default gateway device has two network interfaces, one on the internal network with a non-routable IPv4 address and one on the Internet with a "real" (routable) IPv4 address. When an internal computer wants to send a message to a destination on another autonomous network, it builds an IP datagram (with a TCP or UDP packet and header—we will discuss this shortly) and forwards that datagram to the default gateway because the destination is not local; that is, the network portion of the destination address is not the same as the sender's.

The default gateway removes the source IP and source port number from the datagram and replaces them with its own routable IP address and a port number selected from the available set. It also adds an entry to a table of active connections that it maintains. That table provides a link between the NAT-assigned port number and the true IP/port numbers of the internal computer. Using its valid IP address, the gateway forwards the datagram onto the Internet, where it should be delivered. The receiving computer sends responses back to the gateway's IP address, with the destination port number being the one assigned in the NAT process. When the gateway receives the response, it uses the destination port number to look up the IP address/port number of the respective internal computer and alter the response message to contain those values, and then it forwards the response to that internal computer. This translation process continues until the internal computer terminates the communication session. Of course, the destination computer may also be on an autonomous network with non-routing IP addresses and so its respective gateway would perform NAT as well.

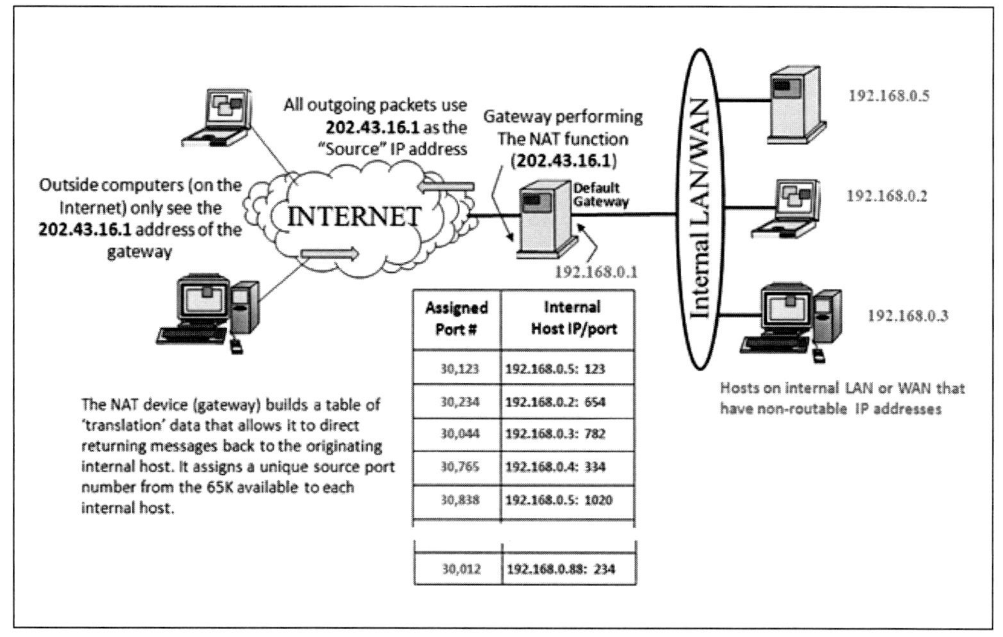

Figure 4-30. Using NAT to Translate Non-Routable IPv4 Addresses

Few systems are ever directly connected to the Internet. Most (like your computers at home) connect to an Internet Service Provider (ISP) that supplies your PCs with a non-routing IP address. That ISP may, in turn, connect to another ISP that eventually connects to the Internet. Thus, several layers of NAT may be taking place as your message traffic traverses the Internet.

All this was done to allow IPv4 to survive until a replacement could be developed. As was mentioned earlier, the replacement is IPv6, a protocol that provides end-user to end-user addressing (each end user will have a unique global address and there are no private or public addresses).

IPv6

IPv6 brings several changes to the inter-networking protocol. The first change that we will address is in the datagram header. Of the 40 octets in the header, IPv6 uses 16 octets (128 bits) each for the source and the destination IP address (3.4×10^{38} addresses or 340,000,000,000,000,000,000,000,000,000,000,000,000 addresses). That should keep the number of free addresses greater than the need for them for quite a while. Figure 4-31 illustrates the 40-octet IPv6 header.

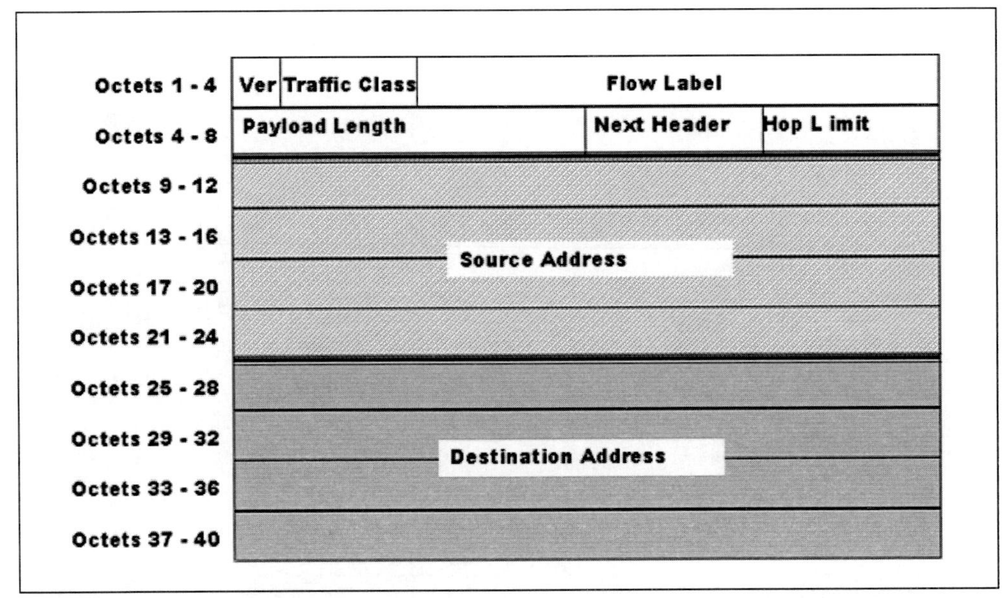

Figure 4-31. IPv6 Datagram Header Structure

Each of the various bits and fields in the header are assigned a meaning. These meanings are discussed next. Note that this base header has a fixed length (40 octets) and is always required (the minimum length of the IPv4 header was 20 octets). Option (extension) headers, if desired or needed, are appended to this base header.

- **Version** (four bits) will obviously be a 6 (binary value) for IPv6 (it was the binary value 4 in the version field of the IPv4 datagram header). During the transition from IPv4 to IPv6—which may take several more years—devices will need to know which IP/header version they are to process. That is why the version octet is the first in both types of IP headers.

- **Traffic Class** (four bits) is a priority system, which states whether the packet can suffer delays or whether it is to be treated as streaming (real-time) traffic. Values from 0 to 7 indicate that the packet may be delayed, while values 8 through 15 indicate real-time traffic. This allows IPv6 to offer enhanced delivery performance for traffic, such as streaming audio and video (e.g., VoIP).

- **Flow Label** (24 bits) is to be used for virtual private networks (VPNs) or other pseudo connections that have specific requirements.

- **Payload Length** (16 bits) indicates the number of bytes in the packet following the base header. The base header (unlike with IPv4) is not included in the length count because the it has a fixed length. However, any option header appended to the base header IS included in this length value.
- **Next Header** (16 bits) indicates which type of option (extension) header, if any, follows this base header (this makes the IPv6 header extensible). If no extension/option header follows, then Next Header indicates what protocol (like TCP) is following the header.
- **Hop Limit** (16 bits), in practice, serves the same purpose as the Time to Live (TTL) field does in IPv4. Although TTL could have been measured in seconds, it was almost always measured in hops (router connections). In IPv6, the Hop Limit states the number of hops that the packet can be relayed.
- **Source Address** (128 bits) is the IPv6 address of the sending device.
- **Destination Address** (128 bits) is the IPv6 address of the receiving device.

Another enhancement that was added with IPv6 is the ability to use cryptographic methods to provide confidentiality and integrity verification, plus source validation of message traffic. This collection of capabilities is grouped together under the name of IP Security (IPsec). We will discuss this in more detail in a later chapter.

Another Layer 3 Scheme: OSI-IP

The OSI Layer 3 protocol is considerably more detailed than the protocols described thus far. The U.S. government made a considerable push to adopt the OSI protocols (Government Open System Interconnection Profile [GOSIP], 1990). However, although every day and in every industry government standards permeate the very fiber of technology, mandating a policy rather than allowing the marketplace (or the users) to determine what will be standard can increase the chance of policy failure and indeed the Federal Information Processing Standards (FIPS) no longer mentioned the policy as of 1995. One might very well recall the competition between the Betamax and VHS videocassette technologies. A number of sources thought Betamax technically superior to VHS. Price, number of suppliers, and superior marketing made VHS the overwhelming favorite, technical superiority or not. OSI routing protocols are well thought out and certainly much more robust than either IPv4 or IPX (although IPV6 makes use of them, particularly the Intermediate System to Intermediate System [IS-IS] standards; IS is a router). They are an open international standard but they did not prevail. TCP/IP has effectively eliminated the competition, except as computer science curiosities.

LAN Layer 4 Overview

Layer 4 is concerned only with end users (a gateway with seven layers is also considered an end user). Intermediate nodes (such as bridges and routers) have no use for Layer 4 or any higher layers. Layer 4 is necessary when Layer 2 is operated in a connectionless manner. In most industrial networks of the three-layer variety, protocols are either connection oriented or the application layer handles the packet sequencing and accountability. With the encroachment of public standards upon the proprietary industrial networks, the most used standards are the TCP/IP protocols.

TCP/IP

As we've seen, TCP/IP, an acronym for Transmission Control Protocol/Internet Protocol, was developed by the Department of Defense's Advanced Research Projects Agency (ARPA) to work across dissimilar computer platforms, ranging from mainframes (then) to micros (now). As routing of LAN data became desirable, proprietary schemes were used to fill the functions of Layers 3 through 7. In order to use an open set of protocols, LAN vendors began using IP. IP interfaced to higher layers using TCP (or UDP, which we will discuss shortly), to the lower layers (notably IEEE 802.3), and to peer, gateway-to-gateway transactions.

IP was designed to be as flexible as possible. It, therefore, offers very little in specific services and is basically an unreliable datagram service. In other words, it only supports a connectionless type service, so minimal error checking (i.e., frame bit checking by the CRCC, fragment reassembly, and delivery-failure timeout) is provided. Although IP does not expect a delivery acknowledgment message to be returned for a transmitted datagram, it may receive failure notifications in the form of an Internet Control Message Protocol (ICMP) message (e.g., the Time to Live hop count was exceeded and the datagram was discarded). ICMP, which is actually an extension of IP, is used by routers or the destination node to report unrecoverable errors to IP.

At the receiving end, a CRCC error (failure to receive all datagram fragments within a reasonable timeframe), will cause a received datagram to be discarded. For some applications this is adequate or even preferable; however, for many it is preferable that a reliable, end-to-end delivery service be available. TCP was developed to provide this class of service and to support connection-oriented communications.

Connection or Connectionless

The difference between a connection-oriented and connectionless service can seem confusing, however, there are good examples in real life. Mailing a letter is a connectionless service. You hand the letter to the post office and hope it gets delivered (best effort for delivery). The postal service will do its best, but delivery is not guaranteed. It is up to you to decide if there is a problem (e.g., several weeks have passed and no return letter) and to take action (e.g., send another letter). Each letter is independent; there is no notion on the part of the postal service that you may be sending further letters. This is a connectionless service. On the other hand, you can also place a phone call to someone. Once the call has gone through (a connection has been established), you can keep the line open and keep talking in both directions until one end or the other decides to terminate the call. The connection remains open, even if you stop talking for a minute or two. This is a connection-oriented service. The three types of service (described previously) are based on these general categories. Type 1 is connectionless, Type 2 is connection oriented, and Type 3 is Type 1 with packet acknowledgment.

TCP, a Transport Layer 4 protocol, supports several reliability mechanisms for ensuring successful message delivery, along with other higher layer functions. TCP/IP has found wide use and is widely adapted to LANs, particularly those with mixed operating system platforms. TCP/IP is built into all versions of UNIX, Linux, Windows, Mac OSX, Solaris, and other operating systems, including some embedded operating system versions.

On LANs, TCP/IP is mostly used layered above Ethernet (IEEE 802.3)-type networks, although it is usable on any network. The primary difference between OSI-compliant standards and TCP/IP types is that the OSI model encompasses and attempts to standardize all possible elements, so a manufacturer knows precisely what to put into a data communications product. TCP/IP, on the other hand, starts out with very few rules as standards and allows add-ons (e.g., new Layer 7 protocols) whenever necessary. Although some add-ons are not universal, therefore not everybody can use them, perhaps not every station needs to use (or is hardware capable of using) a particular add-on. The TCP/IP concept has made it possible for a range of new Internet applications to be developed. All the World Wide Web technology was successfully layered on top of TCP/IP, this has also been done with industrial protocols, such as Modbus/TCP and DNP3.0.

The Transport Control Protocol (TCP) assumes that the lower layers (e.g., IP and Ethernet) only offer unreliable datagram service. This means that the TCP software must provide the transport services of error recovery/retry, packet sequencing, and flow control. TCP is considered to be a transport protocol because its job is to enable message exchange between programs running on different computers that are connected to a common IP-based network.

This brings up an important question: Given that IP protocol and datagram delivery works on an IP address (typically one per computer) but there can be many different programs running on a given computer, all of which may be requesting communication services; how should messages be delivered to the right program? The solution was to assign each program a unique identification number called a *port number* (or just *port*) and to use that additional information to achieve final delivery. Since IP datagrams do not support (do not have a field for defining) port numbers, an added block of data is required. This is part of the information in a TCP header. Figure 4-32 shows the structure and content of a TCP segment header.

Since TCP is connection-oriented, the assumption is that messages will be flowing in both directions as the two endpoint programs exchange them over the connection. A given message may be broken into one or more parts, called *segments* (depending on overall message size), and each segment will be transmitted separately and have its own TCP header. A segment could be a maximum of 64 kb, including the header, or it could be just a few bytes plus the header. At each end of the connection, the local copy of TCP will accept data from the local sending program and buffer it in memory until the sending program requests that all accumulated data be sent (by setting the PuSH [PSH] flag). The PSH flag provides the option of "pushing" data out immediately, rather than waiting for the buffer to be filled. When commanded, the PSH flag in the outgoing TCP packet is set to 1 (true). Upon receiving a packet with the PSH flag set, the other side of the connection knows to immediately forward the segment up to the application.

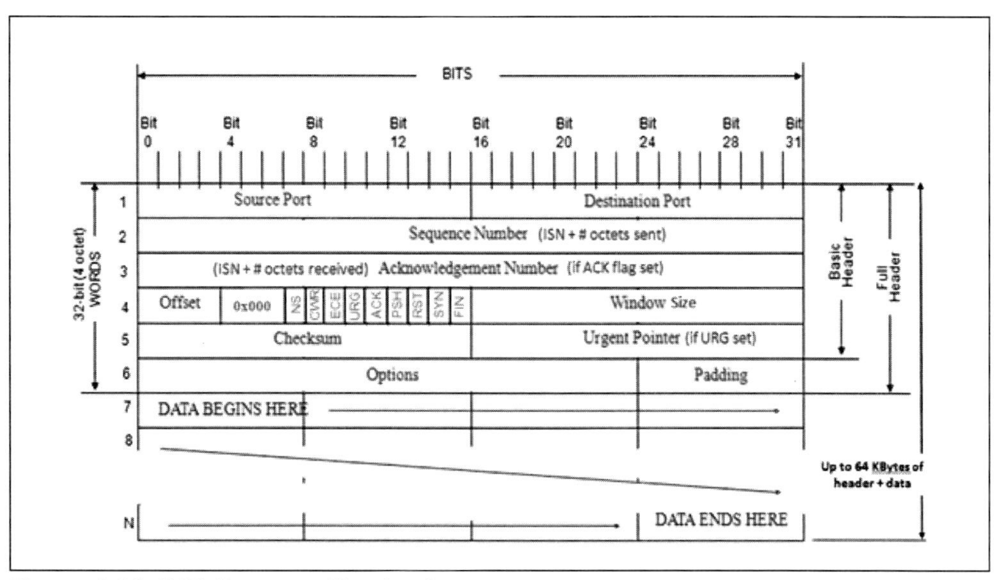

Figure 4-32. TCP Segment Header Structure

When TCP sends bytes of data it expects, at some point in the near future, to get a confirmation that all data were received. The sequence number and acknowledgment number fields are actually a count of bytes sent and a count of bytes received. In addition, each program advises the other program of how much buffer space (it's "receive window") it has available to use in receiving data. Each program can only send up to that many bytes and then it must stop, until the receiver acknowledges that all bytes were received and provides an updated window size.

Like IP, TCP has a few options that can be appended to the header. After options and their associated data are appended, the header is padded to an even 32-bit/4-octet multiple and the data is attached. TCP has several commonly used options that set, for example, the maximum window size allowed (Maximum Segment Size [MSS]), as well as any multiplier to be used when computing or setting a window size scale factor (Windows Scale Option [WSOPT]).

TCP uses message acknowledgments, timeouts, re-transmission of messages, and other means to ensure that messages get delivered completely and without errors, yet this does introduce a lot of overhead and can introduce transmission delays. For some purposes, doing this is unnecessary or even detrimental in that it consumes a non-renewable resource—time. For some applications, IP provides adequate delivery service but IP has no means for supporting port numbers for end-to-end delivery. Therefore, a minimal-function transport layer protocol was devised to add port knowledge: User Datagram Protocol (UDP). UDP is used quite often because it is a connectionless transfer mechanism that introduces little delay and has very low overhead.

UDP and TCP both enable program-to-program message delivery but UDP is connectionless and TCP is connection-oriented. TCP establishes, maintains, and terminates connections between processes; provides an acknowledgment process for reliable message delivery; and performs sequencing (fragment assembly and disassembly) using a duplex (bi-directional) mechanism. UDP allows a higher layer to take responsibility for delivery reliability; it basically appends a source/destination port number to a packet and hands it to IP for delivery. UDP is used in industrial applications, where the higher layers perform the reliability checking and correcting. A UDP header is displayed in figure 4-33, for comparison with the TCP frame in figure 4-32.

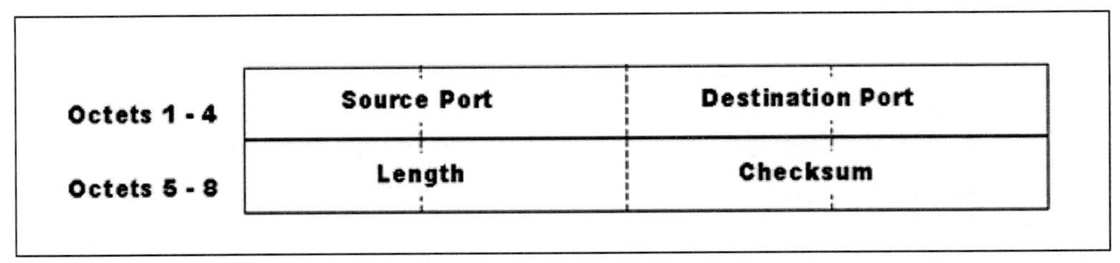

Figure 4-33. UDP Header Structure

The process of receiving an Ethernet frame, examining it using the EtherType code to determine if it contains an IP datagram, then examining the datagram to see if it contains a TCP/UDP or ICMP message (using the IP header's protocol number), and finally delivering that message to the right application program (using the destination port number) is diagrammed in figure 4-34. As the figure notes, some messages are Layer 2 messages (such as LLDP or RSTP) and never go up to higher layers. Some messages (such as ICMP messages) are used at the network layer (Layer 3) to manage connectivity and routing. Other messages end up delivered to specific application programs or to registered services based on their TCP or UDP port numbers. Figure 4-34 is based on the 5-layer IEEE 802 model and not the 7-layer OSI reference model.

Port Numbers

Looking at figure 4-34, it should be noted that the process starts at the bottom and proceeds upward. This diagram represents how a TCP/IP frame informa-

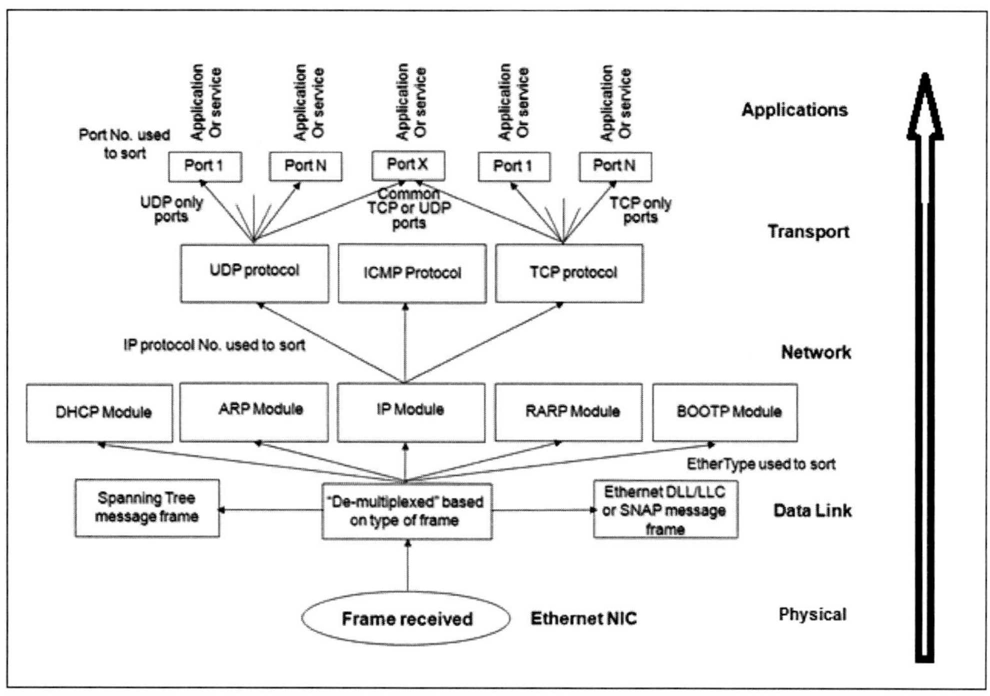

Figure 4-34. Frame Decapsulation and Delivery

tion is acquired by Decapsulation—the opposite of the process that encapsulated the data within the layer information on the transmit side. In figure 4-34, the layer formatting is being stripped off and the information used. An Ethernet Frame is received at the correct MAC address (Physical and Data Link) as long as there are no bit errors (any errors and the frame is dropped and a replacement requested). The EtherType is used to determine which TCP/IP suite module will handle this frame. If it is an IP frame, then where the frame information goes will be determined by the network header (whether it is a UDP, ICMP, or TCP type information). At this point, it is sent to the port address, which is the address of the handler for this type information. As was mentioned, end-to-end message delivery, regardless of whether TCP or UDP is used, requires a means for identifying the specific program to which message data is to be delivered. To save memory, the IETF designated a 16-bit integer number as a means for defining a port number. If you think of an IP address as the street address of an office building, the port number is like an office number within that building. Delivery of a letter requires both! So does delivery of message data. In an IP-based network architecture, programs wanting to communicate will request a port number and be assigned one, either permanently (as in the case of a service that starts up every time a computer is rebooted—for example, a web server or ftp server) or temporarily (as in the case of programs you run when needed and then terminate—for example, a browser or telnet client.) Port numbers range from 0 to 65,534 but fall into three groupings:

- 0–1,023: Well-known port numbers permanently assigned to common services
- 1,024–49,151: Registered port numbers known to be commonly used for a given purpose
- 49,152–65,534: Temporary, reusable port numbers

Note that most programs and services will accept messages on the same port, regardless of whether delivery is by TCP or UDP. Some applications and services require either TCP message delivery or UDP message delivery but not both. Some programs (or services) use two consecutive ports. Table 4-9 shows some examples of well-known port numbers. Table 4-10 shows some examples of registered port numbers. A complete list of both can be found on the Internet at the Wikipedia.

Table 4-9. Examples of Well-Known Port Numbers

Service/Function	TCP port	UDP port
Quote of the day (qotd)	17	17
File transfer protocol (ftp)	20,21	———
Remote host login (Telnet or ssh)	23	23
Time of day (time)	37	37
Domain name server (dns)	53	53
User terminal login authentication (tacacs)	49	49
Trivial file transfer protocol (tftp)	———	69
Distributed document search (gopher)	70	———
System/user status report (finger)	79	———
Hyper text transfer/WWW (http)	80	80
Post office protocol (pop3)	110	———
Kerberos authentication/"ticket" server (kerberos)	88	88
Bootstrap protocol server (bootp or dhcp)	———	67,68

Of course, TCP isn't the only Layer 4 protocol. There is the OSI TP and Novell's SPX (Novell's equivalent of IP protocol is IPX). The OSI Transport layer provides five classes of transport, four of which require connection-mode network service (CONS). The vast majority of OSI installations use connectionless transport (CLNS) and do not (of course) require CONS service. A

Table 4-10. Examples of Registered Port Numbers

Service/Function	TCP port	UDP port
Microsoft DCOM services (not just these ports)	1026,1029	———
SOCKS proxy service	1080	———
Tripwire IDS	1169	1169
OpenVPN	1194	1194
Nessus Security Scanner	1241	1241
IP_{SEC}	1293	1293
IBM/Lotus Notes RPC	1352	———
Microsoft SQL Server	1433	———
Citrix XenApp Thin Client	1494	———
Microsoft P2P Tunneling Protocol	1723	1723
RADIUS authentication server	1812	1812
Adobe Macromedia Flash	1935	———
World of Warcraft Online Gaming	3724	3724

list of the bit sequences is not provided here, due to their complexity and multitude of options, and because the OSI architecture has been effectively eliminated by TCP/IP.

Two features of OSI are worth commenting on: congestion avoidance and flow control window adjustment. The congestion avoidance algorithm allows the Network layer to notify the Transport layer when congestion is detected, for which the Transport layer can reduce its outstanding packet balance. If the Transport layer determines that the network has lost a packet, it can reduce the outstanding packet window to one, then increase the balance by one each time an acknowledgment is received. This process is called *flow control window adjustment*.

TCP has similar mechanisms for flow control and congestion relief. TCP also has a mechanism called *slow start*, whereby messages are initially sent using the minimum segment size and then each successive message is doubled in size until the MSS is reached or the sender is notified of delivery failure. In this second case, the segment size is dropped back to the prior (successful) size and all further message segments will use that size unless subsequent delivery problems arise due to traffic congestion and the segment size has to be reduced again.

As has been mentioned, TCP, being a connection-oriented protocol, establishes a persistent connection (called a *session*) between the two end-point applications that are communicating. The session (at each end) can be in one of a dozen states (e.g., SYN SENT and LAST ACK). Based on the current state, there are only certain permissible (valid) commands permitted and, for some states, only a limited amount of time allowed to remain in that state. (By "commands" we mean the setting of various flags in the TCP header to request/indicate state changes.) Figure 4-35 shows the state transition diagram for TCP and the flag combinations (or other actions) that cause transitions to successive states. When we later discuss security and firewalls, the ability to identify the current state of a TCP session and the allowed/illegal state transitions will be important and are based on this diagram.

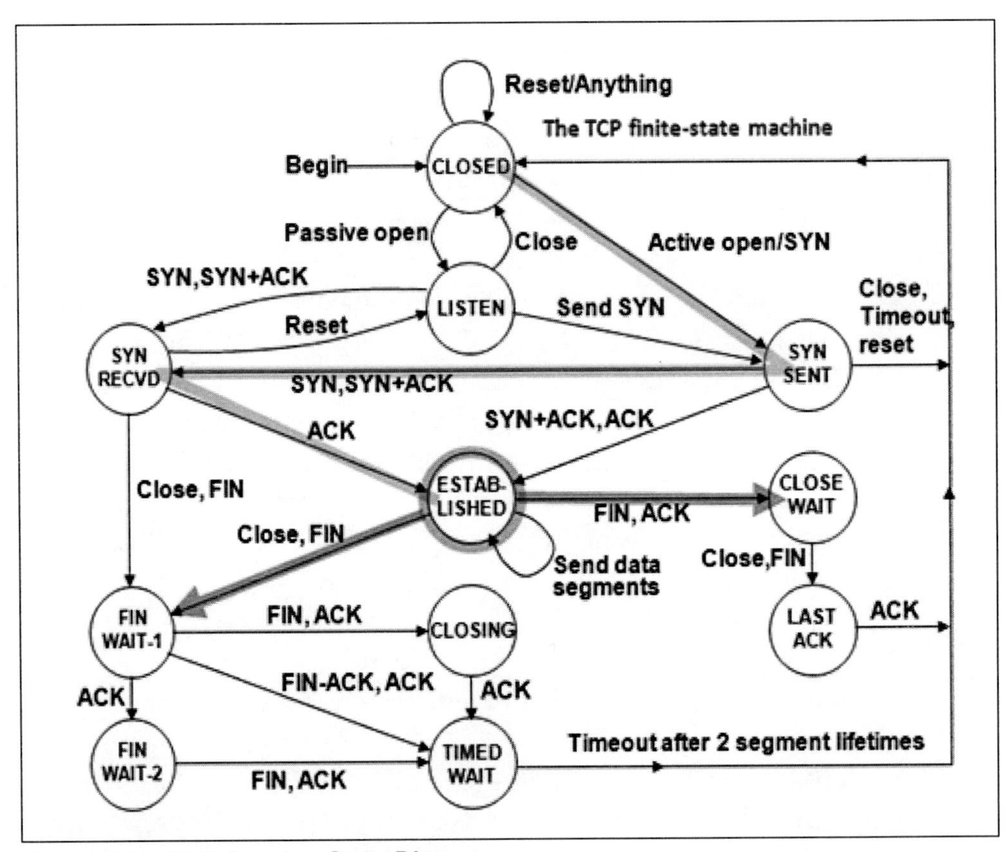

Figure 4-35. TCP Session State Diagram

Each segment of data sent from one application to another application through the TCP session typically includes (in addition to any data being sent) the acknowledgment value and ACK flag to indicate how many octets of data have been received at the other end. Each segment received includes a "window" value advising each application of how many additional octets of data the other application is prepared to receive and accept.

Other flag bits may be set to indicate a requested state transition. (Refer to table 4-11 for a description of the TCP header flags and their meaning and use.) For example, if the application at the other end needs data sent immediately, rather than letting it be buffered, it will set the PSH (push) flag. If the application at the other end (or TCP at the other end) has had a corruption of data, it may set the SYN (synchronize) flag to request a re-synchronization of the octet counts, tracking the data flowing in each direction. If communication problems and data errors are being experienced, the other application (or TCP at the other end) may set the RST (reset) flag to request that the session be reset and restarted. When the application at the other end is done sending data, it sets the FIN (finished) flag to indicate this and to allow the other application to conclude its data transmission and then close down the session.

The flags in the TCP header are critical to understating how TCP works. (Note that there is a different version of TCP for IPv6 because of differences in how the [pseudo] header checksum is computed.)

Table 4-11. TCP Header Flags

NS		ECN-nonce sum, used for concealment protection.
CWR	Congestion Window Reduction	This flag is set by the sending host to indicate that it received a TCP segment with the ECE flag set and had responded in congestion control mechanism.
ECE		ECN-Echo indicates: If the SYN flag is set (1), the TCP peer is ECN capable. If the SYN flag is clear (0), a packet with Congestion Experienced flag in IP header set is received during normal transmission.
URG	Urgent	Indicates that the Urgent pointer field is significant.
ACK	Acknowledge	Indicates that the Acknowledgment field is significant. All packets after the initial SYN packet sent by the client should have this flag set.
PSH	Push	Push function. Asks to push the buffered data to the receiving application.
RST	Reset	Resets the connection.
SYN	Synchronize	Synchronize sequence numbers. Only the first packet sent from each end should have this flag set. Some other flags change meaning based on this flag, and some are only valid for when it is set, and others when it is clear.
FIN	Finished	No more data from sender.

LAN Layer 5: Session

The upper three layers, starting with Layer 5, are usually considered to be data processing more than data communications. These functions are typically found in the network operating system. Integrated and not well structured into finite layers, most network operating systems perform these communications functions as part of their core services and usually through a protocol stack.

LAN Layer 6: Presentation

This function is usually integrated into the operating system but may be accessible for encryption (Crypto API) or for other forms of syntax conversion.

LAN Layer 7: Application

LAN Layer 7 is where the data is interfaced to the application software. There are many utilities (file systems, directory services, e-mail, etc.) or services running that support various applications. These utilities or services are very much part of the operating system or proprietary Layer 7 for industrial LANs. Ethernet can be the Layer 1 and 2 standard and TCP/IP can be the Layer 3 and 4 standard; however, there is a great deal of confusion about how to standardize Layer 7 services. On one side of the commercial LAN world is the UNIX camp, which uses the Network File System (NFS), and on the other side is the Windows camp, which uses the Server Messaging Block (SMB). The industrial Layer 7 arena is no better, with no particular standard prevailing, resulting in confusion. Even if systems are standardized from Layers 1 to 4, it does not mean that they will talk to each other if they have different Layer 7 protocols.

The IEC 61158 standard for fieldbuses defines Application layer standards for each of the service types. Of these types, the most common are FOUNDATION Fieldbus (Type 1) and PROFIBUS (Type 3), which share a common Application layer, called *Fieldbus Messaging Specification (FMS)*. Another commonly used protocol is ControlNet (Type 3) and its Ethernet equivalent, EtherNet/IP, both of which share Common Industrial Protocol (CIP) at the Application layer. Modbus/TCP also has a well-defined set of Modbus commands at the Application layer.

Summary

In this chapter, local area networks were discussed in concept from the vantage point of Layers 1 through 4. The majority of industrial networks use just Layers 1 (Physical), 2 (Data Link), and 7 (Application). However, with the growing emphasis on communicating across the entire enterprise, routing and transport layers are becoming more widely used. Routing and Transport operations, as well as the different access methods, were also discussed, along with the structures of various LAN protocol frames from Layer 1 through Layer 4.

The reader should be aware that—beyond Layer 1—you need detailed knowledge of a particular protocol to achieve more than a conceptual understanding. In most cases these protocols are not adjustable, not even to system programmers, so monitoring and analyzing are the usual procedures for determining irregularities or discontinuities in service.

In addition to 802.3, this chapter also discussed a number of other protocols, including the newer wireless ones under the 802 umbrella. These will be of increasing importance to industrial settings insofar as wireless technologies offer some unique advantages in application. The wealth of information you can gain by analyzing these protocols is reason enough to seek deeper understanding. If you want to study the many topics in this chapter in greater depth, an Internet search engine will provide you with all the data and perspectives you will need.

Bibliography

Note that Internet links may change.

Comer, D. E. *Internetworking with TCP/IP Principles, Protocols and Architectures.* Upper Saddle River: Prentice-Hall, 2000. ISBN 0-13-018380-6.

Deering, S. and R. Hinden. *Request for Comments: 2460 Internet Protocol Version 6 (IPV6) Specification.* December 1998.

International Electrotechnical Commission. http://www.iec.ch.

Institute of Electrical and Electronic Engineers. http://www.ieee.com.

Institute of Electrical and Electronic Engineers. *IEEE 802.2, Logical Link Control.* New York: Wiley-Interscience, 1998.

---. *IEEE 802.3, Local Area Networks and CSMA/CD Access Method.* New York: Wiley-Interscience, 2005.

---. *IEEE 802.4, Local Area Networks: Token-Passing Bus Access Method.* New York: Wiley-Interscience, 1988.

Palmer, M. J. and B. Sinclair. *A Guide to Designing and Implementing Local and Wide Area Networks.* Boston: Course Technology, 1999.

Stallings, W. *Local and Metropolitan Networks.* 6th ed. Upper Saddle River: Prentice Hall, 2000.

Webopedia. http://webopedia.internet.com.

Wikipedia. "IEEE 802." http://en.wikipedia.org/wiki/Category:IEEE_802.

Wikipedia. "Wireless Mesh Networks." http://en.wikipedia.org/wiki/Category:wireless+mesh+networks.

5 Network Software

Introduction

In this chapter we discuss network operating systems and some software ancillary to the core network operating systems. The specific systems selected for discussion here were chosen because they made the best possible illustration of the variety of system types available. Their inclusion does not constitute either author's endorsement. Much of the information about these systems comes from the manufacturers' specifications and maintenance literature, which takes precedence over any system data presented here.

However, the authors have offered comments based on their experience with these systems. Given the unusually contentious atmosphere surrounding discussions of network operating systems and particular types of industrial control systems, the authors hasten to add that the opinions expressed here are their own, not those of any organization (particularly the ISA), and that they have neither received, nor will receive, compensation of any type for their comments for or against any network operating system.

Object-Oriented Programs

To start a chapter on network operating systems, there is a need to build some definitions and discuss the modern goals of software design. Object-oriented programming (OOP) is a good place to begin discussing network operating systems. Much has been written about what object-oriented programs are and are not. Their advantages include: reduced development time, better organization of programming efforts, and reusable code. Today, any modern network system used in either commercial or industrial applications should be object-oriented and have a good "object model" (the structure and interface requirements for objects and classes of objects).

Object-oriented programs have classes of objects. A particular manifestation of a class is called an *instance* and to create such an instance is called *to instantiate*. The concept of "class" can be explained by an example: a class of objects known as Dog. An instance of the class Dog is an object called *Poodle*; another instance of this class is an object called *Doberman*. They have the same main features found in all objects of the class Dog—i.e., four legs, canine teeth, a propensity to bark, and some semblance of a tail—but they are different in detail. If you had an instance of the class Dog (and you could make changes to it), you could make your own object by changing the various properties (size, weight, color, hair style, appendage proportions, etc.) and perhaps have a cocker spaniel. The point is that the class must be defined before objects can be made from it. This object-orientedness enables software to be built out of components, rather than be all original or new each time. An automobile is built from components; so, also, is a complex software program. This concept is the basis for the Component Object Model (COM) and now .NET, which are a legacy and a current object model, respectively.

Distributed control systems (DCS) and supervisory control and data acquisition (SCADA) systems (and more recently, programmable logic controllers [PLCs]) have long used a basic object model concept. In most OOP programming, the objects are created by the program and cease to exist when the program terminates. A special category of objects is called *persistent* objects. Once created, these objects continue to exist until they are intentionally destroyed, which usually means that they are held in some form of database. In a DCS, for example, the creation of regulatory control strategies is typically done by instantiating copies of function blocks from the vendor's library, linking them together into a chain, and downloading them into a controller for execution. These function blocks are objects that have specified functionality, as well as "methods" (things you can tell them to do) and both settable and readable parameters and values. For example, a basic control loop might be built by creating and linking an analog input block, a PID block, and an analog output block.

To be robust, extensible, and scalable, a networked application should have a good object model. In computing, an object is program code and data that can be summoned or called (e.g., a function), may be reused, can have "relations to siblings" and "inheritance with children," and generally makes life easier for programmers. Objects are also the basis for other network application software programs, such as JAVA. Objects make possible faster production of programs, more consistency in program operation, and the use of less code.

An object can be something as simple as the code for a push button that is included as part of a graphical human-machine interface display. Using this code, the programmer could indicate whether the push button is depressed, has been depressed, or is not depressed. The object code could make the button's graphical representation "illuminate" (change color), dictate the alphanumeric text on or around the button (e.g., "RUN" or "OFF"), or locate the button's position on a screen. All of these parameters are easily amended as attributes of an object, and simple data entry, by manual or automated means, can set these values. Objects can also be as complex as a chart recorder or a PID controller.

Microsoft Visual Basic (as well as C# and C++) is an example of a programming language that employs objects. Microsoft used programming based on OOP, so a program can be built entirely of software components (objects) using a COM server (Microsoft Transaction Server for NT; Component Services for Windows 2000). These are now legacy products. Those familiar with Windows know that most applications are made up of an EXE program (the executable) and usually one or more dynamic link libraries (DLLs), and that DLLs were actually objects in the Component Object Model (COM). What Microsoft used to call object linking and embedding has (with exceptions) had its name changed to ActiveX. The difference between an EXE and a DLL is that the EXE is referred to as *in process*, running in same memory space, and the DLL is for "out of process," running in a different memory space. Dynamic link libraries (DLLs) are so-called because they are programs that are brought into available memory when needed. The DLL physical memory location can change from one use to the next and the "links" (memory addresses), referenced by the calling program, are edited by the program loader to point to the library location in physical memory.

Microsoft currently uses .NET (.NET Framework 4.5), a system based in part on Extensible Markup Language (XML) and what was formerly called Simple

Object Access Protocol, now simply called *SOAP*. Microsoft .NET uses a virtual machine to implement the data and structures in XML files; technically, it uses an intermediate language, common language runtime (CLR), and a real-time compiler for the virtual machine. A programmer could compose his object or program snippet in any of the languages that output the CLR, such as Visual Basic, C#, or C++, and they will execute on the virtual machine.

Companies typically use object-oriented programming with Oracle's JAVA (created by Sun Microsystems which was later acquired by Oracle), which assumes that a JAVA virtual machine is on the device receiving the JAVA code. JavaScript is a scripting language which bears no relation to JAVA, other than similarity in name and syntax (it should actually be called *ECMA Script*), and it uses a C++ programming style. JavaScript was developed by an entirely different company and then purchased and named by Sun Systems. Older methodologies, such as COM and Distributed COM (DCOM), are now known as legacy items. An entire set of COM and DCOM objects, called *object linking and embedding* (OLE) for process control (now-OPC Classic), is still used in many industrial applications today. We discuss OPC and its move away from COM/DCOM further toward the end of this chapter. Microsoft's migration away from DCOM (and the underlying remote procedure calls [RPCs]) to their .NET architecture has had a major impact on the evolution and future direction of OPC.

Commercial Systems

Although this is a book about industrial data communications, most of the systems used in the industrial environment had their start in the commercial world. At present, that world is divided into several camps, all of which give the impression that they are "standards-based." Unfortunately, interoperability between them is often difficult to accomplish.

Stand-alone Systems

The various application models were discussed in chapter 2 and the reader is referred back to those models as they are mentioned. Not so long ago, the PC was much like the model shown in figure 5-1: not networked and not connected to anything but the occasional bulletin board. Note how self-contained this so-called one-tier PC model was.

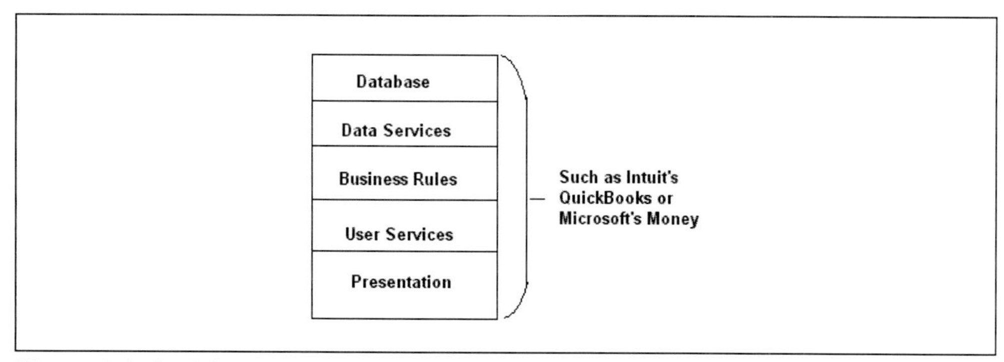

Figure 5-1. Self-Contained (One-Tier) Model

This one-tier model illustrates the way most PC application software was written. A proprietary presentation means (usually a graphical user interface [GUI]) would present the information to the user. User services would interface between the business rules and the presentation, usually by presenting hooks (a programmer's way of saying addresses that handle events) to the resident operating system using the GUI. The business rules would dictate what the user could and could not do with the data. Data services would interface the business rules and the database, which stores the information. This model is how a stand-alone analyzer in the industrial area would operate.

Two-Tier Systems

Two-tier models, such as client-server, are more complex. Before discussing the client-server model, we need to define a few terms and identify the resources a server shares (see table 5-1).

A two-tier system is a client-server model. The figure for the two-tier system may be found in chapter 2 (figure 2-10), as well as a detailed discussion of thin/fat clients and n-tier systems. The reader is referred to those discussions for clarification of the material that follows.

Table 5-1. Server Resources

Servers share resources such as:
Printers/Scanners/Fax
Floppy/Hard drives/Optical drives/Bulk storage
CD-ROMs/DVDs/Flash drives
Backup media (e.g., DAT tape or Blu-Ray)
Peripherals (that allow sharing)
Servers provide network resources and services such as:
Communications
Email
Internet
Web
Printer queue
Database
Backup
File system
Security
Network management

With respect to a client server, thick and thin are not absolutes. There are hybrid systems in which the local PC has some application software but accesses a server for other applications. In the world of computer-based automation, the classic DCS and PLC architectures have been thick client designs in which each workstation or operator console has a full suite of application software and user tools. SCADA systems have made more use of the thin client approach, where there may be a redundant server that supports multiple thin-client operator workstations. A web server/web browser design is a form of thin-client architecture and is being used more often these days in industrial automation systems.

The advantage of the thick client is this: if the primary database server is unavailable and the client is attached to another database server, then the business rules for the client are intact (hopefully the database is a replica of the one that failed, otherwise, problems will result). The thin client, on the other hand, is good for the administrator because there is only one place to upgrade and correct and only one point of control. Thin clients can even be diskless machines that will not operate unless the network is up. Client-server is probably not the model to choose for industrial use, although it has been used in that capacity as noted above.

N-Tier (Three-Tier and Above)

The problem with the two-tier model is that it has no way of distributing functions. To distribute functions it is necessary to go to the three-tier or n-tier model. Since the functions on a three-tier model may be spread out over three or more machines, the three-tier model is just a subset of the n-tier model. The n-tier model from chapter 2 is shown in figure 5-2.

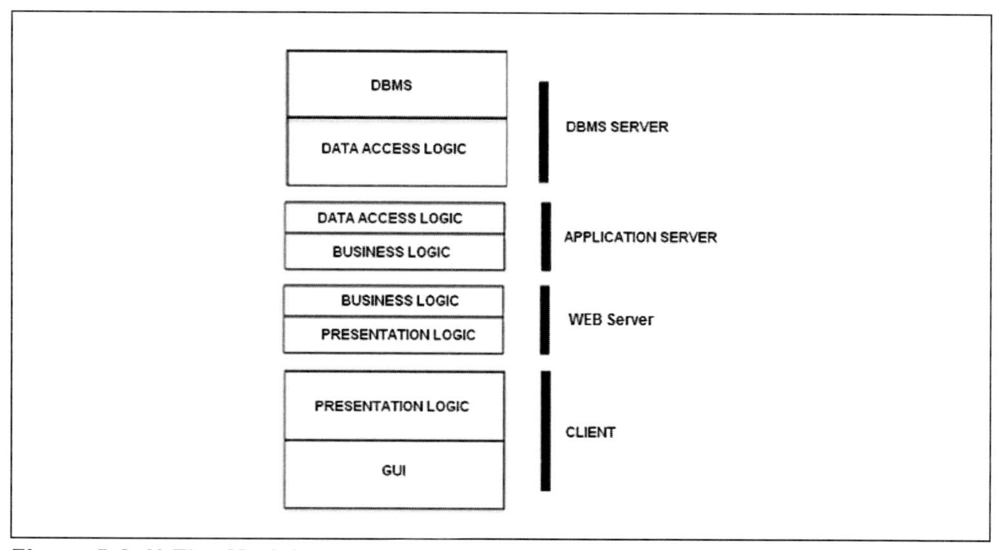

Figure 5-2. N-Tier Model

The n-tier model in figure 5-2 uses an Internet perspective and technologies, such as web servers/browsers, which can and are being used in current industrial automation system designs. The browser can be implemented using an inexpensive, minimal-functionality display device. The process control field has traditionally used the three-tier model for distributed control, using a sole-source proprietary system or mixed stores (such as PLCs, somebody's database, and an integrator's human machine interface [HMI]) to put together a distributed control system. In commercial systems, the integration of the Internet with all business work has meant the widespread use of the n-tier model, which can be used effectively in control systems, even those not attached to the Internet.

One difference between the three-tier and n-tier models is how data is accessed. In the three-tier model, the client interface (typically a proprietary Visual Basic or C++ interface) is operated and requests and receives information through the business rules server. The business rules server may very well use a process like Microsoft's ActiveX Data Objects (ADO) that requests data and receives responses from the database server. The reasons for the business rules are: security and ease of use. The client is not supposed to be able to access the relational database (RDBMS) directly, only through the business rules.

For the n-tier model, the user services may employ the Hypertext Markup Language (HTML) delivered by HTTP or HTTPS (hyper-text transfer protocol or the secure version) for the user interface. The client presentation service is typically a web browser that understands HTML and transfers files and data using HTTP or HTTPS. Examples of web browsers include: Internet Explorer (IE), Opera, Firefox, Pale Moon, Ice Dragon, and Chrome. The browser presentation language (HTML) has been supplemented by eXtensible Markup Language (XML). Microsoft IE 5 (around 1999) through IE 11, Firefox (2004), and Chrome (2008), are some of the browsers that fully support XML, although other browsers have been continually improving their support for it. XML conveys meaning about syntax, rather than simply page layout. In addition, the introduction of dynamic HTML and XHTML, as well as various server/client-side scripting languages, has dramatically expanded web server/client capabilities. One of the present n-tier models has the client presentation services requesting responses (via HTML), such that one of the application web servers will request data via an active server page (which is enforcing the business rules) from the database server that accesses the data and returns it to the client.

The n-tier model generally used today is called the *application model*, which is illustrated in figure 5-3.

In no case can the user services ever directly connect to the data services; this ensures that the business rules are enforced. The direct applicability of the application model to control systems should not be overlooked.

The Internet

As a networked system, the Internet is an n-tier design. We describe it in some detail in chapter 8, "Internetworking," but the key point here is that both it and the technologies it has spawned are either used now or soon will be used in industrial systems. The application model and Internet n-tier technologies have already revolutionized office systems by coming closer to the goal of providing simple, universal, and platform-independent access to corporate

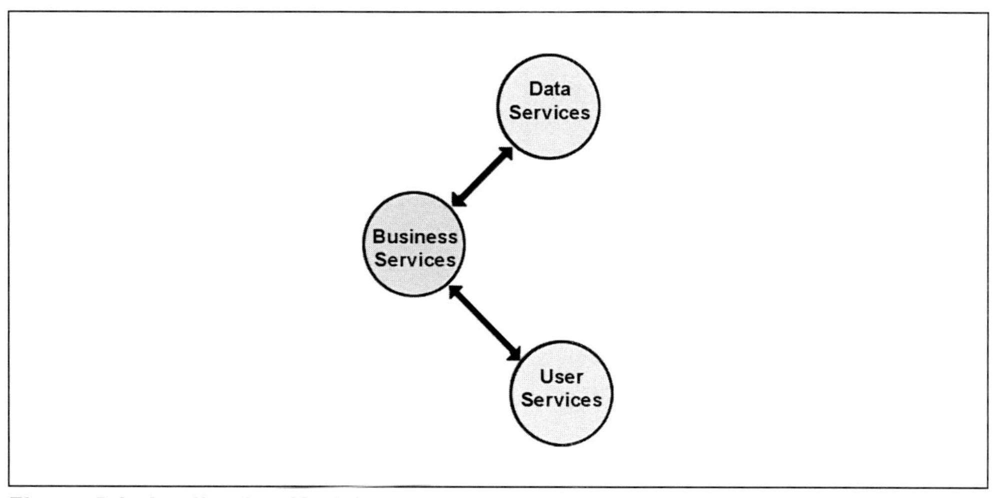

Figure 5-3. Application Model

data. As a medium for connecting geographically separated plants, the Internet offers the tremendous advantage of costing only the services of an ISP for truly high-speed interconnection. As time passes, more vendors are building a web server (as well as wired and wireless Ethernet) into their network-connected products and systems, and offering web access for system administration and maintenance functions. (*Please note here that the authors do not advocate running a control system over the Internet.*) However, Internet technologies have established themselves on company Intranets (the facility network, not the Internet). Also note that there *have* been SCADA applications deployed using the Internet as the WAN technology that connects all the sites together.

Network Operating Systems

Given the previous background and supporting information, this would be a good point to start a discussion of basic network operating programs, known as Network Operating Systems (or NOS). These differ from application programs in that the NOS supplies the basic network services for the applications. An entire book could be devoted to network operating systems, their design, and implementation. There is neither the space nor the requirement for a detailed explanation, so only some salient facts about data communications applications will be discussed along with a few of the more popular network operating systems. If your favorite system is not included, our apologies.

Microsoft Family of Windows Products

Microsoft Windows has the largest market share of desktop systems (over 90%) and has made significant inroads into the server business, particularly in process control. It acquired its share by offering reasonably-priced software that would do 90% of what most people wanted at least 90% of the time through an easy-to-use interface. No one who ever used the *earlier* Windows network versions (3.11/95/98/ME/NT) will say that it always did what he or she wanted, however, it certainly gave reasonable performance for its intended use and price.

PC Operating Systems

Microsoft has offered a myriad of operating systems under the marketing-driven title "Windows" that are different in concept and execution. Windows 3.1 (not used anymore) was a 16-bit system with a GUI glued onto Microsoft DOS. Windows 95 was a hybrid system supporting 16-bit DOS programs (using shared memory and cooperative multitasking) and true 32-bit (Win32) programs (using protected memory and preemptive multitasking). As with anything, trying to be all things to all programs meant that it did not excel at any one of them. Windows 98 and Windows ME were really upgrades to 95. Windows 98 added support for plug-and-play, while ME was multimedia oriented.

Windows NT ("New Technology" – NT 3.1 through 4.0) was good to start with (protected memory, small kernel, preemptive multitasking) and has done nothing but improve the operating system's capabilities (networking as well as operating system performance—at the cost of increased memory and disk space requirements). Windows 2000 limited plug-and-play to those devices that could auto-configure, so plug-and-play was far more successful and less rebooting was required to refresh the registry. W2K Professional was a suitable industrial workstation operating system (OS) for many applications and the server editions (Server, Advanced Server, and Data Center Server) performed reliably, according to many who used these systems (including the authors).

Windows XP was an effort by Microsoft to do two things: Primarily, it wanted to establish a standard code base across all of its operating systems (XP Home, XP Professional, and Server 2003). Secondly, it wanted to move more into subscription services, where software is rented in perpetuity rather than purchased. (At least that was the thought at the time; it seems it has not quite worked out that way yet.) Windows XP Pro (Service Pack 3 [SP3]) was quite probably one of the best clients available (our opinion, remember?) at the time.

Since the last version of this book, Microsoft has introduced a number of operating systems:

Windows Vista, circa 2007, (Starter, Home Basic, Home Premium, Business, Enterprise, Ultimate); Windows 7, circa 2009, (Home Premium, Professional, and Ultimate; others available by licensing); and Windows 8, circa 2012, and 8.1, circa 2013, (Windows 8 and Windows Professional; Enterprise available by licensing).

Most users in industrial data communications circles seemed to be of the opinion that Vista was not quite ready for prime time at the time of its introduction, so they held on to XP. Windows 7 was generally praised by most, who considered the operating system to be a major improvement over Vista due to its increased performance, its more intuitive interface, fewer User Account Control pop-ups, and other improvements. Additionally, Windows 7 offered a compatibility mode for including those older programs that could not quite run under Windows 7. In general, Windows 7 has been implemented in industry (since 2012) to replace Windows XP SP3 (which is no longer supported by Microsoft as of April 2014, meaning no more patches except for embedded systems); however, that replacement is neither total nor complete at this date (2014). According to Wikipedia, "From its formal release (2009) in just 6 months, over 100 million copies had been sold worldwide, increasing to over 630 million licenses by July 2012, and a market share of 47.49% as of February 2014 according to Net Applications, making it the most widely used version of Windows."

Windows 8 introduced major changes to the operating system's platform and user interface to improve its user experience on tablets, including Android and iOS. These changes were based on Microsoft's "Metro" design, with an emphasis on touchscreen input, and Windows Store, an online store for downloading and purchasing new software. Windows 8 added support for USB 3.0 and cloud computing (where data and programs are located in a remote server available through an Internet connection). Additional security features were introduced, such as built-in antivirus software and integration with support for Unified Extensible Firmware Interface (UEFI) Secure Boot on supported devices with UEFI firmware (to prevent malware from infecting the boot process).

Windows 8 had a lukewarm reception. Proponents pointed to its performance improvements, security enhancements, and improved support for touch-screen devices; however, many questioned the new user interface, as it was widely thought to be both confusing and hard to learn without a touchscreen. Its lack of applicability (other than to HMIs) in the industrial arena is thought to be mostly due to interface issues.

On October 2013, Microsoft released Windows 8.1. It addressed some of the criticized aspects of Windows 8 and incorporated additional improvements to the operating system, primarily to the user interface.

Windows 10 is the newest version of the Windows operating system family. Announced in September 2014, it is scheduled for release late in 2015. Windows 10 adds the ability to run Windows Store applications inside a desktop window (rather than demand the whole screen as in Windows 8) and a Windows 7-like Start menu, in place of the Metro Start screen seen in Windows 8. Microsoft also unveiled the concept of the "universal Windows app," allowing applications to be utilized by Windows Phone 8.1 and Xbox One, while sharing a common codebase and allowing user data and licenses for an application to be shared between multiple platforms. It is rumored that Windows 10 will be a free upgrade to Windows 8 users, more in-line with other vendors' distribution methods, as Windows 8.1 is presently free to devices with screens smaller than 6 inches (2014). As to its applicability to modern industrial systems, that will remain to be seen.

Server Operating Systems

A server uses the underlying desktop operating system core base; however, because its purpose is to share resources, it requires different software applications and hardware (depending on the server) than the basic operating system on a desktop. Server tools are added to make the sharing more efficient and easier to access. Centralization of applications on a server(s) allows centralization of security and updates, and perhaps directory services. Microsoft entered the server world with the introduction of NT Server 3.1, a processor-independent, multiprocessing, multi-user operating system.

The server family of the basic operating systems since NT 4.0 is shown in table 5-2.

Table 5-2. Microsoft Server Applications

Application	Description	System
Server 2003	Web, Standard, Enterprise, Data Center, Small Business Server	IA-32 and x86-64
Server 2008	Windows Server 2008 Standard	IA-32 and x86-64
	Windows Server 2008 Enterprise	IA-32 and x86-64
	Windows Server 2008 Datacenter	IA-32 and x86-64
	Windows Web Server 2008	IA-32 and x86-64
	Windows HPC Server 2008	Clustered comp.
	Windows Storage Server 2008	IA-32 and x86-64
	Windows Small Business Server 2008	IA-32 and x86-64
	Windows Server 2008 Foundation	x86-64bit
Server 2008 R2	Foundation	x64 64-bit only
	Standard	x64 64-bit only
	Enterprise	x64 64-bit only
	Datacenter	x64 64-bit only
	Web	x64 64-bit only
	HPC Server	x64 64-bit only
	Windows Storage Server 2008 R2	x64 64-bit only
Server 2012 (& R2)	Foundations	x64 64-bit only
	Essentials	x64 64-bit only
	Standard	x64 64-bit only
	Data Center	x64 64-bit only

Table 5-2 Definitions

x86-64 (also known as x64, x86_64 and AMD64) – 64-bit version of the x86 instruction set, can run either 32-bit or 64-bit server applications

Clustered comp – Two or more connected computers working on the same software actions to increase computing power

IA-32 (Intel Architecture, 32-bit) – x86 versions that support 32-bit (but not 64-bit) computing

x64, 64-bit only – 64-bit version that runs only 64-bit server applications (32-bit applications require emulation)

While not yet announced, Microsoft has stated that their next server (Windows Server 20XX) will thoroughly support Azure (Microsoft's "cloud computing" platform) and, through compatibility with Linux (via Docker Inc.) and other operating systems, it will support their vision of cloud computing.

Regardless of the edition, the code base is essentially the same between the desktop and the server software in XP/Server 2003, Vista/Server 2008, Windows 7/Server 2008R2, and Windows 8/Server 2012.

Cloud Computing

The big push commercially is "cloud" computing, this is an effort to again penetrate the subscription market. Office 365 is a good example of an application living in the cloud; for a monthly fee (quite reasonable) you have an up-to-date, patched version of Office. The problem is that most users of control systems are reluctant to base operating their control systems over the Internet. One means of obtaining the advantages of cloud computing, which is really a thin client scheme, is by using Internet technologies on the Intranet. This means installing a local cloud server, or set of servers, onto an Intranet (company owned network).

Embedded Operating Systems

Windows CE was Microsoft's entry into the small-footprint operating systems market for personal digital assistants (PDA) and sub-notebooks, including tablets. It has found applications in industry because of its low resource requirements (compared to its big brothers) and its ability to interface easily with its kin. Microsoft has made a big marketing push to have Windows CE, or a selective combination of CE objects, on wireless phones, copiers, and office equipment of any description.

Windows CE 5.0 became Windows Embedded CE 6.0 (12/2006), a componentized operating system designed to power small-footprint devices such as:

- TV set-top boxes
- Thin clients
- Digital media adapters
- Voice-over-IP (VoIP) phones
- Navigation devices

- Medical devices
- Portable media players
- Home gateways
- Digital cameras
- Networked digital televisions
- PDAs

Windows-embedded CE 2013 was released in June 2013 and will be supported until October 2023. It supports a number of networking features, including IPv6 (Internet Protocol version 6).

The Microsoft family of embedded operating systems includes:

- Windows Embedded Standard 7, released in 2010, is based on Windows 7 and includes features such as Aero, SuperFetch, ReadyBoost, BitLocker Drive Encryption, Windows Firewall, Windows Defender, address space layout randomization (ASLR), Windows Presentation Foundation, Silverlight 2, and Windows Media Center, among several other packages. It is available in IA-32 and x64 versions.
- Windows Embedded Professional consists of Windows Vista for Embedded Systems and is available for both IA-32 and x64 processors.
- Windows Embedded Industry is for industry devices and was based on Windows XP Embedded. However, an updated Windows Embedded POSReady 7 (based on Windows 7) was released in 2011 and the name of this product was changed to "Windows Embedded Industry." Microsoft released its version to manufacturing as Embedded 8 Industry in April 2013.
- Windows Embedded NAVReady, also called Navigation Ready, is a plug-in component for Windows CE 5.0 that is useful for building portable handheld navigation devices.
- Windows Embedded Automotive (known previously as Microsoft Auto, Windows CE for Automotive, Windows Automotive, and Windows Mobile for Automotive) is an embedded operating system based on Windows CE for use on computer systems in automobiles.

- Windows Embedded Handheld 6.5 (released in January 2011) has compatibility with Windows Mobile 6.5 and is targeted at retailers, delivery companies, and other companies that rely on handheld computing.
- Windows Embedded Server is the same as their desktop server products. Technically there is no difference between the two; however, in order to conform to the license requirements, Windows Embedded Server should be used only for embedded applications and products.

Starting with Windows Vista, Microsoft also added support for a next-generation TCP/IP protocol stack that allows customized message processing between protocol layers using what Microsoft calls a "shim" (user code invoked as messages pass up/down the stack). That stack also permits concurrent operation with both IPv4 and IPv6 (a dual stack architecture.) Microsoft also added support for some of the IPv4–IPv6 tunneling protocols (e.g., 6in4 and Teredo) that allow IPv4 traffic to pass across an IPv6 network (like the Internet itself) and IPv6 traffic to pass across an IPv4 network.

In short, there are many factors affecting the use of Windows operating systems in the industrial environment, not the least of which is the upgrade/replacement time period. Unlike the IT world—where hardware and software are expected to be capitalized, depreciated, and then replaced every 5 years—an industrial automation system, once installed and commissioned, may be expected to operate as is for 10 or more years. The authors have worked with systems that were over 20 years old and still running the same software and hardware as when they were installed. For this reason, it is not unlikely to find automation systems and smart devices in a plant that are running anything from DOS and Windows 95, all the way up to Windows 8.

UNIX

UNIX, which has been around for over 40 years, has had time to mature. Yet because vendor products (Sun's Solaris, IBM's AIX, Apple's OS X, SCO UNIX, et. al.) needed modifications to the kernel in order to do specific tasks that the vendors deemed necessary, UNIX became a multi-flavored operating system that, in many cases, was not source-code compatible between vendors. UNIX vendors are far less numerous now because many have consolidated and partnered in the past few years, resulting in fewer language variants. The bulk of their competition comes from a UNIX clone, Linux. UNIX has suffered through the years from each vendor's lack of sufficient market share, preventing mass-marketing of their software. Hence, UNIX was (and is) more expensive than the Windows equivalent. To make matters worse (from a marketing standpoint), UNIX originally was, and in its core still is, a command-line-based system (not a GUI system). Although many front ends have been applied to UNIX to give it a graphical user interface, they do not have the low cost or popularity of Microsoft Windows. This is particularly the case within industrial applications where the tools, the ease of use, the cost of Windows 2003/2008/2012 servers, and the different distributions of Linux have greatly eroded what was once a UNIX-only bastion.

Linux

Linux is an open-source clone of UNIX (FreeBSD - Berkeley Software Distribution is open-source UNIX). Open-source does not necessarily mean free, only that the source code is available. In most of its distributions, Linux has a good reputation as an application server and its total cost of ownership rivals or, in some cases, betters that of Windows. It is in the application software field that Linux has had marketing problems. While there are many good Linux programs that will do at least some of what the users want, most programming firms, which are in business to make money from programming services, do not want to give away software or intellectual property rights, as is required by some of the open-source Linux license agreements. While the application programs will run on Linux, the application programs themselves may not be open-source (or free). As a result, the use of Linux has been somewhat limited (when compared to the adoption of Windows applications) in industry. But times are changing and open-source is gaining viability, even in industrial markets.

There are a number of problems with using open source programs, particularly with concern to both financial and operational liability of an open-source industrial program. Whom do you call if the program does not perform correctly, particularly in safety-related areas? Another concern is that, even with greater availability, many of the Linux programs may be challenging for a nonprogrammer to install. They are certainly more difficult for the average operator to install, particularly when compared with Windows' ease of use. On the other hand, for new product development, where specialized hardware and interfaces may be needed and where real-time response is a requirement, Linux may be a better platform of choice. The hardware abstraction layers and task-switching mechanisms in Windows do not provide a real-time response capability much better than fractions of a second.

In spite of the challenges it has faced, Linux closes the gap with each subsequent release. A number of usable GUIs are available for Linux, including some that look very similar to the Windows XP GUI. Now the major competitive issue in the open-source camp is whose distribution is better. Sun has thrown its weight behind Red Hat, while Microsoft (yes, you read that correctly) is now assisting with a competitive distribution: Novell's SuSE (originally a German software distribution company). Apparently, Novell will do the open-source things while Microsoft will make Windows (particularly the servers) work better with Linux—a win-win for users, if it works out. In 2011, Microsoft and SUSE, an independent business unit of The Attachmate Group Inc., announced a 4-year extension of the groundbreaking agreement struck nearly 5 years before, between Microsoft and Novell, for broad collaboration on Windows and Linux interoperability and support. This relationship will extend through 1 January 2016, with Microsoft committed to invest $100 million in new SUSE Linux Enterprise certificates for customers receiving Linux support from SUSE.

Another popular release of Linux-based systems is Ubuntu, which is applicable wherever Linux computers can be used. Ubuntu is composed of many software packages, of which the majority are free. Free software (as stated by *Wikipedia*) gives users the "freedom to study, adapt/modify, and distribute it." Ubuntu can also run proprietary software. The Ubuntu Desktop is built around the Unity GUI. Ubuntu comes installed with a wide range of software that includes LibreOffice, Firefox, Empathy, and Transmission. Programs are mostly free but there are also priced products, including applications and magazines. Ubuntu can also run many programs designed for Microsoft Windows, such as Microsoft Office, through Wine (a Windows environment emulator) or by using a Virtual Machine, such as VirtualBox or VMware Workstation, although these virtual machines require a copy of the operating system license. Ubuntu can be installed onto the hard disk from within the Live CD environment.

For increased security, the *sudo* tool is used to assign temporary privileges for performing administrative tasks, allowing the root account to remain locked and thereby preventing users from making catastrophic system changes or opening security holes.

NOTE Sudo is a program for the Linux operating system that allows users to run programs with the security privileges of another user (normally the superuser, or root). The name is a concatenation of "substitute user do."

Ubuntu can close its own network ports using its own firewall and it compiles its packages using GCC (Gnu C Compiler), enabling particular features that greatly increase security with a minimal performance cost. Since Linux is free, it can be configured to run in a diskless environment and it includes or supports other free packages, such as Apache web server, MySQL relational database, and PHP scripting. It has become a popular choice as an embedded operating system for specialized industrial devices, including: analyzers, condition monitoring systems, emission monitoring systems, and smart RTUs.

Windows CE (and the embedded systems that follow from it) and Linux are real-time operating systems, meaning that the operating system can respond very quickly and make a context switch to applicable support functions at the microsecond level. Thus, although the popular desktop Windows has made significant inroads into the server and workstation portions of many industrial automation systems, the real-time process controllers of such systems tend to still use either vendor-proprietary operating systems, Windows CE, or, more recently, Linux variations. There is also a high-security version of Red Hat Linux, called Security Enhanced Linux (SELinux) that is aimed at applications where a hardened, cyber-secure platform is essential, such as in enterprise firewalls and network intrusion detection systems (NIDS).

Other companies make network operating systems (e.g., Apple, which uses UNIX as the core of its OS-X operating system); however, Windows, UNIX, and Linux are the current "Big Three." Naturally, any company in the industrial networking business would like to build off of an already-successful system because writing an operating system is difficult and tedious at best, particularly when graphics and interfaces to other systems are involved. Yet most industrial system vendors of proprietary systems did just that from the beginning of industrial networked systems (circa 1960s): they designed, programmed, and implemented their own operating system (usually a derivative of UNIX) and developed all the applications, interface screens, and graphics. Some still try to develop new systems—see chapter 6, "Industrial Networks and Fieldbuses."

In chapter 6, you will find some vendors still trying to reinvent the wheel and others capitalizing on successful network operating systems. This proprietary operating system development issue is not as much a factor in the operator workstations and servers for such systems, Windows is well accepted for those elements. It is far more common to see a vendor-proprietary operating system in the process controllers of a DCS system, in the RTUs of a SCADA system, and in the main processor of a PLC, even if the operating system is derived from some other source, such as Linux or another proprietary commercial offering (e.g., VxWorks, pSOS, FlexOS, Windows CE, or QNX).

Windows, UNIX, and Linux manage OSI Layers 5 through 7 somewhat differently. Most layers handle communications in a protocol stack but Layers 5 through 7 are bound (by a process referred to as *binding*) from the stack to addresses and locations that are generally part of the operating system. All three operating systems claim to be compatible with network standards and all have differing degrees of compliance. All offer many features not needed in an industrial network operating system, yet they leave out a few features which must be added through other software programs or perhaps through hardware additions. Perhaps the most-missed feature, at least in desktop Windows, is a real-time dispatcher that would limit the uncertainty over execution times (but that in itself is also a problem with the asynchronous PC).

Protocols Used by Vendors

Each network operating system vendor has a preferred set of protocols. Each will supply the TCP/IP protocol stack and, in general, the entire suite of protocols. Windows (in NT, 95, 98, and ME versions) supplied its own implementation of other vendors' protocols, such as Apple's AppleTalk, Netware's IPX/SPX, and IBM's DLC, in addition to NetBIOS Extended User Interface (NetBEUI). Windows XP can use other protocols but is designed for TCP/IP, as are all current versions of Windows (TCP/IP is required if one is going to run Active Directory services). Novell version 5.2 offers TCP/IP as its native connectivity, and Apple's OS-X is a flavor of UNIX, so it offers TCP/IP as a native protocol.

Microsoft's NetBEUI

Before the Internet and TCP/IP became as popular as they are today, Microsoft was selling software to run on your IBM PC: DOS and eventually Windows 3.1. In that same timeframe, Xerox, IBM, and Digital joined forces to create "Ethernet" (actually a Xerox-owned trademark). People wanted to connect PCs together using Ethernet and Microsoft had to either devise a means for this to happen or allow companies like Novell to dominate in that niche. Microsoft developed a protocol (NetBIOS) to run on top of Ethernet to provide basic messaging services (similar to UDP datagrams and TCP sessions) and to name resolution services (similar to DNS). They added another application layer on top of NetBIOS, called Server Message Blocks (SMBs), to enable file and printer sharing among PCs. To enable application programs running in different PCs to communicate with each other, Microsoft invented a support library of functions that could be called from user applications.

NetBEUI is both a protocol extension of NetBIOS and an application programming interface with a naming convention and a way of interfacing network hardware and network software. NetBIOS (or NBT) transports data. Older Windows systems ran NetBIOS over IEEE 802.2 (Ethernet) and Novell's IPX/SPX using the NetBIOS Frames (NBF) and NetBIOS over IPX/SPX (NBX) protocols, respectively. In modern networks, NetBIOS normally runs over TCP/IP via the NetBIOS over TCP/IP (NBT) protocol.

Although NetBEUI is now a legacy networking protocol, for small networks (less than 100 nodes), NetBEUI is as close as you could once come to a plug-and-play network. There is very little setup—you only have to assign a unique name (a Microsoft NetBIOS name, not a domain name as with the Internet) to each piece of equipment on the network. Identical names are not allowed, as they would confuse the protocol. Using each node's unique computer name, NetBEUI is actually a Layer 4 function providing end-user packet sequencing and recovery, and interfacing with the LLC of Layer 2.

What is missing here is Layer 3. NetBIOS is *not routable* (it only works among computers on a common LAN). It has no provision for routing because it was intended for small networks and uses Ethernet MAC addresses; for that reason, and because it uses acknowledgment windows, NetBIOS is not designed for use with a WAN. However, there is a workaround: to route NetBIOS/BEUI packets, they must be encapsulated in a routable protocol. The interfacing ends must also be spoofed or the time-to-respond will expire on the LANs. Spoofing means generating a response as a result of networking that fools the software into thinking it received the correct response in the correct time.

Unless you have a legacy network, there is now no justification for the use of NetBEUI, as TCP/IP networks are typically now just as plug-and-play as NetBEUI ever was. Remember, since almost all the network diagnostic and troubleshooting tools came from one vendor, very few firms marketed to that vendor's customers. Today, Microsoft has a TCP/IP version of NetBIOS, called NBT (NetBIOS over TCP), that uses SMB messages to enable resource sharing among Windows PCs. Up to and including Vista (but not later versions), these communication functions are usually enabled by default and offer a juicy target for hackers due to inherent cyber vulnerabilities. The solution was to lock down the computer; however, this had to be done by the user and, in many cases, all the user did was to ensure that the PC would run. Later versions of Windows come already in lockdown and the user has to open ports and features (which is why the operating system asks questions during installation), plus the software itself has been hardened. Windows PCs automatically attempt to create a "workgroup" (network neighborhood) when connected on a common LAN. (PCs will revert to being broadcast nodes and will try to resolve for themselves an IP address and network name.) They will auto-configure and assign themselves the correct global IP (provided a network connection is provided) or the correct local link address (no network connection) if they are running IPv6. The user can create a Windows "Domain" for larger collections of computers and establish a domain controller and a Windows Internet Name Server (WINS, although Windows 2003 was the last Server version to support WINS). However, Microsoft suggests going with Internet standards, such as DNS, and using Microsoft Active Directory (AD) on the (Windows) domain controller. Whether you use AD or not, a DNS is a must because Windows networking will only support TCP/IP (unless you like to manually update host files on every workstation).

Regardless, Windows networking still requires the computer name (NetBIOS name), as well as the IP and MAC address, to be unique. Table 5-3 lists the major protocols unique to Windows networks.

Table 5-3. Windows Networking vs. OSI Layer

Name	OSI layer	Description
Redirector	7	Directs requests for network resources to the appropriate server and makes network resources seem to be a local resource
SMB	6/7	The redirector helps implement the SMB messages which makes the resource (server) appear to be local to the requesting device (client)
NetBIOS	5	Controls the sessions between computers and maintains connections
NetBEUI	3/4	Provides data transportation. This protocol may sometimes be called the NetBIOS Frame (NBF) protocol.
NDIS and NIC Driver	2	NDIS allows several adapter drivers to use any number of transport protocols. The NIC driver is the driver software for the network card.

All Windows nodes use parts of the Microsoft networking protocol, primarily consisting of SMB and NetBIOS. The Server Message Block (SMB) protocol is a function of both Layer 6 and 7. It uses the NetBIOS computer name. Microsoft states that the name must be in uppercase letters and can be up to 15 characters in length. There are a number of variants of the SMB, so the first transmitted block will contain the "negotiate protocol" command, which will then negotiate SMB features.

Microsoft can use TCP/IP to transmit Windows networking functions by using NetBIOS over TCP/IP (NBT), where connection-oriented (TCP) and connectionless (UDP) communication are both supported. NetBIOS over TCP/IP is described by product names such as NBT and NetBT. NetBIOS names are used, even though NetBEUI is not used. Windows supports several name-address translation functions that allow for the determination of an IP address from a NetBIOS name. As mentioned above, Microsoft previously supported a WINS server function but now recommends using conventional DNS services in place of WINS.

NetBIOS supports both broadcasts and multicasting and supports three distinct services:

- Naming Services (port 137)
- Session Services (port 139)
- Datagram Delivery Services (port 138)

With SMB used over TCP/IP, NetBIOS names must be used. Microsoft requires that NetBIOS names be in upper case, especially when presented to servers as the "CALLED NAME."

The take-away from this is: if you use Windows networking features (e.g., to see all nodes in the Network Neighborhood—so called in early versions of Windows and now called just *Network*), then you are using NetBIOS over TCP. Modern Windows operating systems are fully capable of using straight TCP/IP, with no NetBIOS computer name.

There are, of course, many variants of the SMB protocol. SMB 2.0 is currently used and has very few (compared to the original 100+) commands. These commands are listed and described in table 5-4.

Table 5-4. SMB 2.0 Commands

Field Name	CMD	Description
SMB2negotiate	0000	Protocol negotiation
SMB2session_setup	0001	Authentication start
SMB2logoff	0002	Authentication end
SMB2tree_connect	0003	Share access connect
SMB2tree_disconnect	0004	Share access disconnect
SMB2query_info	0010	File access locate
SMB2set_info	0011	File access set location
SMB2cancel	000C	File access quit
SMB2query_directory	000E	Directory locate
SMB2change_notify	000F	Directory change
SMB2oplock_break	0012	Cache coherency
SMB2echo	000D	Simple messaging

Some of the protocol variants include particular SMBs for Windows 2003/2008/2012 and for SAMBA, which is an interface between Windows networking and UNIX computers. (Using SAMBA, a UNIX computer can appear on a Windows network and communicate.) SAMBA was developed in the UNIX world to spoof SMB messages and to allow a UNIX system's files and directories to appear in a Windows network neighborhood and be shared (and vice-versa). (The name SAMBA is derived from the use of SMB and the idea of getting the two operating systems to "dance" together.)

Common Internet File System

The TCP/IP suite offers FTP for file transfers; however, it is somewhat limited when it comes to file sharing for applications and browsers. The Common Internet File System (CIFS) is a variant of the SMB protocol that Microsoft proposed (1996) to the Internet Engineering Task Force (IETF) as an open standard. Although some non-Microsoft users have complained that CIFS is not as open as they would like, a surprisingly large number of non-Microsoft SMB variants are available. Since Microsoft-based clients are a large number of the Internet hosts (nodes), it is highly likely that CIFS will reign and that Sun's Network File System (NFS, the UNIX version of SMB) will be used strictly for UNIX-only answers.

Netware's IPX/SPX Suite

Until version 5.0 and up, Novell used its own IPX, derived from Xerox's XNS protocols (see table 5-5 for the Netware components). Note that to talk to Netware from Windows required that there be a device or software programming to translate the different Layer 7 protocols, even if both were running IPX/SPX. This is only a historical reference, as IPX is now legacy software.

Table 5-5. Novell OSI Layer Implementation

Abbr.	Name	OSI Layer	Brief Description
SAP	Service Advertising Protocol	5-7	Allows clients to identify servers and network services
NCP	Netware Core Protocol	5-7	Opens/closes files, reads/writes data blocks to open files, lists directory, and provides high-level connection services
SPX	Sequenced Packet Exchange	4	Transport for connection-oriented applications
IPX	Internetwork Packet Exchange	3	Routing protocol
RIP	Router Information Protocol	3	Provides routing information about networks to routers

TCP/IP Suite

There is no UNIX suite since, for the most part, UNIX has used the TCP/IP suite. Sun originated, and most UNIX machines use, the NFS. This Layer 7 protocol performs the same functions as Microsoft SMB and the Novell NCP. Although there are other programs in the suite, table 5-6 lists several of the most common Layer 3 through 7 protocols. Note that because of the flexibility of TCP/IP to allow new Layer 7 (Application) protocols to be added, there has been a constant stream of new protocols for things such as voice over IP (VoIP), instant messaging (IM), and video streaming. IPv6 brought new capabilities to the Internet, which some of these new application protocols make use of to support their enhanced functionality.

Table 5-6. Internet OSI Layer Implementation

Abbr.	Name	OSI Layer	Brief Description
HTTP	Hypertext Transfer Protocol	6-7	Used in web communications
SNMP	Simple Network Management Protocol	6-7	Used to manage network nodes and devices
FTP	File Transfer Protocol	5-7	Transfers files from one network location to another
Telnet	Telecommunications Network	5-7	Has a workstation emulate a dumb terminal
SMTP	Simple Mail Transport Protocol	6	A standard for email
RPC	Remote Procedure Call	5	Enables a computer to execute services on a remote computer (usually a server)
TCP	Transport Control Protocol	4	Ensures end-to-end data reliability
UDP	User Datagram Protocol	4	A connectionless service with no acknowledgment
DNS	Domain Naming Service	4	A distributed database of IP to computer names
IP	Internet Protocol	3	Routing protocol
ICMP	Internet Control Message Protocol	3	Network error reporting
RIP	Routing Information Protocol	3	Provides routing information about networks to routers
OSPF	Open Shortest Path First	3	Provides routing information about networks to routers
ARP	Address Resolution Protocol	3	Resolves IP addresses to network name address

It should be apparent that if standard Layer 1 through 4 protocols are used, then the only differences are in Layers 5 through 7. For industrial applications with the three-layer model (1, 2 and 7), there is a similar problem. Even though the different network operating systems have a common standards-based set of protocols, they still cannot talk directly to each other (although there are often gateways that can interconnect such different systems, at least for a specific set of functions such as email exchange and file access). This is illustrated in figure 5-4. Although the authors took a little liberty with the layers that SMB, NCP, and the TCP/IP upper-layer protocols actually cover, figure 5-4 should leave no doubt about the current problems posed by using mixed platforms.

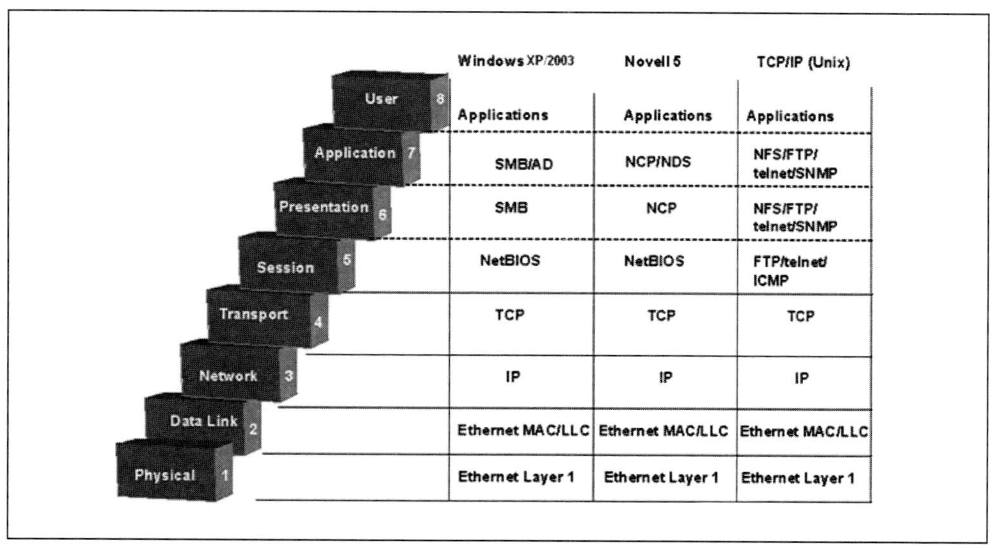

Figure 5-4. Network Operating Systems and the OSI model

Directory Services

A network operating system is really only good by itself on a small peer-to-peer system, or perhaps a very small server-based system. Identifying all the nodes on a large network, let alone setting a policy that has to be originated on each node and updated on each node when required, is work intensive—not to mention tedious, prone to error, and inefficient.

Network operating systems will establish a set of services to handle network management and networks will have their own, or some other vendor's, directory service. "Directory service" has several slightly different interpretations. In this case, we mean a centralized administrative software on a server of some type that provides network management, provides node updating and patching, handles address allocations, provides location/node information for distributed services, and handles user and application authentication functions. Here we will describe the directory service that is included in Windows Server operating systems and is probably one of the most installed: Microsoft's Active Directory (AD). AD is a directory service implemented by Microsoft for Windows domain networks that makes use of Lightweight Directory Access Protocol (LDAP) versions 2 and 3 (Microsoft's version of Kerberos), as well as DNS.

Microsoft Active Directory

An AD domain controller authenticates and authorizes all users and computers in a Windows domain network, assigning and enforcing security policies for all computers and installing or updating software. When a user logs into any computer that is part of a Windows domain (single sign-on [SSO]), AD checks the submitted password and determines the user's privileges based on predefined configuration settings for the group to which the user belongs. The AD server can "push down" (modify settings on all computers in the domain) a group policy object (GPO), which is a set of local access policy settings, that will be used for this particular user on the system into which the user has logged on.

Root Tree

AD uses the LDAP structures. An LDAP directory tree is a hierarchical structure of organizations, domains, trees, groups, and individual units. At the top of the AD structure is the *root tree,* which holds all the objects, organizational units (OUs), domains, and attributes in its hierarchy.

Forests

Under that root tree may be other trees that replicate the root tree for reliability and availability purposes. Subsets of the main root tree may be physically distributed around a corporate WAN to enable a measure of fault tolerance and to provide faster response. Changes to user accounts and other configuration information can flow up and back down between and amongst trees to ensure data synchronization. The collection of trees (connected to the root of the domains or the highest part of the hierarchy) that contains all the domain's OUs, objects, security policies, and attributes is called a *forest.*

AD defines a multi-level tree-structured database organization (where the term "forest" comes from) as illustrated in figure 5-5. While a forest must only contain one tree, a large, world-wide organization might end up with a large, logically and physically distributed forest (i.e., one that has many trees). A structure like the one shown in figure 5-5 might be used by organizations with more than one operating company.

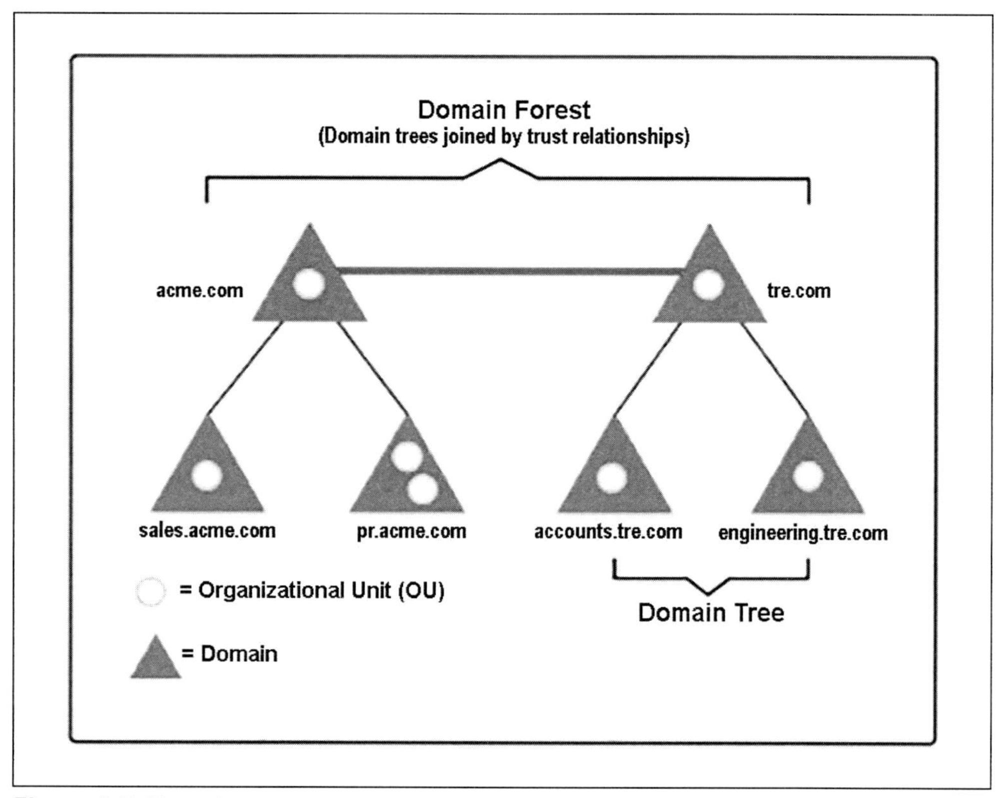

Figure 5-5. Forest

A structure might also be designed with multiple forests; however, these would be designed for very specific reasons and are not common.

Trust Relationships

Trust Relationships are important elements in an AD environment, as they enable forests, and the domains they represent, to communicate with one another and pass credentials. Within a single forest, there are implicit two-way transitive trusts (a two-way relationship is automatically created between parent and child domains in a Microsoft AD forest). By default, two-way transitive trusts (defined in the table 5-7) are automatically created when a new domain is added to a domain tree using the Active Directory Installation Wizard.

Table 5-7. Microsoft AD Trust Definitions

Trust type	Transitivity	Direction	Description
External	Nontransi-tive	One-way or two-way	Use external trusts to provide access to resources located on a Windows NT 4.0 domain or a domain located in a separate forest that is not joined by a forest trust.
Realm	Transitive or nontransitive	One-way or two way	Use realm trusts to form a trust relationship between a non-Windows Kerberos realm and a Windows Server 2003 domain.
Forest	Transitive	One-way or two-way	Use forest trusts to share resources between forests. If a forest trust is a two-way trust, authentication requests made in either forest can reach the other forest.
Shortcut	Transitive	One-way or two way	Use shortcut trusts to improve user logon times between two domains within a Windows Server forest. This is useful when two domains are separated by two domain trees.

Organizational Units

An Organizational Unit (OU) is a container which gives a domain its hierarchy and structure. It is used for ease of administration and to create an AD structure in the company's geographic or organizational terms. An OU can contain other OUs, allowing for creating a multi-level structure. There are three primary reasons for creating an OU:

1. Organizational structure
2. Security rights
3. Delegated administration

Domain Naming System

AD is integrated with DNS and requires DNS to be present to function, as Microsoft no longer supports nor recommends using their WINS server. DNS is built into Windows Server 2003 and all newer systems. An example of a third party DNS infrastructure and server is Berkeley Internet Name Domain (BIND), in a UNIX/Linux network environment. Use of Window's DNS service is recommended as it is integrated into Windows and provides the easiest to use functionality. AD uses DNS to name domains, computers, and servers, as well as to locate services.

AD (or any Directory Service) Conclusion

Is AD an "IT-only" technology? Today, there are large networks of Windows computers distributed over (multi-) plant/corporate networks. It is probable that the IT organization has established one or more centralized AD servers to manage user access rights, push out "policy" objects, and administer company PCs.

Automation systems are now (in many cases) and if not now may eventually be tied into one of these AD servers and then be dependent on its functioning and availability. *Automation technologists need to be aware of the functions of an AD architecture and of its role in a centrally-administered architecture so that they can understand how it will impact operations, security, and reliability.*

An Application Object Model: OPC

The evolution and development of computer-based automation technologies, smart devices, and subsystems had, by the 1990s, produced a lot of proprietary designs and a lack of interoperability. Making two different products communicate and share data was often a programming challenge for the user. In 1996, a group of industrial user, integrator, and developer vendors came together to create the OPC standard to enable data exchanges between products and systems from differing vendors. It was originally focused on enabling communications between PC-based HMI packages and PLCs, but OPC gained traction and expanded into several variations and is now supported by many vendors. OPC is typically implemented by adding "OPC Gateways" to systems that cannot directly support OPC, having these gateways communicate in the native protocol of the system, and then acting as either or both an OPC client and OPC server to other systems and devices and other OPC gateways. Figure 5-6 illustrates a simplified example of a typical OPC implementation.

Object linking and embedding (OLE) for process control (original OPC) was based on Microsoft's Component Object Model (COM) and Distributed Component Object Model (DCOM), which uses Windows proprietary remote procedure call capabilities. As these are legacy protocols (as mentioned, Microsoft is replacing them with its .NET architecture), the OPC Foundation is transitioning this functionality into the OPC UA for Unified Architecture, an open system based on network standards that will not require a Windows platform to function.

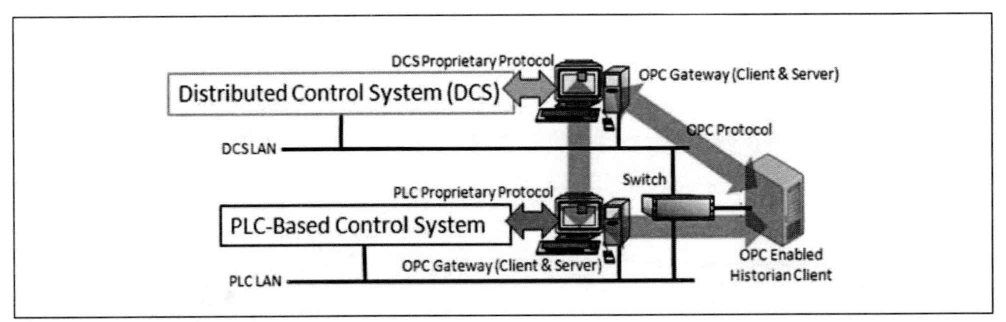

Figure 5-6. Typical OPC Implementation

The object set started out being called *OLE* and is now under the umbrella name of ActiveX technologies. OPC consists of a standard set of interfaces, properties, and methods for use in the process control industries. COM itself is a binary standard, meaning it is generic and not beholden to any particular development language. COM could, in theory, be supported on platforms other than Microsoft (who now considers it to be legacy code), but, in reality, classic OPC has invariably required a Windows platform. It was/is not uncommon to see stacks of PCs being used as OPC gateways for the purpose of allowing systems from different vendors to exchange data with each other.

The goal of OPC (and particularly OPC UA) is plug-and-play for process control, where only one set of drivers for a device has to be written and may be reused, where only one software toolkit is required for development, and where the configuring of software and hardware is automatic. For example, if a PLC collects data from its I/O, the PLC then becomes an OPC server for those OPC clients that want the PLC's collected data. Application developers can write in whatever language they deem appropriate. OPC offers a number of benefits to users:

- Lower system integration costs
- Ease of integration (plug-and-play)
- Auto-configuration of tags
- Elimination of proprietary lock
- Access to data by every level of the hierarchy

An OPC client is a system or device that wants to receive data from an OPC server. A system might want to receive data and also provide data to others and, thus, be both a client and a server. Figure 5-7 illustrates the concept of

OPC. With OPC UA it should be possible to implement OPC directly in smart devices, even if they do not use a Windows operating system.

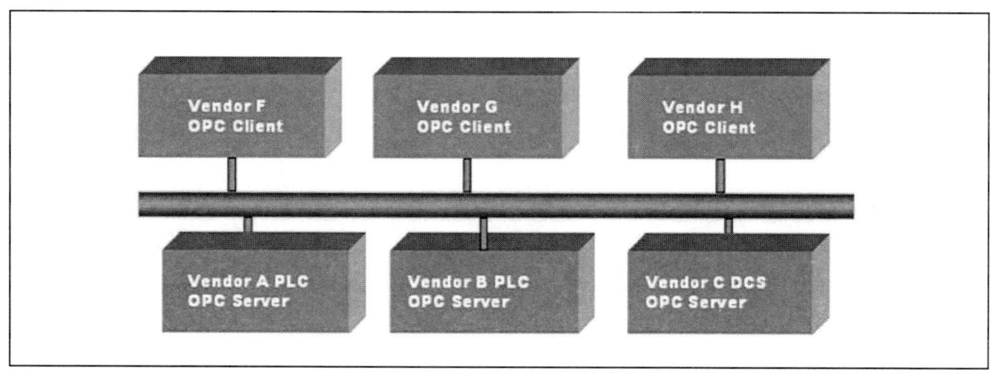

Figure 5-7. OPC Concepts

Other application object models have been promoted by this organization or that organization. However, OPC has obtained a critical mass of users and vendors and was one possible answer to the "Who is doing Layer 7?" problem, since OPC runs on top of (whoever's) Layer 7 and implements the application model described previously.

The OPC Foundation has support from many automation integrators and suppliers. Microsoft has hosted the OPC Foundation's annual meeting but has otherwise kept a low profile, supplying technical support and briefings on software upgrades that might affect OPC. OPC initially merely supported the exchange of time or quality tagged process measurement values (OPC/DA-Data Access), but over time, OPC was expanded to include several variations aimed at the exchange of other types of information, such as alarms and events, batch recipe information, and historical trend data.

Things change. As mentioned above, OPC now offers a unified architecture (UA) that is designed to use open standards and those OPC programming models that are currently in use. This UA is developed in a layered manner so technology changes, such as switching from COM, will not require rewrites. It is also designed to be forward looking and backward compatible, preserving users' investments in the previous issues of OPC.

Three different code bases will be used depending on the user equipment:

1. *ANSI C/C+*, which meets specific requirements for embedded systems
2. *A* portable JAVA implementation
3. *Microsoft .NET*, where the Microsoft .NET 3.0 (or higher) platform will be used

As a workaround to bring classic OPC to interoperate with OPC UA, OPC Express was developed. Running a device as a Classic OPC Server, OPC Express can wrap (encapsulate) the server's data in XML and send it to OPC.NET clients. Figure 5-8 illustrates a possible scenario where OPC.NET clients can access data on a Classic OPC Server. WFC means Windows Foundation Classes—the WFC library is a set of packages. These packages include one or more classes that organize the methods and capabilities into useful groups, particularly with respect to communications.

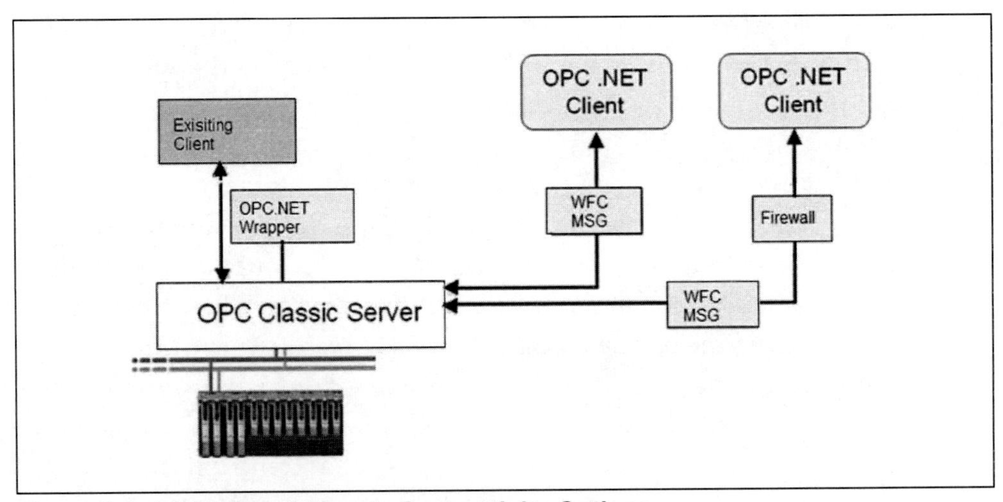

Figure 5-8. OPC UA and Classic Connectivity Options

While some UNIX flavors will run COM/DCOM (reluctantly), the first incarnation of OPC was essentially a Windows-based suite; the unified architecture overcomes that limiting feature and runs on many architectures. Microsoft's .NET has delivered the most complete feature set but .NET, itself, is capable of being run on many different architectures and operating systems because it is standards-based.

Figure 5-9 shows the parts that comprise the OPC Unified Architecture Specification.

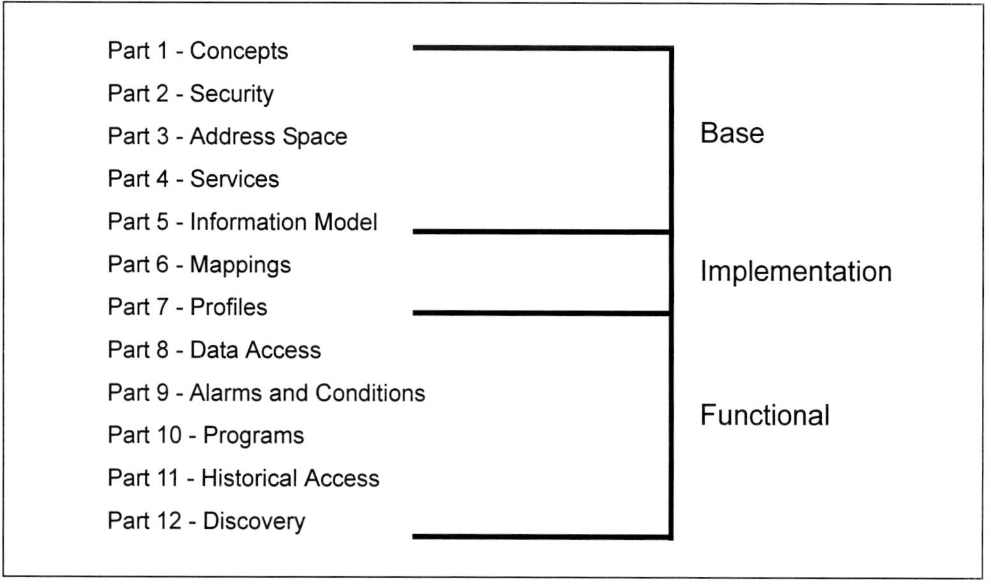

Figure 5-9. Unified Architecture Specification

The Base parts documents describe the basic concepts, design, and abstract interfaces for all OPC-UA applications, regardless of functionality. These parts are completely independent of the underlying implementation infrastructure.

The Implementation parts describe how to implement OPC UA. They detail how the abstract specifications layer onto existing technologies or accepted standards, such as TCP, XML, SOAP, HTTP, and WS-* specifications, as well as web security applications. The Profiles part actually falls under multiple categories but can be generally classified as Implementation. Since OPC UA covers such a wide range of functionality, a more elaborate system than that used for Classic OPC was needed to ensure proper levels of compliance. The Profiles part outlines the multiple levels of functionality that an OPC-UA product can implement. This gives clients and servers much greater granularity in determining the expected level of functionality each product will offer.

The Functional parts are what really give the unified architecture its name. These parts encapsulate the essence of the classic OPC specifications, namely: OPC Data Access (DA), OPC Historical Data Access (HAD), OPC Alarms & Events (OPC A&E), and OPC Complex Data.

Additional OPC-UA parts are anticipated for OPC Batch and OPC Data eXchange (OPC DX). These parts do not introduce any new services or concepts beyond the Base parts. They are primarily information model extensions and define how to use the Base parts to deal with a common functional model in a standard way.

This multiple parts model is designed to allow other key Functional Documents or Companion Specifications to evolve. At present, there are active collaborations amongst groups developing standards such as EDDL (Electronic Device Description Language), ISA S95 and S88, and MIMOSA.

EDDL is the language in which Electronic Device Descriptions (EDDs) are written, at least those that are an integral part of FOUNDATION Fieldbus, PROFIBUS, and PROFINET, as well as 4–20 mA/HART, WirelessHART, and OPC UA. These are the entities that developed the EDDL. Members of both the HART Communication Foundation and Fieldbus Foundation approved a merger proposed by their respective boards through a voting process that concluded on 30 August 2014. This completed a year-long study and diligence period by a dedicated team of volunteers representing each of the foundations. The new corporation, called FieldComm Group, will be led by a board of directors composed of representatives of the collective companies from the current boards of each foundation according to the HART Communication web page, October 2014).

The EDD file contains a description of all the I/O signals, internal variables, and other information in the intelligent device, their data type, block/slot, index, sub-index, and range etc. The EDD file for each device type and version can be loaded onto a modern networked control system and all the device measurement, control, and feedback signals can be automatically integrated into the control system database.

EDDs (EDD files) can be used by control systems to automatically configure their OPC servers to make all data in underlying intelligent devices (using a mix of HART, FOUNDATION Fieldbus, PROFIBUS, or WirelessHART) from different manufacturers available from the same server to any OPC client. This includes not just bringing process variables to plant historians and statistical process control, but also the ability to bring device diagnostics to Safety Instrumented Systems (SISs) or messaging software that sends email to service personnel anywhere or sends text to mobile phones.

System integration based on OPC is about multi-vendor interoperability and easy access to data. EDDs make the vast amount of setup information and diagnostic data in intelligent devices accessible to software, enabling automatic configuration of the OPC server without manual data mapping, which in turn makes system integration using OPC even faster and easier.

This solution for OPC servers owes much of its ease of use to the open standard configuration power of EDDs. EDDs save numerous man-hours of OPC server configuration and speed up project completion.

EDD is platform independent, much like OPC UA, and is text-based, suitable for a web services-oriented architecture. Devices are described with text-based files for functions including diagnostics, configuration, and calibration. OPC, together with the EDDs, allows the data in the devices to be accessed and it does this without affecting the reliability and integrity of the control system, a key user requirement, as no device drivers need to be added to the system to make it work. EDD is the only device integration solution that is declarative (declarative programming is often defined as programming that describes the computations to be performed—not necessarily the instructions on how to do it—and the mathematical/logic rationale used). Presently, no other software achieves a comparable set of results.

Conclusions to Chapter 5

As we have seen, communicating between dissimilar platforms is almost as difficult in the commercial world (which includes the business and administrative networks at an industrial site) as it is in the industrial world. The problems the two spheres share grow, as most industrial networks standardize around Ethernet (Layer 1 and 2) and TCP/IP (Layer 3 and 4), leaving the operating system to determine Layers 5 through 7. Even with this Ethernet/TCP/IP consolidation (ongoing), if you wished to communicate between different platforms (Layer 7s) using LAN technologies, somewhere along the line you would find a gateway, a computer with two network protocol stacks, to convert between this Layer 7 and that Layer 7.

However, if you will recall, the n-tier method was not terribly dependent upon platform, only upon the capabilities of the presentation software (the browser). After all, a browser looking for a website generally cannot tell whether the site is running on Microsoft Internet Information Server (IIS) over Windows Server 2012 or on an Apache Web Server over Linux. As XML begins to be used much more widely, platforms will not matter as much as the presentation program. XML is capable of inter-processor communications, particularly when coupled with SOAP. Whether running on the Internet or over an Intranet, the techniques are the same. It should be noted that Microsoft Office 2007/10/13 (also Office365, a widely used suite of productivity tools including word processor and spreadsheet) is XML enabled. However, although Office 2007 (and newer versions) comes in a multitude of feature formats, to be cost effective for its target audiences, the underlying code base is the same. It is quite probable that other programs vying for the same market space will also be XML enabled.

One of the priorities that could be construed from the previous discussions on Directory Services, particularly applying to large corporate IT groups, is the goal to enforce "policies" (user access rights) from a central AD facility (which they manage, of course). These central AD facilities link EMS/SCADA systems and plant automation systems across the corporate WAN, so that they can push down security policy changes and account setting updates, such as adding and deleting user accounts.

WARNING: IT-centralized AD can bring about a safety/vulnerability problem. To wit: loss or compromise of the corporate AD server(s) or of network connectivity between the AD server and the system requesting authentication can potentially disable directory and authentication functions if the AD design is not adequately robust and is not adequately distributed in the automation areas. The solution described in IEC/ANSI/ISA-62443 is to install a firewall with restrictive access control lists (ACLs), which can be configured to allow or prevent these actions. The security necessary will have to be determined by both the Business IT and the Control IT personnel.

Summary

This chapter discussed some of the models used in both industrial and commercial networks, as well as some of the vendor products used in industrial networks. Since both of the feature sets and details are apt to change during this book's life span, the level of detail provided here is sufficient. In fact, this chapter could be summed up as follows: For network operating software, Layers 1 and 2 are Ethernet, Layers 3 and 4 are TCP/IP, and whichever operating system you choose will supply Layers 5, 6, and 7. You will be using either the two-tier, three-tier, or n-tier model, or all of them, and your application software will run at the User layer.

Bibliography

Note that Internet links, although accurate at the time of writing, may change over time.

Microsoft Corporation. *MSDN Library.* Microsoft, July 2000.

———. Corporate website. http://www.microsoft.com.

Morneau, K. *MCSD Guide to Microsoft Solution Architectures.* Boston: Course Technology, 1999.

Novell Inc. Corporate website. http://www.novell.com/linux.

OPC Foundation. Organization website. http://www.opcfoundation.org.

Palmer, M. J. and B. Sinclair. *A Guide to Designing and Implementing and Local and Wide Area Networks.* Boston: Course Technology, 1999.

Red Hat Inc. Corporate website. http://www.redhat.com.

Wikipedia. References to Windows Embedded. http://www.wikipedia.com.

Wikipedia. References to Microsoft Windows 2003/Vista/7/8/2010 and server families. http://www.wikipedia.com.

Ubuntu. http://www.ubuntu.com.

6 Industrial Networks and Fieldbuses

Before launching into the requirements of an industrial network, the term *industrial network* should be defined: An industrial network is three or more devices connected through a shared media or distribution in an industrial environment, including the media, infrastructure, protocols, and all other peripherals and actions necessary for functional communications, particularly between industrial devices (author's definition).

For just fieldbuses alone, a simple perusal of the Synergetic Fieldbus Comparison Chart (www.synergetic.com) lists 18 different fieldbuses, excluding two types approved in IEC 61158 (the international fieldbus standard that regulates around eight non-interoperable protocols). If you perform an Internet search on "industrial fieldbus," you will find over 138 different protocols and systems (on a multitude of sites) that are used in industrial applications. Many entries are from user groups and many entries point to one or another of the popular protocols. We start this chapter with much the same disclaimer as the previous one: Of the numerous industrial networks available, the authors have selected for discussion those that illustrate a type, have a large installed base, and/or offer the authors sufficient technical information (as opposed to marketing features). Again, the authors accepted no payments, fees, or quid pro quos for describing one vendor's product over another's, and, as always, the authors' views and descriptions here are their own and do not represent official endorsement by ISA (International Society of Automation) or anyone but the authors. We do not have the space to explain each and every industrial network in this chapter, so we have selected the following as examples of types that best illustrate key principles.

Industrial Network Requirements

All networks, commercial (the business side of an industrial facility or just commercial IT) and industrial, share some common requirements:

1. They must provide effective performance for resources used.
2. They must offer multilevel security.
3. They must be cost-effective.
4. They must be standards based.
5. They must provide reliable transmission.
6. They must offer ease of access.
7. They must provide ease of use.

Industrial networks (those in the industrial environment) have these extra requirements:

1. They must offer predictable throughput.
2. They must provide predictable scheduling.
3. Their downtime rates must be extremely low (zero downtime is preferred).
4. They must be able to operate in environments hostile to equipment.
5. They must be scalable from one to many.
6. They must be operable by other than communication specialists.
7. They must be maintainable by other than communication specialists.

All designs are compromises, so some of the common requirements (e.g., being standards-based or cost-effective) may have to give way to ensure meeting key industrial requirements, such as low downtime and so on. Let's take a closer look at the seven requirements that apply specifically to industrial networks (referred to in this book and other reference materials as industrial systems, industrial control systems, industrial automation control systems (IACSs), local area networks (LANs), and data highways).

Predictable Throughput

Industrial requirements are such that network response times must fall within a specific window of time or an error can result. This functionality is called *determinism*. In general, the window must be wide enough so that the response can be acted on in real time; that is, the controlling device must issue an output in time to affect the control or alarm before a process's operation becomes unstable.

Predictable Scheduling

Time-sensitive operations, such as computing a PID algorithm (consisting of proportional, integral, and derivative action—or, for old timers: proportional, reset, and rate), must provide a definitive window of time for data input or the results will be incorrect.

Extremely Low Downtime

If the control system is down, the process is down and the company is losing money. Downtime requirements in modern industry are such that a system must be up (must be available) above the "five nines" (99.999%), downtime must be less than 0.001 percent of total system availability (uptime). This may only be achieved by using distributed control, redundancy, fail-over clustering, or other techniques that will reduce system non-availability under any circumstances.

Operation in Hostile Environments

Industrial operations are usually performed in extremes of temperature, vibration, chemical atmosphere, and electrical interference, as well as in areas that lack cleanliness. Only commercial equipment that has been industrially hardened or encased in an industrial enclosure can withstand this treatment. Fortunately, much of the data communications equipment in industrial settings can be located in the far more protected control room, where the environment is more closely monitored and maintained for human activities.

Scalable from One to Many

Scalable (from one to many) means the design doesn't lose its efficiency when the network is expanded by an order of magnitude (10 times) or more.

Although additional resources will be required, the overall design must accommodate very small to very large network configurations, with no loss in features and timing.

Operable by Non-specialists

Most operators know their process, however, asking them to learn the intricacies of communications is probably beyond their desires and training. In a properly designed industrial control system, the communications technology will be totally transparent to the operator. He or she shouldn't have to determine anything network related, except how to report an alarm to the correct party.

Maintainable by Non-specialists

Maintaining industrial control systems has become an involved process. The days of electronic component repair are gone. Now, locating a problem in a unit requires programming, diagnosing, and analyzing systems. If the problem is communications related, the process control technician generally does not have the background, training, equipment, or skills to quickly and efficiently locate it. A well-designed system has sufficient diagnostics, so that a control technician can find the errored unit (malfunctioning device, circuit, or signal set) and replace it, without having to suffer through ROM revision levels, protocol switch settings, and missing software components.

In many cases, the effort required just to meet the last two requirements drives an industrial system's cost far above a commercial system offering similar performance. One method originally developed to address these specialized requirements was to provide a proprietary turnkey approach: the vendor-proprietary distributed control system (DCS). This evolved into the open DCS. As of the early 2000s, most (if not all) DCS vendors had migrated onto high-speed, switched Ethernet as the replacement for their earlier proprietary local area network (LAN) designs.

Distributed Control Systems

A distributed control system (DCS) has been previously defined but the basic meaning is repeated here: a system in which the control functions are shared among separate intelligent devices. In DCSs, the intelligence is located in various nodes that are connected by some media and networked together, enabling the nodes to perform peer-to-peer communications and enabling each node to perform a portion of the overall set of tasks. As mentioned above, for several years the DCS was the province of proprietary systems. The amount of capital needed to develop a system, work out the bugs, market, install, and provision such a system usually necessitated that the vendor company be well-established. The number of potential customers would be few, and so the cost could not be broadly amortized, resulting in relatively high prices for these systems.

At the low-cost end of industrial control systems, the programmable logic controller (PLC), a device developed to replace relay logic, enjoyed significant success in industrial applications. Here was a digital logic device that, as it grew in capability, acquired increasing intelligence in various nodes, was connected by a data highway (really a LAN), performed peer-to-peer communications, and performed a portion of the overall tasks at each node. This PLC system was not called a DCS, however, as DCS had come to mean devices capable of performing continuous process control actions, rather than just on/off or two-state control. Then, analog I/O and the PID control strategy were added to the PLC. A facility of networked PLCs using a third-party human-machine interface (HMI) and performing both discrete and analog control is called a *DCS*. The difference is simply that a PLC/GUI setup on an I/O point basis will have considerably lower initial costs, because using market-available devices, with their much lower research and development recovery costs, means a lower price. On the other hand, the traditional DCS vendors have done a better job of integrating the system-wide configuration and development tools, in addition to providing bumpless fail-over redundancy at the controller level. DCSs also offer tools that have the ability to make configuration and control changes to running controllers without the need to reboot or halt execution. This was not a capability generally found in PLCs until the latest generation of software (circa 2006) and in some fieldbuses.

Although PLCs are, by and large, proprietary devices, the Physical layer on most PLC networks today is a variation on the Ethernet standards. PLCs are used primarily for discrete manufacturing, offering only the PID control strategy for handling continuous processes. Process DCSs, on the other hand, offer advanced strategies. Other than the Physical layer, PLCs have undergone little standardization. Industrial communications—all three layers (Physical, Data Link, and Application)—must meet many rigorous requirements, most of which we have encountered already. It's a sound assumption that proprietary systems meet or exceed these requirements. It remains to be seen how the proprietary-versus-open networks battle will turn out; at present, open networks appears to be the encompassing trend.

Process Automation Controllers

Analog I/O in a DCS primarily refers to continuous processing. A new generation of more advanced PLCs has come onto the market and are being called *Process Automation Controllers* (PACs). A typical PAC can serve as a simple single-loop controller or as a multi-loop controller, with complete control and logic functions for small unit batch or continuous processes. These controllers, when used with IEC 61131-3 programming tools and function block libraries, provide advanced process control capabilities and have been well accepted for industrial batch (as opposed to continuous) process automation. PACs evolved from the older standalone digital controller, which could control one loop, to controllers that now have multiple input-multiple output capabilities.

Proprietary DCS: A Brief Look

Most instrument manufacturers still sell proprietary DCSs. These systems perform PID control for several loops and offer advanced control strategies such as cascade, feedforward, ratio, and other multivariable control strategies. These proprietary DCSs also supply trending, optimization, nice graphics, diagnostics, operator stations, and an engineering workstation for configuration and changes. Their communications methods were originally closely held. The first of these DCSs, like the Honeywell TDC2000 and EMC Controls Emcon-D and D/3, had proprietary local area networking schemes, as well as proprietary operating systems in both the workstations and operator consoles, particularly in the distributed controllers. It was in the next generation of DCS products in the mid-1980s where systems based on some version of UNIX (and eventually even Microsoft Windows starting with Windows NT) began to appear. These early systems used vendor-proprietary LANs to link all of their components. This was typically a token-passing scheme or a modified master/slave arrangement, connected as a logical or physical ring, that met the manufacturer's performance standards. These proprietary DCSs were turnkey operations. Once a given vendor's DCS had been selected, you were locked in—at least for major components (and a great number of the minor ones too). Figure 6-1 shows a simplified architecture for a typical second generation (1980s) era DCS.

What was the industrial company to use as an external interface for these proprietary DCSs? When provided (it usually was an option), the external inter-

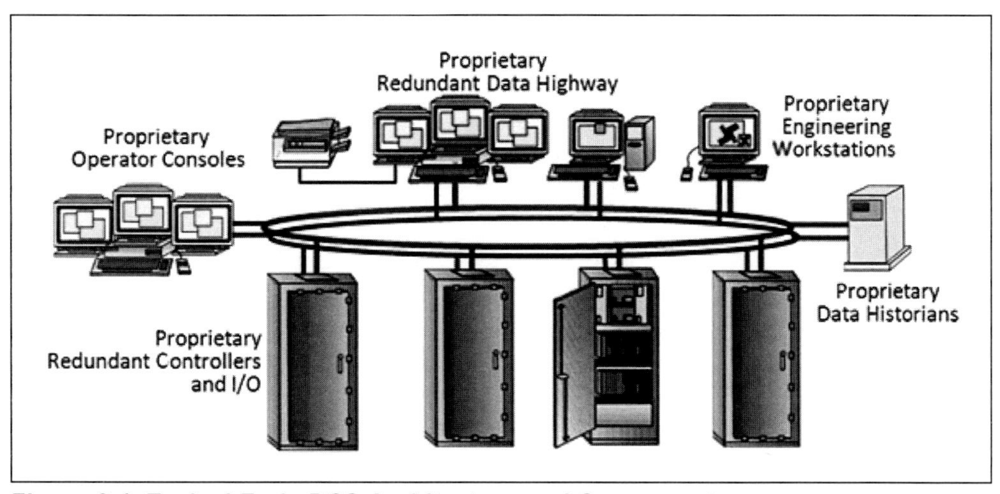

Figure 6-1. Typical Early DCS Architecture and Components

face would normally be compliant with either EIA-232 or EIA-422 and the software for communications would typically be proprietary. Just having a standard communications port on a processor or two does not fit the definition of an open architecture. However, most of the communications were still proprietary. Time has passed and many of those proprietary systems that remain closed are neither being purchased, nor added to. They are being replaced either by more open systems or by standards-based systems. Conventional DCS technology has lost a large percentage of the industrial automation market share to PLC-based offerings, except in specific market segments (e.g., refining and power generation) where the "bumpless" (a.k.a. "hot standby") redundancy, advanced control features, and live configuration change capabilities of a DCS are seen as essential.

Vendors who were interested in making their proprietary systems profitable, had to take several actions to remain competitive. First, they had to become truly more open. For a vendor, this is a dangerous step because the system and all its components are vulnerable to shifts in marketplace preferences. If one chooses an open platform or a standards-based system and the platform does not have market acceptance or the standard is superseded, it would damage the viability of that system. One method vendors undertook to ensure market acceptability was to base their systems on widely accepted platforms. Most of the old-line DCS manufacturers now base their systems on Microsoft Windows (except for their controllers) and on Ethernet-TCP/IP for their LANs. Microsoft Windows networking offers the advantage of a native GUI (not one added on or developed through many man-hours of programming), an excellent security model (although minor implementation bugs have occurred, each patch makes it a little less vulnerable), and reduced development costs, while providing an easy interface to office systems software. Having a host of third-party vendors offering components at any level allows for quick and reliable development.

Another vendor method for becoming more open has been to embrace the standards by offering Ethernet and TCP/IP to the outside world and using various recognized (by standards committees or user groups) standard buses and protocols internally. The DCS manufacturer has, in effect, become a systems integrator, even though part of the systems integrated are designed, developed, and manufactured by the DCS manufacturer itself.

Programmable Logic Controllers

A PLC using analog I/O and a PID algorithm performs continuous process control and, as such, is a loop controller. Also offered is the PLC mainstay: discrete control. PLCs vary widely in their capabilities and this is certainly not the place to list them all. Of interest is communications between and with the PLC. As with the early DCS vendors, each PLC vendor developed their own proprietary scheme for communicating with their PLCs. Two such schemes predominated: Allen-Bradley PLCs used variants of the Data Highway (a token passing scheme) and almost every other PLC vendor made use of Modicon's Modbus protocol (a poll-response scheme). We will discuss both in the following section. Today, both of these communication schemes have been migrated onto an Ethernet-TCP/IP platform as Application layer (OSI Layer 7) protocols and most PLC manufacturers offer Ethernet connectivity. In the last decade, various vendors, industry groups, and standards organizations have developed a range of competing Ethernet-TCP/IP-based protocols for PLCs and other types of smart instrumentation. In the next section we will be examining several of them.

Selected Industrial Systems: Allen-Bradley and Modicon

The topology of Allen-Bradley's Data Highway (DH), Data Highway Plus (DH+), and Data Highway 485 (DH-485) is illustrated in figure 6-2. The Data Highway was used for peer-to-peer transmission between controllers and is a local area network for up to 255 nodes (Octal 377, the 256th node, cannot be used as an address because it is all 1s). The Data Highway bus operated at 57.6 Kbps (kilobits per second) and could have a trunk line of up to 10,000 ft.

(3.048 km) at the 57.6 Kbps data rate, with drop lines from the trunk of no more than 100 ft. Data Highway used a modified token-passing scheme known as *floating master*, which provided a bid scheme for controllers who wished to be master; the controllers bid for the bus based on their priority and need to use the bus.

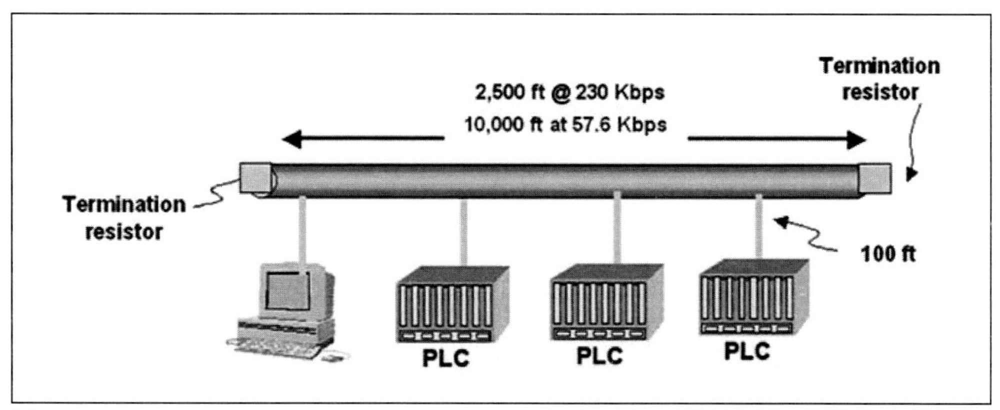

Figure 6-2. Data Highway/Data Highway Plus/Data Highway 485 Topology

Data Highway Plus was a token-passing bus that allowed up to 64 nodes (0–77 Octal) on one highway, with speeds up to 230 Kbps. This is not a field device connection, rather a processor-to-processor connection for PLC2, 3, and 5 series devices (the model number of the A/B PLC line, 2 being the oldest and 5 the more current, with quite a few PLC5s still in use as of 2014). Data Highway Plus allows a PC with a KT card (Allen Bradley PC interface card) and appropriate software to be used as a programming or operator terminal. The primary differences between the Data Highway and the Data Highway Plus versions are that the latter is optimized for the lesser number of nodes and allows online programming. Data Highway uses ±14 volts peak-to-peak signaling, while Data Highway Plus uses ±7 volts peak-to-peak signaling. Both use Manchester encoding.

The Data Link layer frame contains the flags and a 2-byte CRCC (cyclic redundancy check character). The message structure is assembled by the Application layer software and then is encapsulated in the Data Link layer frame. Figure 6-3 illustrates how the three layers (1, 2, and 7) are implemented using the two different Data Highway versions. Note that the Data Link layer is divided into the IEEE MAC and LLC sublayers, here the HDLC frame is the LLC and the Floating Bus Master of Token Passing MAC is the MAC.

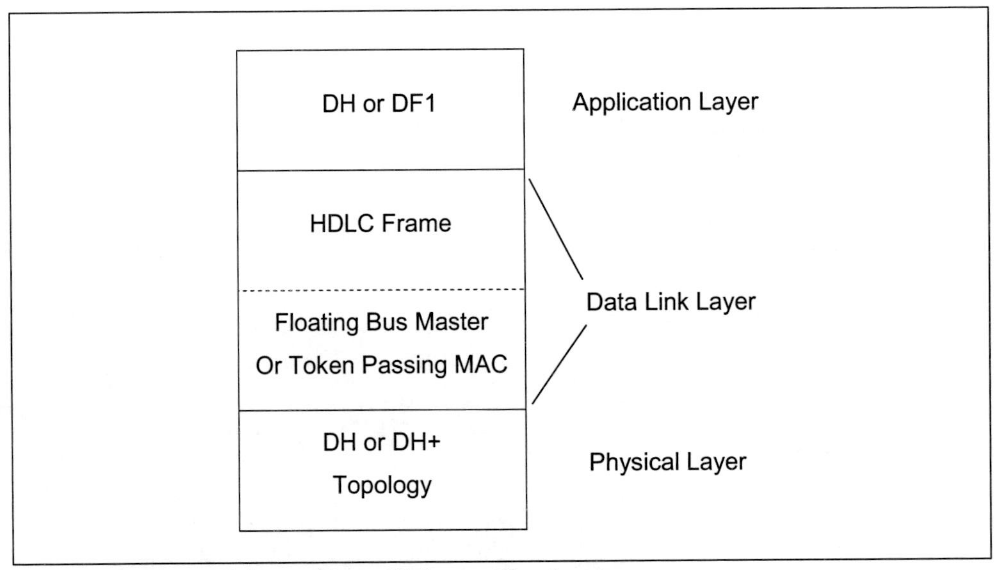

Figure 6-3. Data Highway/Data Highway Plus Layers

Figure 6-4 illustrates the message format for Data Highway and Data Highway Plus. The Application layer assembles the message octets (DST [Destination] Address through Data) sending the formatted octets to Layer 2 where the Data Link layer computes the CRCC, adds it to the frame, and then frames the packet data unit (PDU) with the start and stop flags. Zero insertion is used to ensure that there are no more than five 0s in a row, unless the Data Link layer decides it is time for a flag. This is a synchronous protocol, meaning that system timing is dependent on bit transitions and each bit must be accounted for. At the receive end, the flags are used to determine the frame length and the CRCC is computed independently on the message and compared to the transmitted CRCC. If a match is made, the PDU has successfully been received; if there is no match, then error correction (retransmission of the errored packet) is required.

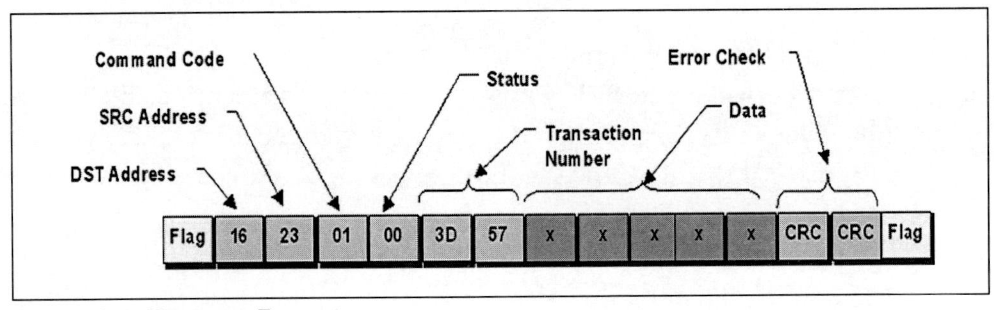

Figure 6-4. Message Format

DF1 is Allen-Bradley's asynchronous protocol for communicating to devices that don't support synchronous communication hardware. For example, it would be used to communicate between a PLC and a PC. It is byte-oriented, rather than bit-oriented, so it can be implemented using standard UART/COM port hardware, and it uses ASCII control characters; the control structure is similar to IBM's Bi-Sync. All control characters are preceded by a data link escape (DLE) character. Some of the control character pairs are as follows:

- DLE STX Start of text frame
- DLE ETX End of text frame
- DLE ACK Positive acknowledgment
- DLE NAK Negative acknowledgment
- DLE DLE Data (which may have embedded control codes)

It requires very little hardware and logic to convert DF1 frames into Data Highway Plus frames (and the reverse), as they have essentially the same fields and structure.

Data Highway 485 uses the EIA-485 standard and will support up to 32 nodes. It has a data rate of up to 115 Kbps (although the EIA-485 maximum is 10 Mbps [megabits per second] for 15 m; 50 ft.). EIA-485 refers to the electrical characteristics, rather than how the network is set up. The network setup is according to Allen-Bradley's definition of full duplex (DF1) or half-duplex (polled), where (logically) the DF1 consists of 2 two-way paths and the half-duplex consists of 1 two-way path.

Modbus RTU

The transmission protocol used by most other PLC manufacturers, in addition to their own proprietary method, is the Modbus Remote Terminal Unit (RTU) protocol. Its OSI implementation is illustrated in figure 6-5. This is a master/slave polling arrangement in which there is one master and one or more slaves. It is a query-response protocol; that is, only the master can query—the slave cannot initiate, only respond. Many manufacturers use Modbus RTU as an accepted standard and the protocol can perform the basic host-to-PLC communications.

The Modbus protocol originally had two modes: an ASCII mode and the RTU mode. The ASCII mode made debugging easier, as the message fields were converted to printable ASCII characters for transmission. However, the conversion doubled the message length, which reduced protocol efficiency. Today only the Modbus RTU version exists.

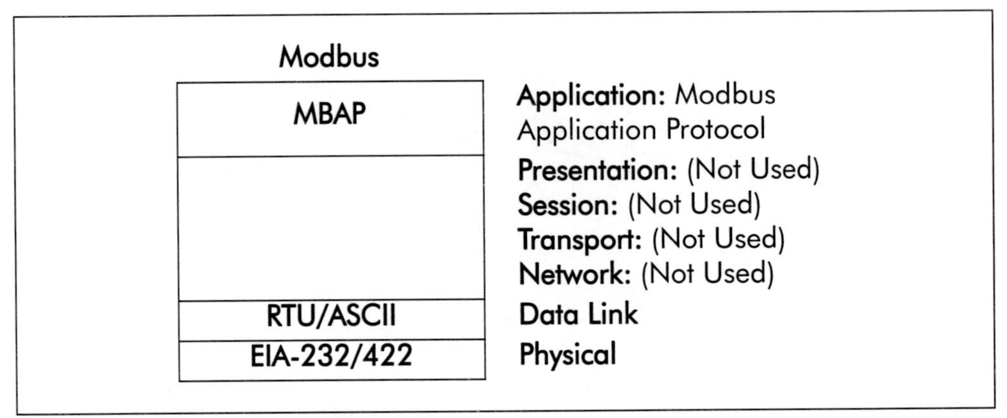

Figure 6-5. Modbus RTU/ASCII

There are potentially 246 RTU addresses (1–247) available. Address 0 is an all-stations (broadcast) address. Speed depends on the devices attached, but it typically goes up to 19.2 Kbps and covers distances up to 1200 m (4000 ft.). However, higher data rates are achievable over shorter distances. The slowest device sets the data rate that all must use in the RT protocol.

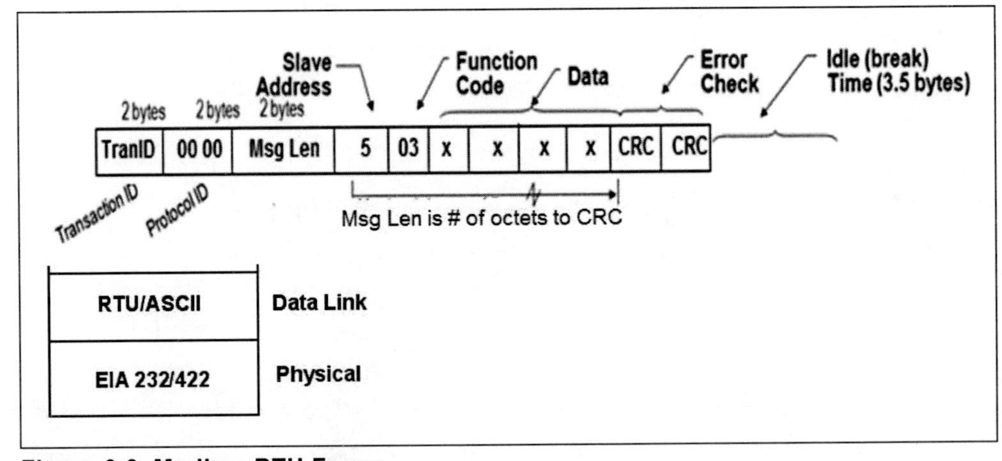

Figure 6-6. Modbus RTU Frame

Figure 6-6 illustrates the makeup of the PDU. The ModBus Application Protocol (MBAP) assembles the slave address through the data bits and hands it to the Data Link layer, which computes and tacks on the CRCC. The MAC and Physical layer use idle time as start and stop delimiters, using greater than 3.5 bit times (a bit being the smallest time unit of information, a 1 or a 0) to indicate that it is time for a new message. Only the master initiates these messages; this timing merely indicates to the slave when a message is to start or stop. If transmission is dropped for longer than 1.5 bit times, the message will be received as a fragment and an error will be generated. Earlier versions of Modbus used different, less effective, versions of the frame check code, but today the standard defines using the CRC-16 check code.

Although the Modbus protocol and data representation method may be more limited than the manufacturer's proprietary code, it provides essential communications and many integrators use it as their primary mode of communications between PLCs and host computers (PCs or otherwise), as well as for a whole range of intelligent devices.

Note that both PLC schemes, Data Highway Plus (includes DH-485) and Modbus, use standards-based Physical layers but are different in the Data Link layer (Media Access and Logical Link Control). Most PLC manufacturers also offer Ethernet or IEEE 802.3 as a standard method of communicating from the master processor to other processors, workstations, operator stations, and gateways. Although this does away with incompatibilities between both Physical and Media Access Command (MAC) layers, a method must still be in place to address the various modules on the PLC, registers, and I/O. Data Highway and Modbus framing/formatting are often employed. Data Highway (DH), Data Highway Plus (DH+), and Data Highway 485 (DH-485) all have different Layers 1 and 2, with Layer 7 remaining the same. ModBusRT and ModBus Plus have different Layers 1 and 2, with the same Layer 7 (ModBus Application layer).

Selected Industrial Networks

In this section, the following competing industrial networks, protocols, and the underlying standards will be explained, some in more detail than others:

1. HART
2. DeviceNet

3. ControlNet
4. EtherNet/IP
5. LonWorks
6. AS-i
7. P-Net
8. PROFIBUS/PROFINET
9. FOUNDATION Fieldbus
10. Ethernet and TCP/IP

HART

Highway Addressable Remote Transducer (HART), originally designed by Rosemount for their smart transmitters, has gained popularity and is used in applications ranging from transmitters to any entity in a two-wire control loop.

Figure 6-7 is a block diagram of a two-wire loop that has a handheld communicator connected to it. The connection location is not technically important as long as a 250-ohm resistance (minimum) exists in the loop between the communicator and the power source. All versions (HART 1 through 7) provide a 1200 bps (bits per second) data rate over the shielded twisted-pair cable that is used in two-wire loop connections. As the HART protocol is half-duplex, most work (configuration or lengthy changes) is performed off line and then uploaded from the communicator or other master device, where the configuration was stored into the field device. The 4–20 mA signal is varied ±0.5 mA at either 1,200 Hz (1) or 2,200 Hz (0). This AC waveform signaling on the loop will average out to a net zero DC signal, so the analog 4–20 mA signal will not be impacted. Most process controllers will limit the input rate of change to less than 3 Hz, so that signaling can go on without affecting the process. Changing ranges or outputs will bump the process, so these changes are performed with the loop in manual.

The HART protocol communicates by using the handheld communicator or a computer that has the appropriate modem card. Rosemount licensed the HART protocol to other vendors and eventually assigned intellectual property rights to the HART Communications Foundation; it is now an "open pro-

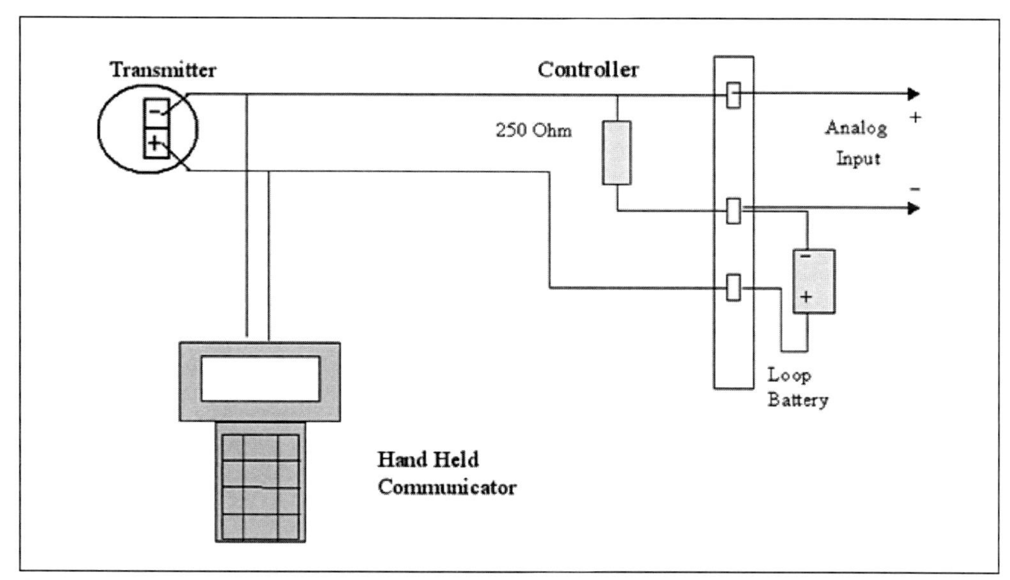

Figure 6-7. HART Instrument with Handheld

prietary" protocol. At the time this was written (2014), more than 200 vendors were using the HART protocol to communicate, not just with smart transmitters, but with valves, chart recorders, and other devices in a two-wire loop. Many HART instruments have advanced diagnostics (the same ones available with FOUNDATION Fieldbus or PROFIBUS) but they are not called by the master (either the master doesn't know or use the diagnostics in a particular application). HART is a slow multi-drop (there can be more than one HART loop device); all signals are digital with the exception of the one variable assigned as the 4-20 signal. (It may be of interest that the HART Communication Foundation and the FIELDBUS Foundation merged in 2014 and the foundation is now known as FieldComm Group).

The HART standard has gone through several revisions and, as of this writing, is now at version 7. Each revision has added features and functionality, yet the additions have not changed any prior functionality, so backwards compatibility is preserved. In version 6, the ability to have peer-to-peer communications was added, thus permitting a pair of HART devices on a common cable pair to directly communicate. Table 6-1 shows a list of functions and features beginning at version 5 and going up to version 7.

Table 6-1. HART Version Functional Differences

HART Feature Summary	Revision		
Feature	**5**	**6**	**7**
PV with Status	✔	✔	✔
Device Status	✔	✔	✔
Broadcast Messaging	✔	✔	✔
Device Configuration	✔	✔	✔
4–20 mA Analog Loop Check	✔	✔	✔
Multi-Variable Reads	✔	✔	✔
PV with status	✔	✔	✔
32 Character Tag		✔	✔
All Variables with Status		✔	✔
Digital Loop Check		✔	✔
Enhanced Multi-Variable Support		✔	✔
Local Interface Lock		✔	✔
Manual ID of Device by Host		✔	✔
Peer-to-Peer Messages		✔	✔
Visual ID of Device		✔	✔
Report by Exception			✔
Synchronized Sampling			✔
Time- or Condition-based Alerts			✔
Time Stamp			✔
PV Trends			✔
Wireless Co-Existence			✔
Wireless Diagnostics			✔
Wireless Mesh and Star Topologies			✔
Wireless Message Routing			✔
Wireless Security			✔

As of version 7, there is now a wireless, radio-based version of the protocol called *wirelessHART* that uses IEEE 802.15.4-compliant radio technology. This enables users to create a wireless mesh network in which devices can forward messages from other devices to ensure that they get to the final destination. The wireless version also added security features and functions.

Of course, the number of devices that can communicate on 1 two-wire loop, be it on the measurement or the control loop side, is limited. It is important to understand that even though the loop may include a device with multiple data elements, only one value can be represented by the 4–20 mA signal. Multivariable transmitters illustrate this limitation. Although these transmitters may measure process temperature, absolute pressure, and differential pressure, as well as use their intelligence to compute compensated flow, the 4–20 mA signal can only represent one of the values. One remedy for this is the HART splitter. Typically mounted in the control room, it communicates digitally with the transmitter to obtain each variable and produces (using the supplied loop battery) a 4–20 mA signal for each variable.

It should also be understood that only slowly-changing process signals can be managed with the splitter, if control is desired. This is because a 1,200 bps transmission has a bit time of 0.83 milliseconds and a whole HART frame may take nearly 50 octet times, based on frame size and running half-duplex, which means that considerable time must elapse between measurements. The actual time (which is for the maximum possible bit count) is certainly shorter than that stated at 0.83 milliseconds/bit but the caution still stands regarding real-time control based on the digital signals.

The HART design also supports a digital multi-drop mode, where up to 15 devices (increased to 63 in revision 6, primarily for wireless applications) can share a common twisted pair cable. Using the multi-drop configuration, an analog 4-mA signal still appears as the 4 mA signal that supplies the device power, but all process variables in each device are only accessible via polling or exception reporting using HART link layer message frames. The protocol is master/slave, with up to two masters allowed (handheld and computer-driven interface). However, the masters cannot intercommunicate. Figure 6-8 illustrates a HART frame. A DCS controller or PLC requires a HART communication card in order to use the HART digital multi-drop mode.

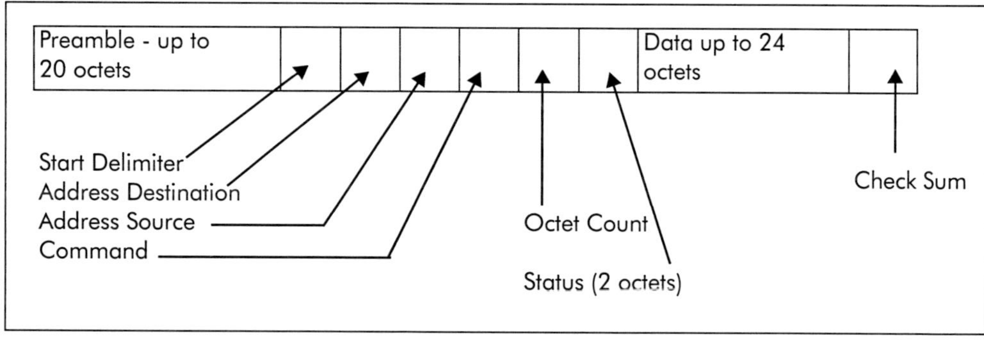

Figure 6-8. HART Protocol Frame

The *preamble* in figure 6-8 is a variable length of an FF (all 1s) octet. Up to 20 preambles may be sent. However, doing so could have deleterious effects on throughput, so each device informs the communicator of its minimum required number of preambles. Some devices can send what is determined (by the Electronic Device Description [EDD]) to be the probable number of preambles required for synchronization, as the number of preamble octets most modern devices require is significantly less than 20. If a response is obtained, throughput has been increased. If there is no response, then the maximum of 20 preambles can be sent. Character parity, as well as a checksum, is derived at the transmitter. The same derivation is performed at the receiver and the frame is accepted if the checksums match. This process provides adequate error detection for the speed and the medium.

HART is primarily used to communicate with smart instruments in an otherwise dumb two-wire loop; it is considered too slow for closed-loop control with many control and measurement points and is therefore not normally used for that purpose. However, nothing prohibits HART from being used (as FOUNDATION Fieldbus is used) as a supervisory system, after all configuration has been accomplished, such a system would include all the new HART transmitters, with PID function blocks included in the transmitters. Supervisory (set point) control would overcome the problem of transmitting the process value and determining control constants at some other location. By doing PID locally at the transmitter, only the set point needs to be digitally transmitted.

Smart positioners with HART communications circuitry are used on valves that are classified as smart valves. They not only locate the valve stem position accurately and keep track of the number of strokes, some models can even transmit the inlet and outlet pressures, as well as the process fluids temperature. This amount of operational information from the smart valve, in addition to that provided by the field transmitters, was not available in the past outside of a small pilot plant or laboratory and should greatly enhance process optimization, safety, and maintenance.

Today, there is a large installed base of HART instruments with ever-increasing amounts of configurability and processing power. This type of network will therefore be around for some time, or at least until the demise of the two-wire loop (the authors offer no prophecies as to that date). Adding HART devices to a two-wire loop means that only the devices and an interface have to be added. Many DCS can communicate with HART instruments directly (digitally, using HART interface cards) over the two-wire loop. Doing so offers the benefit of exploiting the wiring that is there and represents an incremental, rather than a radical, change. At one time (early 1980s), people wondered about the benefits of a smart instrument; today, we know the benefits: increased reliability, self-diagnostics, greater accuracy, precision, and a cost similar to or less than conventional instruments.

WirelessHART uses a time-synchronized, self-organizing, and self-healing mesh architecture and operates in the 2.4 GHz ISM band using IEEE 802.15.4 standard radios. The standard was initiated and developed by the HART Communications Foundation (HCF) in early 2004 under a company- and platform-neutral umbrella.

WirelessHART was approved and introduced to the market in September 2007. That same month, the Fieldbus Foundation, PROFIBUS Trade Organization, and HCF announced a group to develop specifications for a wireless gateway common interface. When the WirelessHART standard was completed, the HCF offered ISA an unrestricted, royalty-free copyright license, allowing the ISA100 committee access to the WirelessHART standard.

Backward compatibility with the HART "user layer" allows transparent adaptation of HART compatible systems and configuration tools to integrate new wireless networks and their devices, as well as continued use of previously proven configuration and system-integration work practices. In April 2010, WirelessHART was approved by the International Electrotechnical Commission (IEC) as IEC 62591.

DeviceNet

DeviceNet is a low-level network that provides connections between simple industrial devices (sensors and actuators) and higher-level devices, such as PLCs and computers. It provides master/slave and peer-to-peer capabilities using the producer-consumer model, which allows devices to share information on a cyclical or event-driven basis. DeviceNet provides interoperability through open/sealed-type device connectors, diagnostic indicators, and device profiles. The DeviceNet network runs on three different cable types, referred to as *thick*, *thin*, and *flat*. The thick and thin cables are round and vary in the amount of current they can carry, while the flat cable is unshielded and has one pair for power, one pair for data, and a mechanical key to ensure proper connection.

DeviceNet is based on the controller area network (CAN) standard, which has found significant use in the automotive industry, linking automotive computers together. (Figure 6-9 illustrates DeviceNet installation concepts.) DeviceNet uses a modified Carrier Sense Multiple Access Collision Avoidance (CSMA/CA) system with arbitration, so there are no collisions. A chip that implements the CAN protocol (CAN chip) is inexpensive, in the $1 to $10 range (2014).

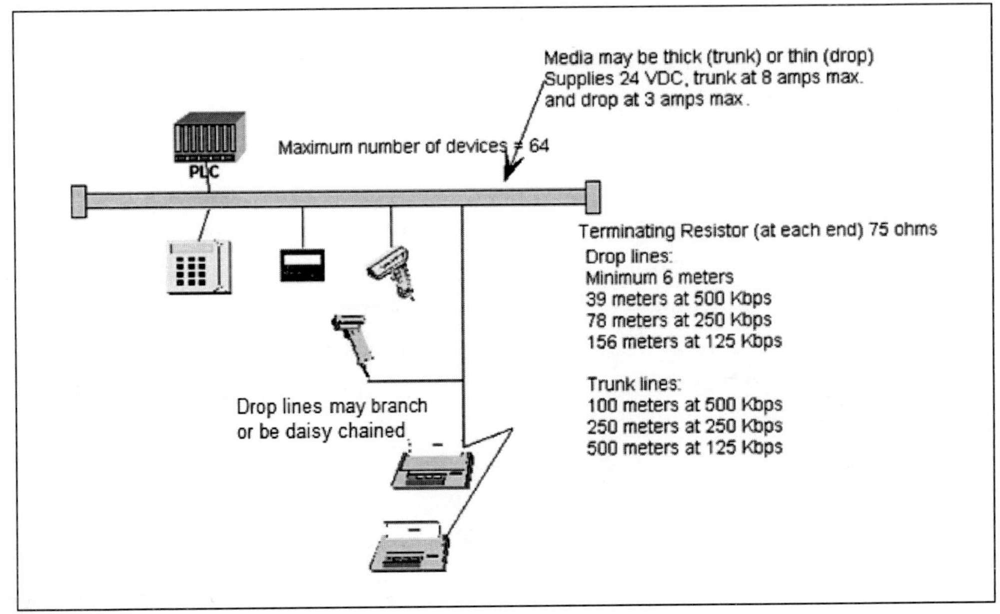

Figure 6-9. DeviceNet Topology

Arbitration of the DeviceNet bus prevents collisions from occurring and guarantees that one of the messages will be completed. The DeviceNet PDU has an Identifier field that contains a unique pattern of bits that, at configuration time, will determine the priority of the device; the lower the identifier, the higher its priority. Zero states have precedence over 1 states. It works like Ethernet: if, after determining that the line is idle, two devices attempt to transmit at the same time, the node that is transmitting a 1 from its identifier stops transmitting if it hears a 0 and the other node continues. This is a bitwise arbitration, and the one with the first 0 has the bus.

ControlNet

ControlNet is an open network that supports simultaneous I/O and explicit messaging. Explicit messaging is a command used for Common Industrial

Protocol (CIP) communications (including DeviceNet). The user is not aware of implicit messages, such as communications commands for remote I/O. Explicit messages are implemented intentionally by the user for a specific purpose. CIP implements the producer-consumer model of communications and supports multiple masters, peer-to-peer, and broadcast communications. ControlNet operates at 5 Mbps, is deterministic, and is intended for controller-to-device, controller-to-controller, and controller-to-external systems that provide seamless integration between DeviceNet and Ethernet. With the correct module, ControlNet interfaces quite well with FOUNDATION Fieldbus. It is approved for installation in intrinsically safe (IS) environments. Figure 6-10 shows a conceptual diagram of ControlNet topology. In this example, the PLCs and the engineering workstation (represented by the computer) are all nodes.

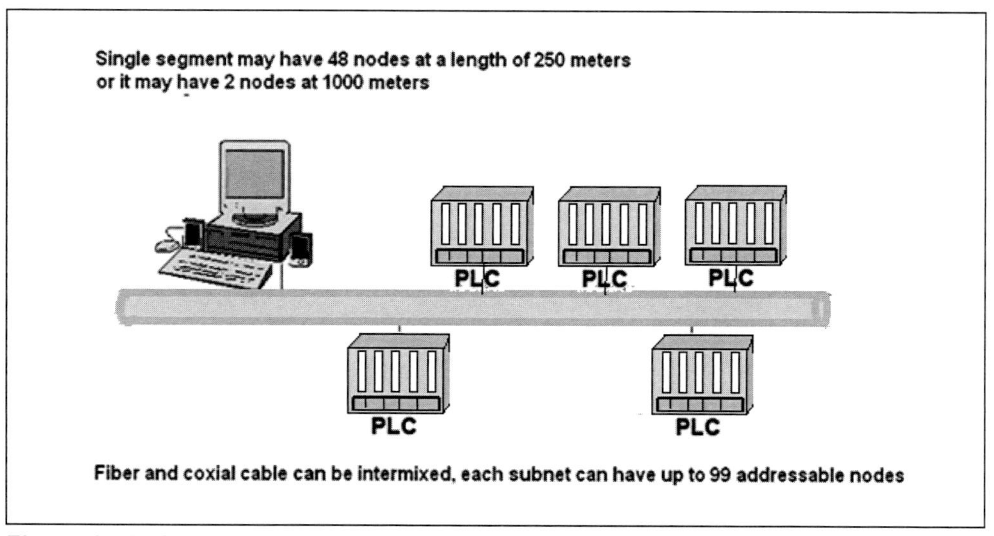

Figure 6-10. ControlNet Topology

ControlNet was developed by Allen-Bradley but since July 1997, it has been controlled by the ControlNet International organization of vendors and users.

EtherNet/IP

By piggybacking on established Layer 1, 2, 3, and 4 protocols (notably Ethernet and TCP/IP), EtherNet/IP has standardized its first four layers. (The "IP" in EtherNet/IP stands for Industrial Protocol and not Internet Protocol.) Figure 6-11 illustrates the relationship between OSI and EtherNet/IP. EtherNet/IP uses the Dynamic Host Configuration Protocol (DHCP) and Address Conflict Detection (ACD) to allow devices to be added and removed from a network. EtherNet/IP supports three types of messages: implicit, explicit, and unconnected (until a connection is made, the unconnected buffer [default 10] is used to originate the connection). Connections can be cached to avoid using the unconnected buffer and the set up times each time a message is sent. All devices on an EtherNet/IP network present their data to the network as a series of data values, called *attributes,* which are grouped with similar data values into sets of attributes, called *objects*. There are EtherNet/IP required objects—Identity, TCP, and Router—that every device must have; the EtherNet/IP specification defines those objects. There are also EtherNet/IP application objects that have the data for specific devices. For example, an EtherNet/IP drive device has a motor object.

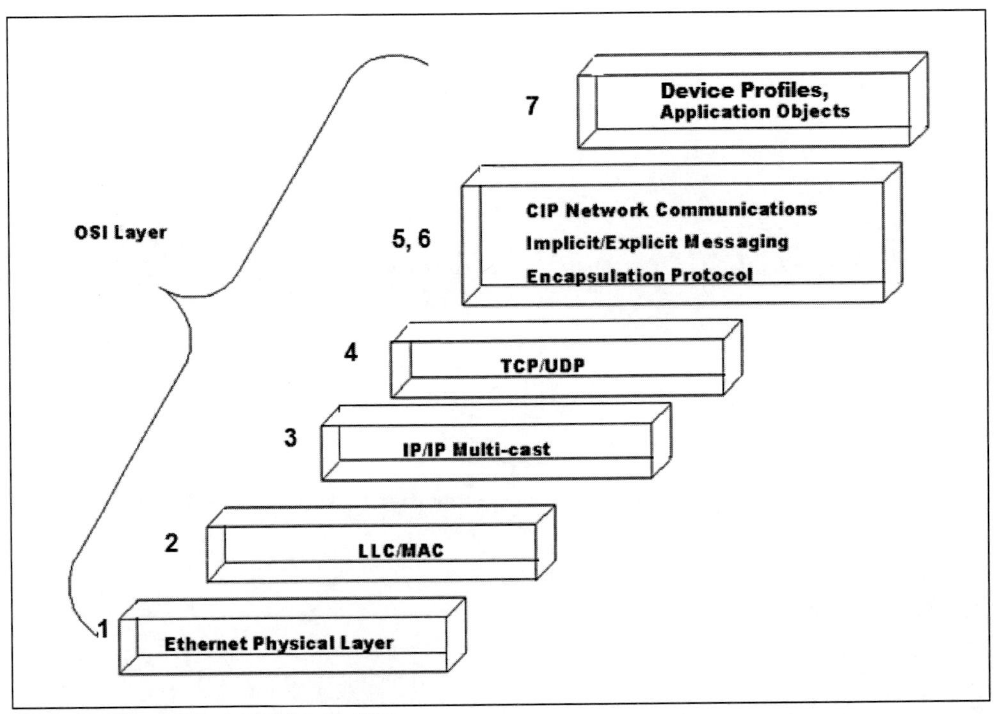

Figure 6-11. OSI and EtherNet/IP

EtherNet/IP enabled devices all have the same set of EtherNet/IP application objects. There are two kinds of messages that are transferred between an EtherNet/IP scanner device (opens connections and initiates data transfers) and an EtherNet/IP adapter device (provides data to the scanner). They are explicit messages (asynchronous, as needed) and I/O messages (data delivery messages that are continuously transferred).

EtherNet/IP is part of CIP, the Common Industrial Protocol. CIP defines the object structure and specifies the message transfer. CIP protocol used over CANbus is called *DeviceNet*. CIP protocol over Ethernet is *EtherNet/IP*. Using the same Application layer in DeviceNet and ControlNet (Layer 7, along with Layers 5 and 6) allows the CIP to be used on several different Rockwell industrial platforms. Layers 1 and 2 are standard Ethernet, and Layers 3 and 4 are standard TCP/IP. The DeviceNet/ControlNet PDU is encapsulated in the TCP/IP payload.

One of the main differences between ControlNet and EtherNet/IP is that ControlNet's Schedule Segment and Schedule Object features are not used in EtherNet/IP. Table 6-2 compares some of the three systems' salient features.

Table 6-2. Comparative Features Table

Feature	DeviceNet	ControlNet	EtherNet/IP
Topology	Passive trunk line	Passive trunk line	Active Star
Determinism	High	High	High (full duplex switched)
Layer 7	CIP	CIP/Routing	CIP Routing
Standards Body	ODVA	ControlNet Int.	IEC 61158
Data rate	125-500 Kbps	5 Mbps	10/100 Mbps
Max Segment Length	500 m	1 km	100 m
Max Network Length	4 km with repeaters	30 km	Unlimited
Maximum Nodes	64	99	Unlimited
Intrinsic Safety	No	Yes	No

Encapsulation is a normal part of TCP/IP, which regards the CIP protocol as data. There is a 24-bit header after the TCP header. However, in the TCP header the target port is set up as 44818 (decimal, or AF12 hex).

LonWorks

LonWorks is based on the EIA-709 standard. It uses a differential Manchester signaling system and a modified CSMA media access called *predictive p-Persistent*, which improves performance in heavily loaded buses. LonWorks uses a special IC—the Neuron Chip—and is a full implementation of the seven-layer OSI model. LonWorks uses the domain model (hierarchical) for addressing four major message types: acknowledged, request/response, unacknowledged, and unacknowledged repeated. Rather than being file oriented, the

LonWorks protocol is message oriented, using short sensor and control messages. Its data rate approaches 500 messages per second, which is approximately 1.25 Mbps. Each LON node has an ID of 48 bits (six octets). The nodes may be addressed by node, or by a domain, or sub domains. A LonWorks network can have a maximum of 2^{48} domains (a domain is a 0 to 6 octet number). Within each domain there can be 255 subnets and each subnet can have 127 nodes. Variables are defined for each node (up to 255). LonWorks has a credible performance, according to the manufacturer, particularly in building automation, its major market. It is an open standard (or considered as such) and might be a formidable competitor for other fieldbus products, except that it does not yet have a large market share for process automation (2014). Figure 6-12 illustrates the relationship between LonWorks and the OSI model.

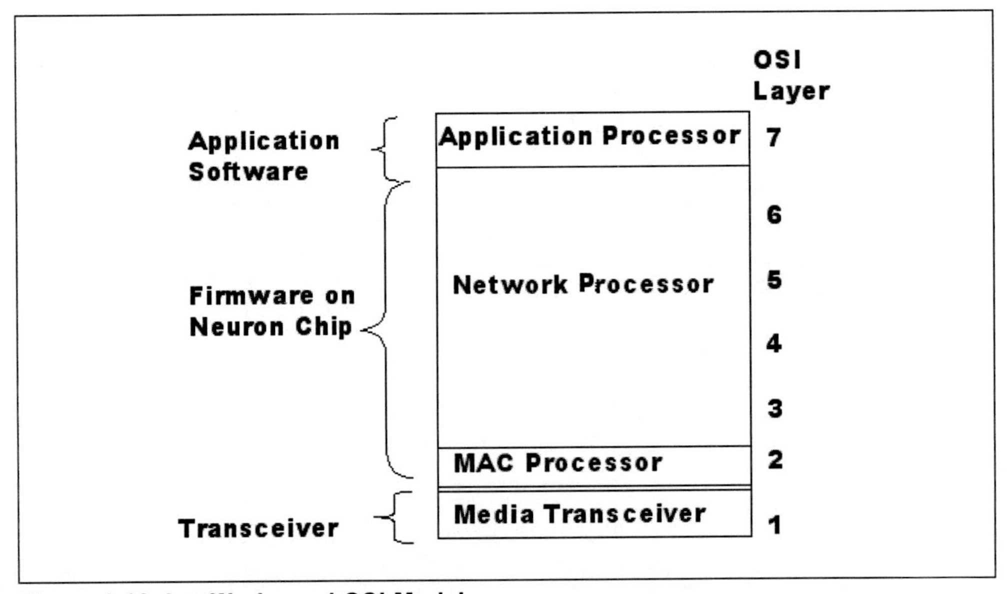

Figure 6-12. LonWorks and OSI Model

The p-Persistent CSMA uses a randomized slot selection and a priority set of slots that is determined at installation time. This is actually a collision-avoidance scheme, rather than a collision-detection scheme. No collisions are possible during the priority time slots and a predictable round-trip time can be made for high-priority messages. A collision-detection scheme is optionally available for lower-priority messages. The packet format in figure 6-13 demonstrates how the data is moved from one address to another address. There are addresses for unicast (one-to-one) and multicast (one-to-many). The messages may be acknowledged or unacknowledged, with the latter conserving bandwidth.

Domain ID	Source Address	Destination Address	Network Value Selector	Network Value

Figure 6-13. LonWorks Packet Format

LonWorks has found acceptance as a sensor network, a device network, and a field network. LonWorks Networks Services (LNS) supports interoperable tools, Windows (2000 up) HMI, Component Software (ActiveX) objects, and local, remote, and Internet access. The EIA-852 standard covers LonWorks on Ethernet. Echlon, the manufacturer of the Neuron chip used in LonWorks, has made their communication software source code available so other manufacturers can implement the source code without the need to embed a Neuron chip.

AS-i

The Actuator Sensor Interface (AS-i) was developed for the European market in 1993 and brought to the United States in 1996. It is a low-level technology that complements, rather than replaces, PLCs and fieldbuses. AS-i consists primarily of an untwisted, unshielded two-wire cable that connects devices with the AS-i chip. The chip has no processor and therefore needs no programming. It handles bit data only and is fast, secure, and built for the industrial environment. Figure 6-14 illustrates how AS-i fits into the hierarchy of buses.

In AS-i, the master can be the controller, or it can be controlled by being coupled into a fieldbus. The two-wire cable carries the data and power and connects using a "piercing" technology, which makes for easy-to-install connectors. AS-i has up to IP67 protection: Ingress Protection Marking where 6 = No ingress of dust, complete protection against contact (dust tight) and 7 = Protected against water immersion for 30 minutes at a depth of 1 meter. The AS-i protocol is simple: master/slave. It is performed on a ring or cyclical basis, where a poll is sent one at a time to each registered slave. The published time for complete rotation with 31 slaves (the maximum permissible is 31 slaves and one master) is 5 milliseconds. It should be noted that even the AS-i master operates under its own firmware and no programming is required of the user. With only 31 slaves, it is still possible to achieve 248 binary signals

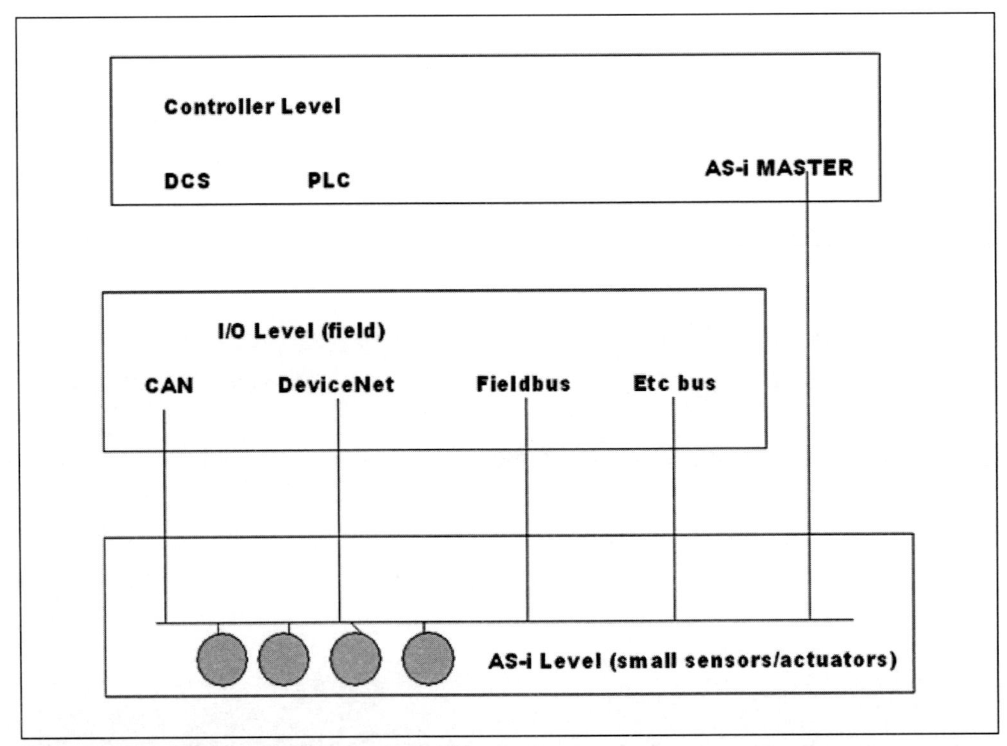

Figure 6-14. AS-i in the Hierarchy of Buses

per network. The maximum length without repeaters is 100 m (330 ft.) and two repeaters can be used to get the maximum length of 300 m (1000 ft.). The AS-i network may have almost any topology the user wishes to create: star, daisy chain, tree, or branch.

All AS-i connectors use coupling modules at their base (see figure 6-15). On top of the coupling module is placed one of the following:

- A tap cover
- A passive distributor
- An application module

P-Net

P-Net is based on the EIA-485 standard. Its topology is shown in figure 6-16. It has an allowable cable length of 1,200 m (4000 ft.) without repeaters. P-Net signals at a 78,600 bps rate. Using a technique called *parallel* (most refer to it as duplex), it will receive the acknowledgment from the receiving device as it is completing the transmission of the frame to that device. This "pipelining" speeds up data transfer and the P-Net people claim it is as fast as any 500 Kbps system.

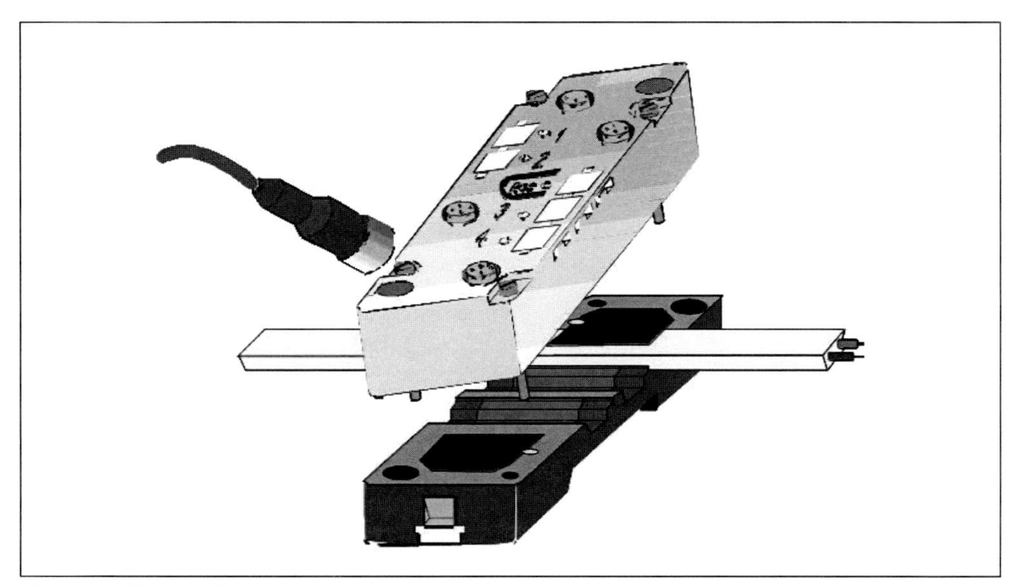

Figure 6-15. AS-i Connector

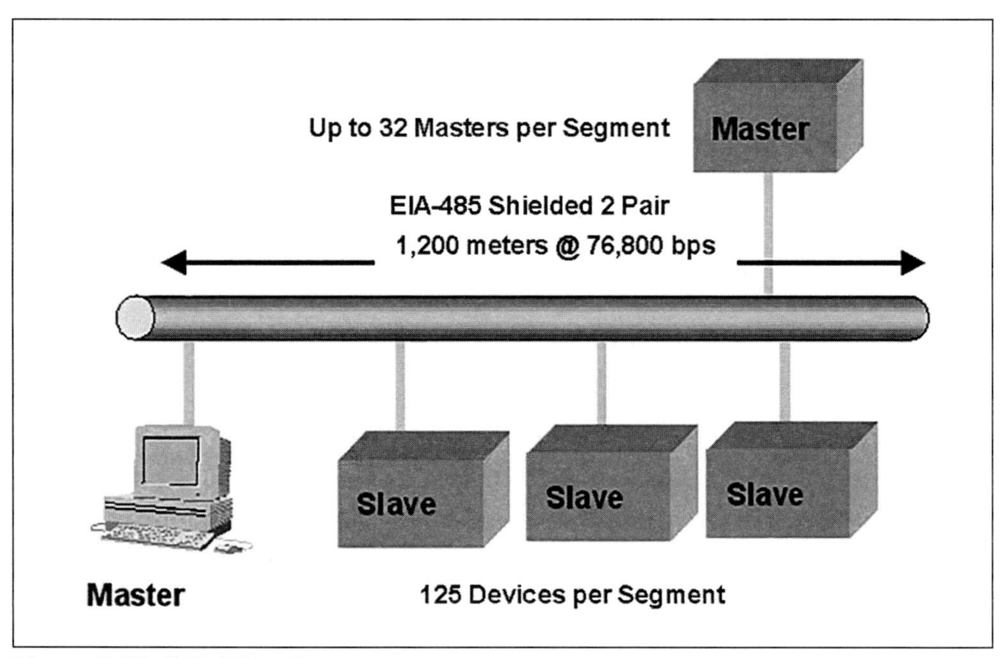

Figure 6-16. P-Net Topology

P-Net's marketing people say it uses a "virtual token," but it is actually a counter-controlled bus mastering technique—a form of polling where the master is assigned an ID number and the bit counter increments a unit at a time; when the master number is up, the master can transmit. It accords 10 bits to each master after the minimum of 40 bus bit idle times are counted. As an example, assume Master 3 (not shown in figure 6-16) had been transmitting and receiving. After the bus has been idle for 40 bit times, the bit counter is incremented by one and Master 4 (not shown) may have the bus. If Master 4 has nothing to say (or isn't even there), the bit counter keeps counting until it reaches 50 bit idle times, plus one. At this point, Master 1 (the computer) can have the bus. If Master 1 has nothing to say, then when the bit counter reaches 60 bit idle times plus one, the second Master (the box labeled Master in figure 6-16) can transmit. If Master 2 is so inclined, the bit counter will be reset to zero until Master 4 has completed its transmission. Since no data is transferred over the bus, this process is deemed an efficient way to handle "token passing." P-Net is object oriented and uses OLE2 (Object Linking and Embedding) as a way for the Application layer to access physical objects via a "virtual object." P-Net has Visual Basic for Applications as a programming structure.

PROFIBUS/PROFINET

PROFIBUS is actually three different protocols. PROFIBUS FMS (Fieldbus Messaging Specification) is a general-purpose solution for peer-to-peer communications tasks. PROFIBUS DP is a high-speed data communications network for factory automation, while PROFIBUS PA is for the process automation market. The Physical layer has three transmission technologies available: EIA-485 two-wire copper cable; fiber optic, when electromagnetic compatibility (EMC) protection is required; and IEC 61158-2, a two-wire copper cable with provision for providing power over the bus. The Physical layer options are outlined in table 6-3.

Table 6-3. PROFIBUS Physical Layer Options

EIA-485	Fiber Optic	IEC 61158-2
Asynchronous 9.6 Kbps to 12 Mbps	Asynchronous 9.6 Kbps to 12 Mbps	Synchronous 31.25 Kbps
Shielded twisted pair	Mono-mode, multi-mode plastic, PCS/HCS fibers	Shielded twisted pair IS and bus power options
32 stations per segment 126 stations maximum		10-32 per segment (depending on power) maximum of 127 stations
Distance: 12 Mbps 100 m 1.5 Mbps 200 m 187.5 Kbps (or less) 1 Km		Up to 1.9 km depending on power requirements
Repeaters allowed	Extendable to 100 km	No repeaters

PROFIBUS differentiates between master and slave devices for access and uses a technique similar to a floating bus master: the current master is determined by a bidding process and the mastership rotates among those designated as masters. The current master being the one who obtained the bid. The slaves are always polled, whether there is just one master or multiple masters. The only passing of command is performed between the masters. Although the physical layers differ, Layer 2 is the same across all three PROFIBUS implementations. FMS and DP use the same cable and their signals can be combined. PROFIBUS PA is dedicated to the process industry and uses a different physical media technology than DP/FMS. PROFIBUS will support bus, tree, or star topologies; the tree topology is preferred.

The PROFIBUS Data Link layer uses virtual field devices (VFD). It was intended to replace the Enhanced Performance Architecture – Manufacturing Automation Protocol (EPA MAP), using the Manufacturing Messaging Service (MMS) as an application utility. Unfortunately, MAP did not succeed in the marketplace and it is now considered a legacy technology. Although the MMS was evaluated by the fieldbus (ISA50) committees, they ended up defining the User layer instead.

PROFIBUS (in all of its implementations) is a German standard (DIN 19245) and an IEC 61158 standard. It has been successful in the discrete manufacturing segment and has a large installed base. It is worth noting that PROFINET/ PROFIBUS organizations, FOUNDATION Fieldbus, HART Communications Foundation, and OPC collaborated to produce the Electronic Device Description Language (EDDL), so the profiles (the device attributes) are common over a large base of equipment.

PROFINET is a continuation of the PROFIBUS system. It is an object-oriented approach that uses standardized interfaces and technologies, such as TCP/IP. PROFINET is an attempt to unify the hierarchy from the sensor to the enterprise network. By using market standards, such as XML, Ethernet TCP/IP, and OPC, it provides a set of open, transparent, and integrated communications protocols. PROFINET is an attempt to move PROFIBUS from a distributed I/O system to a distributed intelligence system in which the intelligence is in the field devices rather than centralized in a control facility or cabinet.

FOUNDATION Fieldbus

FOUNDATION Fieldbus is designed to interconnect field instruments using a distributed intelligence system that puts the controller back in the field. As such, it must have low node costs, offer extremely reliable operation, be relatively simple to operate, and above all be a real-time system. FOUNDATION Fieldbus handles data in two different modes, depending on whether the data is cyclic (operational data: traffic and algorithms) or acyclic (background data: configuration and diagnostics). Cyclic traffic is generally low volume and time critical, while acyclic is just the opposite, high volume and not time critical.

Figure 6-17 illustrates the FOUNDATION Fieldbus OSI model. Note that it uses Layers 1, 2, and 7 of the OSI model but it actually has four layers. The fourth layer is the User layer, which is defined in the specification. However, stating that it has four layers is not totally accurate. Since FOUNDATION Fieldbus is segment routable, the Applications layer handles the majority of the Layer 3 functions for the H1 fieldbus.

The Physical Layer

The Physical layer connects the node to the media and provides for activation, deactivation, and maintenance on a bit basis (octets or byte patterns are not recognized). The SP50 standard, as implemented by the Fieldbus Foundation, defines types of media, signals, speeds, and topology, including the number of nodes and node power.

Continuous process control requires a moderate speed; speed requirements dictate power consumption. The higher the speed, the more power is

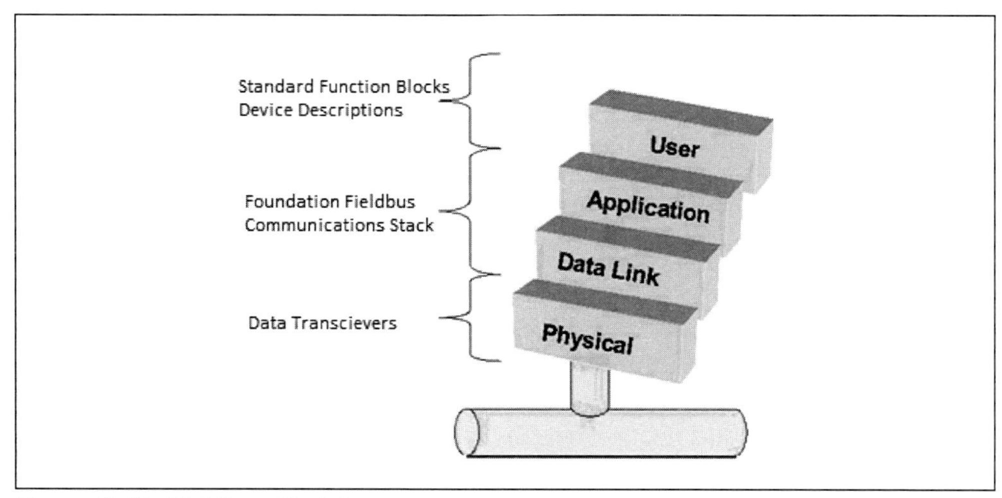

Figure 6-17. Fieldbus Model

required. Large power demands are inconsistent with the intrinsic safety (IS) requirements typically found in continuous processing. The designers of FOUNDATION Fieldbus decided on a moderately high-speed bus for intrinsic and field instrument work (H1) and originally selected two higher speeds for non-intrinsic areas (H2). However, technology and good business sense eliminated H2, as seen in the next paragraph. Each had its own advantages. It should be noted that all devices on one bus had to have the same speed and options for the media. Nevertheless, powered and non-powered devices could be mixed—and on H1—intrinsically safe and non-IS devices may be interconnected.

The first physical medium characterized in the FOUNDATION Fieldbus standard is copper. Standards are currently being developed for fiber-optic media. The H1 standard transmission rate is 31.25 Kbps (the same as PROFIBUS PA). It takes a long time to generate a standard and then populate it with actual equipment. Wisely, the Fieldbus Foundation abandoned the effort to develop a 1- and 2.5-Mbps H2 and embraced instead a proven, workable, cost-effective commercial standard: HSE, or high speed Ethernet, which is full-duplex switched at 100 Mbps and, therefore, is both deterministic and fast. HSE connects fieldbus segments together, connects a fieldbus to other networks, and ties operator consoles together. There is also an increasing number of smart instruments and devices that directly connect to a HSE LAN via an available switch port.

H1 transmission (31.25 Kbps) is a synchronous, half-duplex serial signal that uses Manchester encoding and is, therefore, self-clocking. Manchester encoding places the transmitters' output state into transitions (rather than plateaus), timing can then be recovered at the receive device(s). Preamble, start, and stop delimiters are not Manchester encoded and, thus, are instantly recognized as such.

Signal

With current modulation, the devices can sense the voltage drop across the terminating resistors (100 ohms each). For H1, modulated current (Manchester encoded) is 15 to 20 mA peak to peak and the typical sensitivity of the receiver is 150 mV.

Topology

In FOUNDATION Fieldbus, bus or tree topology is supported by the 31.25 Kbps scheme. A bus has a trunk cable with two terminators. Spurs attach to the bus by way of a coupler. A spur may contain more than one device and it may have an active coupler to extend its length. (The difference between a spur and a drop is that a spur may have multiple devices attached, whereas a drop traditionally only has one device.) The trunk (bus) may have an active repeater to extend its length. An installation guide follows:

1. One fieldbus segment can have between 2 and 32 devices, however the power supply (which has an impedance matching device) and the termination resistor each count as one device, therefore 30 is the maximum number of devices on a non-intrinsically safe segment. Alternately, the segment can have between 2 and 6 powered devices for an IS bus, of which between 1 and 4 of the devices are in the hazardous area. Finally, the SP50 standard specifies up to 12 bus-powered devices in non-intrinsically safe segments. A system can have more than the number of listed devices because devices are calculated to draw 9 mA (±1 mA), so if the devices in fact draw less, the system can have more devices. The IS areas were calculated for a 19 Vdc (output) barrier providing 40–60 mA. The total of 12 bus-powered devices is based on the assumption of a source of 20 Vdc.

2. The total cable length cannot be more than 1,900 m (6,300 ft.). This not the geographic distance but the total length of all spurs and trunks.

3. There can be no more than four active couplers or repeaters.

4. Maximum propagation between any two devices on the segment cannot exceed 20 nominal bit times.

5. Devices can be connected or disconnected without interrupting operations of the bus; any errors caused by connection or disconnection must be detected and corrected.

6. The failure of any communications element (except for such failures as jabber, short circuits, or low impedance) cannot interfere with other transactions for more than 1 ms (millisecond).

7. Connectors must be uniquely marked, with physically guided connectors, in order to maintain correct polarization.

8. Attenuation factors for different topological arrangements must be determined to ensure that a particular topology will meet the signal power budget.

9. For redundant media, each cable has to meet all of the network rules and there can be no non-redundant segment between two redundant segments. Repeaters must be redundant and, if they are transmitting on more than one channel, there can be no more than 5 bit times difference in propagation time for any two devices on any two channels.

Figure 6-18 illustrates connections in an H1/H2 (HSE) network. The high-speed backbone is at 100 Mbps and can connect to standard Ethernet equipment. This architecture allows for the use of standard low-cost wire and fiber-optic media with fault-tolerant communications and linking devices.

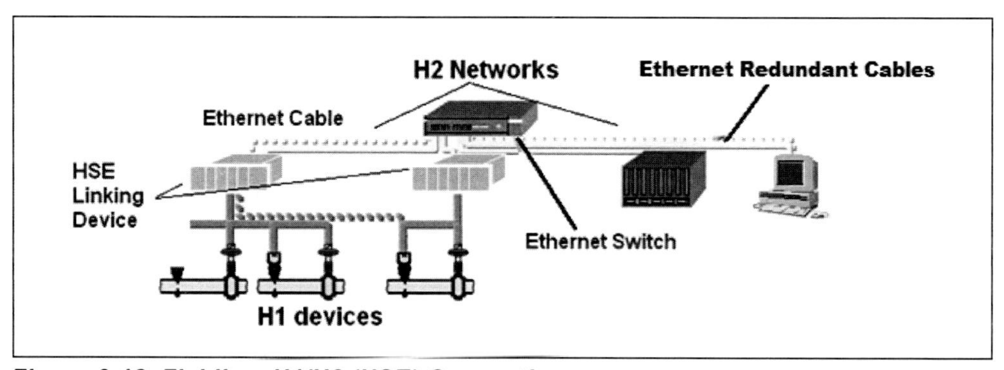

Figure 6-18. Fieldbus H1/H2 (HSE) Connections

Figure 6-19 illustrates the salient requirements for a repeated network.

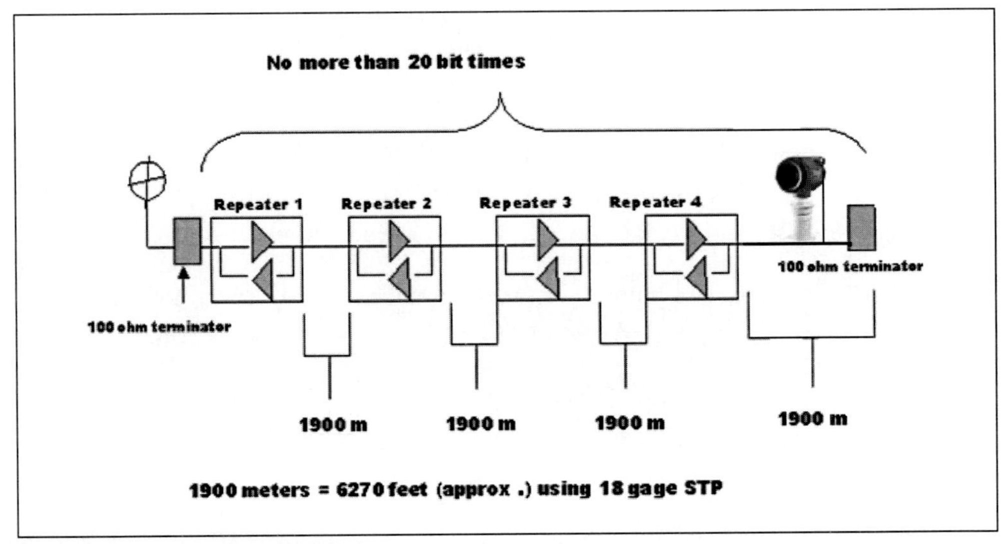

Figure 6-19. Fieldbus (H1) Repeaters

In addition, each node must have a hardware anti-jabber or jabber-inhibiting, self-interrupt capability, which allows transmission from the jabbering node to last no longer than between 120 and 240 ms.

Data Link Layer

As specified in IEEE 802.2, the fieldbus Data Link layer, is divided into two (actually three, one is an intermediate layer) sublayers: Fieldbus Media Access Control (lower) and Fieldbus Data Link Control (upper).

For access control, FOUNDATION Fieldbus modifies the token-passing bus arrangement of rotating the token and, instead, opts to use a method similar to the Data Highway Plus floating bus master. Here, the master is called a *Link Active Scheduler (LAS)* and uses two different tokens, specifically the *delegate* token and the *reply* token. Figure 6-20 illustrates the delegate token.

In figure 6-20, the designated LAS (Station A) passes the token to Station B. As long as the token is in the delegated station's possession (a prescribed time that depends on network loading, configuration, etc.), the station may transmit and request replies from other stations. At the prescribed time, it must return the token to the LAS, which then apportions it out to the next station. Figure 6-21 illustrates the reply token. In figure 6-21, the LAS (Station A) passes the token to Station B. Station B may have one initiated transmission (to one or more stations) and then it must return the token to the LAS.

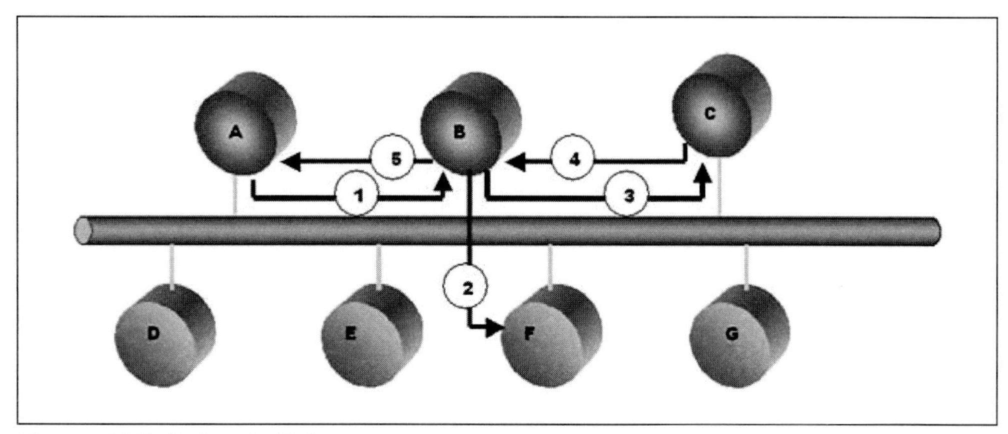

Figure 6-20. Delegate Token

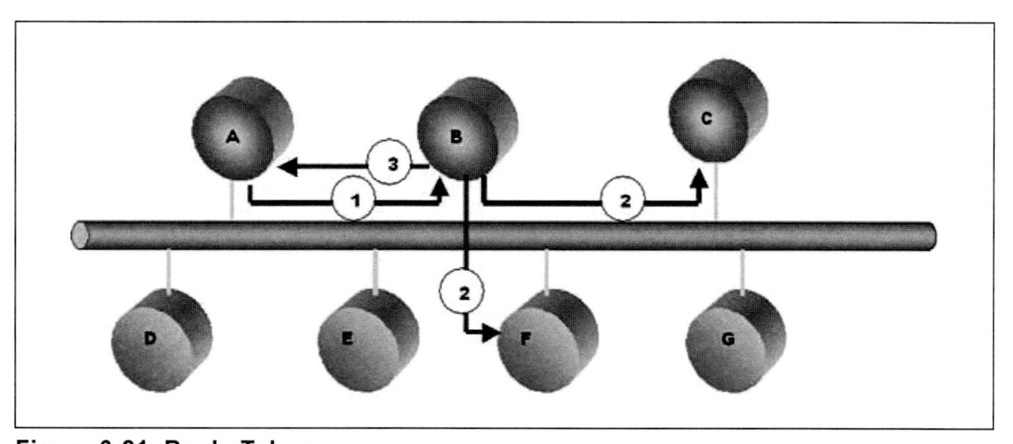

Figure 6-21. Reply Token

Fieldbus Layer 2 Frame

Figure 6-22 illustrates a fieldbus frame. Note that it has a preamble, start delimiter, destination DLL PCI, data block, frame check sequence (CRCC), and a stop delimiter—the same format that has been observed in almost all protocols since LAP-B. The data (all bits except those in the preamble and delimiters are called *data bits*, not just those in the data block) is placed on the wire using Manchester encoding. The data is a valid Manchester-coded signal; the preamble and start/stop delimiters are "nonvalid" (not data) signals and are easily discerned from the data signals.

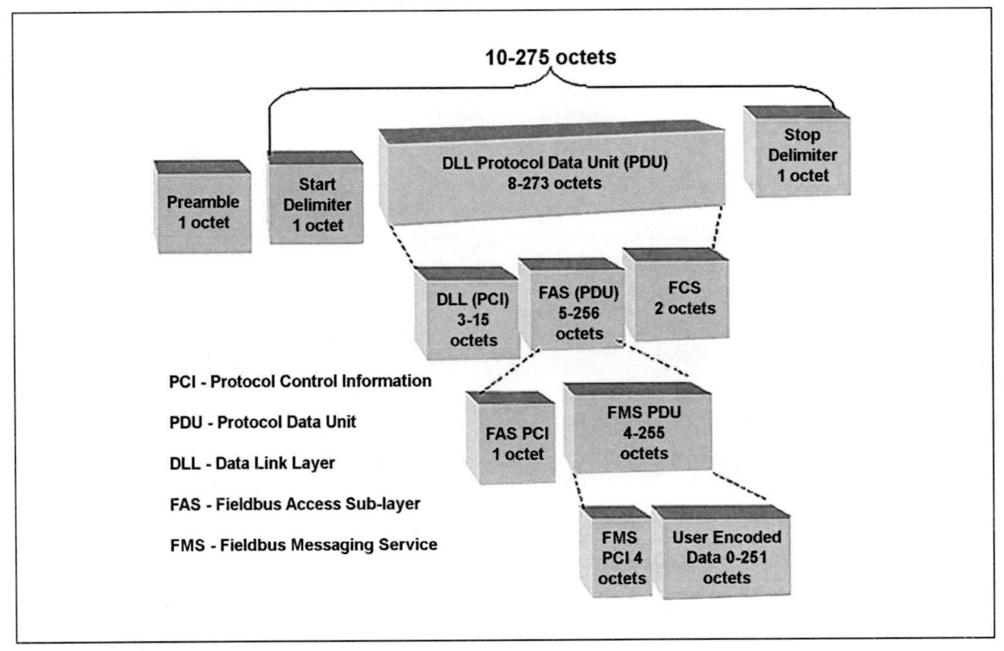

Figure 6-22. Fieldbus Frame

Application and User Layers

The higher layers are implemented in object-oriented design (OOD), in which objects are used rather than lines of linear code. This greatly increases programming productivity, reduces the level of abstraction required, and allows the product to be visually oriented, an essential benefit for users given FOUNDATION Fieldbus's intrinsic complexity.

In FOUNDATION Fieldbus, background processes (e.g., configuration, messages) are acyclic messages and use a form of the client-server model. Recall that a server is any node that shares its resources and a client is any node that uses a server's resources. For the cyclic processes (those that are time dependent, such as a PID algorithm), FOUNDATION Fieldbus uses a publisher-subscriber model. This model is derived from the more frequently used producer-consumer model. The main difference between the two models is that publisher-subscriber normally requires the subscriber to subscribe to a certain event controlled or reported by the publisher, then any time the event occurs, the publisher sends notifications to all those that subscribed. This is called a *push* process, since the publisher pushes out the information without the subscriber having to request this action. In producer-consumer models, the producer typically makes the information available on the network through multicast and those that require it, consume it. This is also called a *push* operation. When a device has to request information from a server whenever it needs the latest value, then the process is usually called a *pull* process.

FOUNDATION Fieldbus builds applications using function blocks. From the User layer perspective, any device is more than just parts, it is a parameterized network node. The User layer sees the nodes as virtual field devices (VFD), which are the interfaces between the communications protocol and the function block. A node may have one or more VFDs.

Both system and network management rely on the tightly coupled Applications/User layer.

System management is concerned with the following five issues:

- **Device Tag Assignment:** Before a device is placed on the network, it must have an electronic device tag assigned to it that is unique to that device and represents its physical attributes (e.g., A0-FIT-71245 is a flow indicating transmitter). This is done using standard facility guidelines for the tag.
- **Station Address Assignment:** Station addresses are assigned by a plug-and-play method. A new device has a default address and, assuming that its device tag is not identical to another device tag in the network—which shouldn't happen if facility guidelines are followed, the network assigns an address to the device.
- **Clock Synchronization:** Clocks are synchronized from a master clock, so all devices maintain the same real time.
- **Scheduling of Application Processes**: Scheduling is required to ensure there is neither dead time while waiting for execution, nor variance in timing delays. The priority is as follows: (1) function block (like PID or AI) execution that is scheduled in the field device, (2) communications (operational traffic) that is scheduled in the ALS (master), and (3) background (acyclic) communications.
- **Function Block Binding:** Function blocks are distributed throughout the network. There is no centralized location for control. In effect, the network is the controller. Communications link the function blocks. When a new block is entered into a system, it must be bound. That is, its stack and the other devices that are interconnected with it must have the service and destination access point addresses (offsets from the

block address) available and stored throughout the system devices that will use the information. In other words, once a block is instantiated, it then goes through a binding process that the software uses to identify function blocks, determine methods and properties, and connect the device.

Figure 6-23 illustrates the concepts of the VFD and function block in terms of the OSI layers for the communications stack.

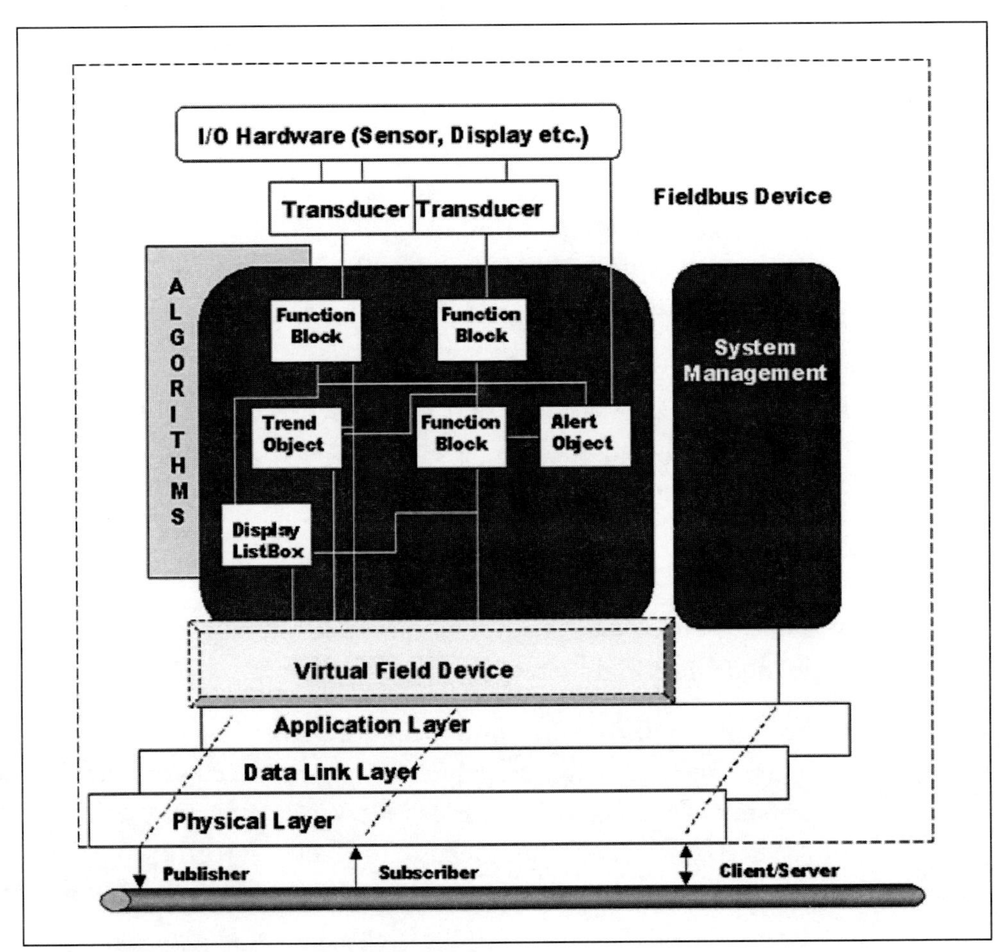

Figure 6-23. VFD and Function Block

Function blocks contain an algorithm and a set of parameters (a parameter block) for processing inputs and producing outputs. The function block is an object, an abstraction of data and software. It exists only in the software model but is useful in enabling humans to organize and recognize the purpose of the function block and its relationship to other objects and function blocks. Just like a physical control system, each block has a unique tag, assigned by the user. The block tag and the parameter tag uniquely identify all parameters. A sample-and-hold system is used, in which an instantaneous picture is taken of all inputs. This keeps them from changing while they are being processed by the algorithm for that block. The outputs are then updated and published to the network.

To build a system, it is necessary to assign tags and to select a control strategy by selecting function blocks, linking them, and setting the parameters. Initially, there were 10 standard function blocks but many have since been added.

The 10 standard basic blocks are:

1. Analog Input block (AI)
2. Discrete Input (DI)
3. PID Control block (PID)
4. PD Control block (PD)
5. Analog Output block (AO)
6. Discrete Output (DO)
7. Control Selector (CS)
8. Ratio (RA)
9. Bias (B)
10. Manual Loader (ML)

Some of the added blocks are as follows:

- Analog Alarm
- Discrete Alarm
- Splitter
- Dead Time
- Lead/Lag

- Pulse Input
- Calculation Block
- Set Point Generator
- Integrator
- Step Control
- Output Signal Selector
- Complex Analog Output
- Complex Digital Output
- Device Control
- Arithmetic
- Analog HMI
- Discrete HMI

As FOUNDATION Fieldbus advances, even more blocks will undoubtedly be added.

A pressure transmitter will likely contain an AI and a PID function block. An equivalent of the I/P (Current to Pressure transducer) would be a Fieldbus to Pneumatic (or F/P) block that contains an AO and perhaps a PID block. A large number of devices exist and they contain a large number of function blocks. The control system and strategies employed are all in how they are connected. Figure 6-24 is a block diagram of a PID function block.

The function block inputs and outputs are standardized for all vendors. However, it is up to the vendor to determine how it produces the output from the inputs; that is what distinguishes one vendor's product from another. Note that this method of ensuring compatibility makes all devices (at least functionally) interoperable and interchangeable, although some will no doubt perform better than others under certain conditions in certain applications. This fact will allow vendors to develop a brand name that is associated with a certain level of performance and cost effectiveness.

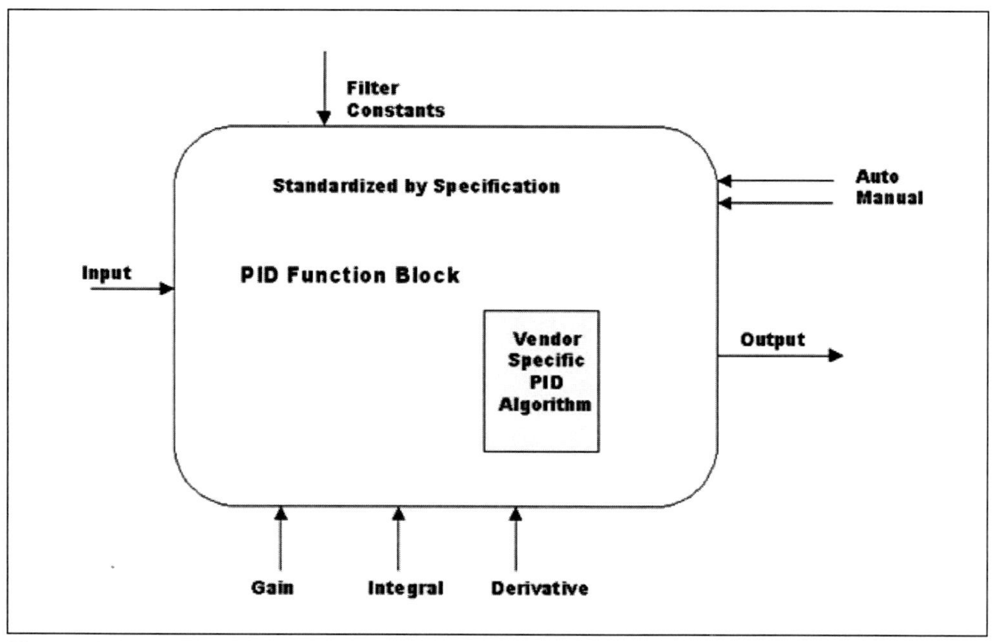

Figure 6-24. PID Function Block Diagram

Hierarchy of Buses

The H1 fieldbus is intended to connect to field devices. The higher-speed mode FOUNDATION Fieldbus (HSE) for High Speed Ethernet (100 MBps) is intended for intersegment or internetwork connectivity, however it may also contain devices other than those intended for interconnectivity such as measurement or control devices. Figure 6-25 shows the hierarchy of buses.

The Fieldbus Foundation, a nonprofit, vendor-supported organization, tests protocol stacks and devices to ensure that they conform to the specifications. If you purchase a device that has its seal of conformance, you can be sure it will interoperate and meet the specifications.

Fieldbus Summary

The advantages of FOUNDATION Fieldbus are as follows:

- It is designed for process control.
- It is a real-time system with node-synchronized clocks.
- It has an open system standard.

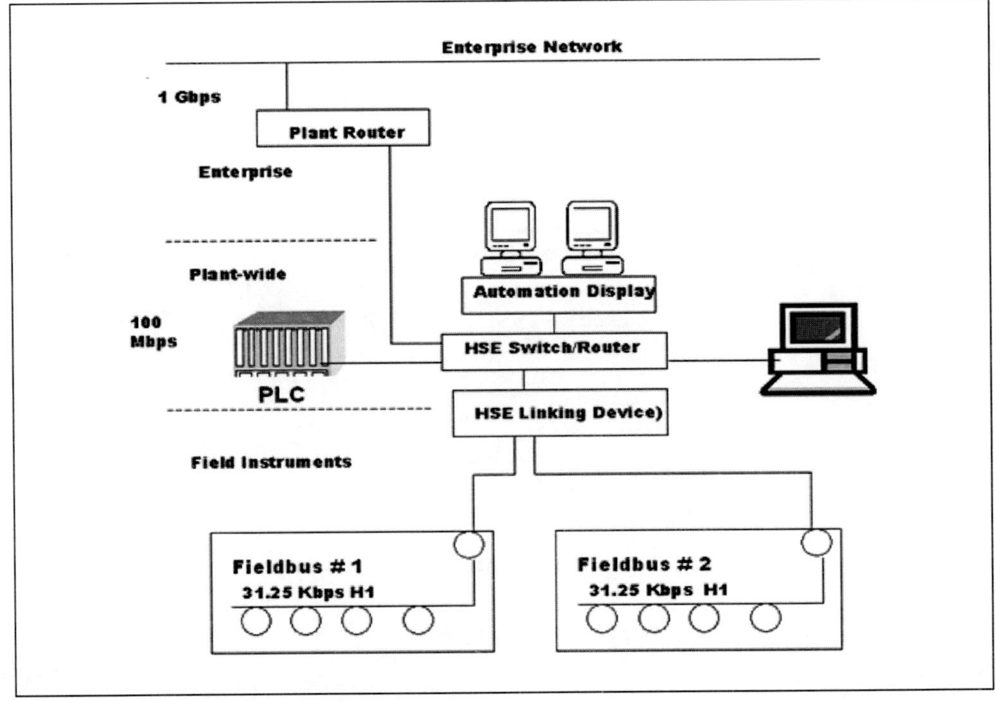

Figure 6-25. Hierarchy of Buses

- Compared to conventional devices, a FOUNDATION Fieldbus device has:
  - High information flow
  - Better accuracy
  - Higher reliability
  - Easier calibration and maintenance
- It offers interoperability and interchangeability.
- It provides good control-loop performance

ISA-100.11a

ISA-100.11a, *Wireless Systems for Industrial Automation: Process Control and Related Applications,* is a wireless networking standard developed by ISA. The ISA100 committee was formed in 2005 to establish standards, and related information, that define procedures for implementing, deploying, configuring, and operating wireless systems at the field level in the automation and control environment. The committee is made up of automation professionals from over 240 companies worldwide and represents all stakeholders (end users, wireless suppliers, system integrators, etc.). In 2009, the ISA Automation Standards Compliance Institute established the ISA100 Wireless Compliance Institute, which owns the ISA100 Compliant certification scheme and provides independent testing of ISA100-based products to ensure conformance to the ISA100 standard.

The standard was approved by the committee in May 2009, by ANSI in 2011, and the IEC will publish the standard as IEC 62734. The goal behind the standard is to provide reliable and secure wireless operation for monitoring, alerting, supervisory control, open loop-control, and closed loop-control applications. It defines the protocol suite, system management, gateways, and security specifications for wireless connectivity with limited power consumption devices. ISA-100.11a/IEC 62734 utilizes Internet Protocol version 6 (IPv6), adheres to the OSI model, and uses object technology. In addition, the standard fully supports the ETSI EN 300 328 v1.8.1 European Union specification taking effect in 2015.

ISA-100.11a systems are deterministic. All devices in a wireless subnet communicate at a pre-arranged time and pre-selected frequency using time division multiple access (TDMA), with the length of a time slot being configurable. When transmitting once a second, the maximum packet time is 4 milliseconds. This is a burst method of transmission using a pre-arranged time (and a 0.004% duty cycle; i.e., 4 milliseconds out of 1000 milliseconds), which increases it ability to override interference. This is primarily for short slots. Longer transmissions can have a long time slot and CSMA/CA is used. Both schemes are used concurrently. The radio scheme itself is IEEE 802.14.5, dividing the usable 2.4 GHz spectrum into 16 channels. It then uses spectrum management (using an awareness of other devices transmitting in the 2.4 GHz spectrum) when assigning slots and uses selective frequency hopping to reduce interference.

The ISA-100.11a system uses a two-layer security (at the Data Link and Transport layers) with authentication required (IEEE 802.1x) on hop-to-hop (Layer 2) and end-to-end (Layer 4) messages. Additionally, the standard employs the latest 128-bit key Advanced Encryption Standard (AES) encryption/decryption. The standard was designed to accommodate multiple protocols at the device and host level, while allowing applications to run interoperably over a common network structure. Figure 6.26 illustrates a single plant deployment of ISA-100.11a.

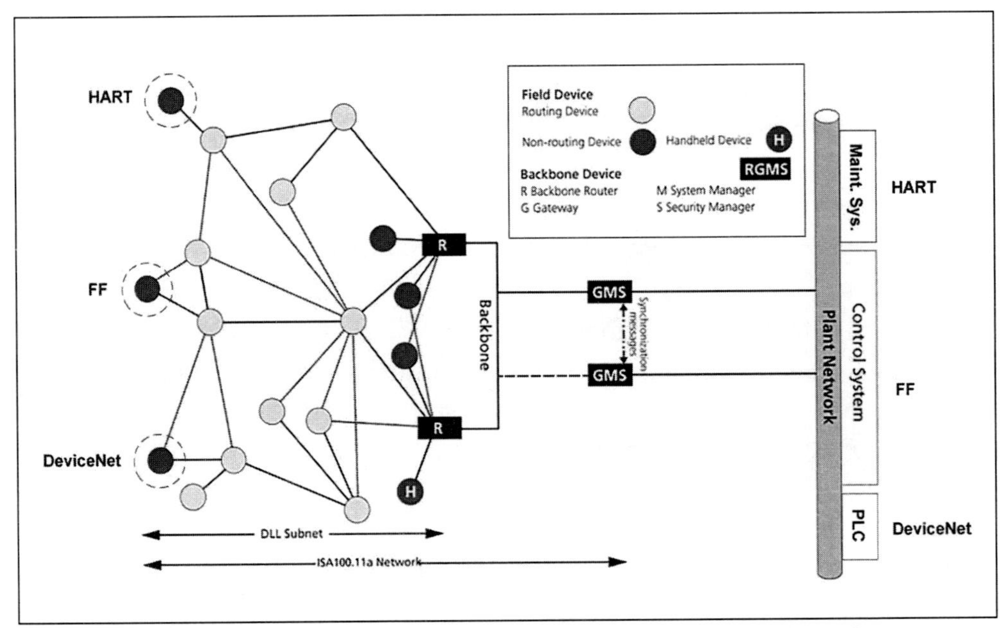

Figure 6-26. Typical Single Plant Deployment of ISA-100.11a (Modified from an original ISA100 Wireless Compliance Institute PowerPoint presentation.)

Ethernet-TCP/IP

The commercial version of Ethernet was explained in chapter 4, in connection with LANs. Here, we will just explain why commercial Ethernet, along with TCP/IP, has made deep inroads into industrial applications and automation products.

First and foremost, it is the least expensive per-node, per-bit-per-second network available today.

Second, it has been proved in installations of many different sizes and under many different conditions.

One of the first arguments against the adoption of Ethernet in industrial settings (from the heydays of MAP/TOP) is that Ethernet is not deterministic: there is no guarantee as to when a node can get on the network. That is true of a half-duplex, hubbed network running CSMA/CD. Yet, that does not mean "never," nor that there is no way to achieve reliable (guaranteed) round trip times. Ethernet (of the shared media, hubbed variety) is based on probabilities.

The statistical probability of a collision or multiple collisions occurring during the time to originate a transmission has to do with both the nature of the transmissions and the loading of the network with a shared medium.

Another of the arguments for a deterministic network is that Ethernet cannot provide an accurate time slice at each time interval to properly calculate PID control. On a heavily loaded network (10 Mbps) the uncertainty may, under some conditions, cause problems on a fast responding loop.

One of the many myths surrounding Ethernet is that heavy loading would bring the network down due to collisions and exponential time-outs. In *Measured Capacity of an Ethernet: Myth and Reality*, Boggs, Mogul and Kent (1995) showed that, in practice, Ethernet delays are linear and range from 2 ms for a lightly loaded network to 30 ms for a heavily loaded network (and this was a 10 Mbps network).

Loading a measurement and control network is obviously different from that of an office network, yet the solution is the same: Do not heavily load the network. Restrict the number of devices in each collision domain, so traffic does not peak at over 10 percent of capacity. The fact that this loading is not a real problem should be understood, as most PLC vendors have included an Ethernet port since at least 1988.

To truly make an Ethernet network deterministic, it is only necessary to employ full-duplex switches. Because each node thinks it has full bandwidth (in fact, twice as much as a half-duplex network) and always has the bus, collisions do not occur and there are no time-outs. This is the option specified for FOUNDATION Fieldbus in the HSE version.

The use of Ethernet switches, rather than hubs, allows IEEE 802(q) and 802(p) standards to be implemented. Both of these standards are quality-of-service (QoS) rules and the switch assigns priorities to different traffic.

Then there is the solution that many proprietary systems use. They determine timing in Layer 7 by time-stamping, dispatching, or otherwise allowing Layer 7 to prioritize and decide communications, rather than first come, first serve. In their "Ethernet Rules Closed Loop System" article for *InTech* magazine, Edison and Cole (1998) demonstrated for Hewlett-Packard a protocol on top of Ethernet that held timing uncertainties to 200 nanoseconds, small enough for all but the most demanding processes.

The problem with Ethernet and, indeed, with employing TCP/IP above it, is that while it would be possible to "ping" everybody on the network (ping is a low-level utility that ensures connectivity over a TCP/IP network), there may still be a problem with communication unless there is a standard Layer 7, which there is not (as of 2014).

Another problem is commercial hardware that does not meet industrial specifications. The hardware that does is not "standard," although installing most of the distribution in the control room alleviates the industrial requirements somewhat. Since the popularity of Ethernet in the industrial setting has continued to increase, a number of hardware manufacturers have released hardened Ethernet products, often applying the undefined term *Industrial Ethernet*. These include industrial switches that use a ring for redundancy, water- and vibration-proof connectors, and copper-to-fiber converters to avoid having unshielded twisted-pair cable on the plant floor. Indeed, just about any component of an Ethernet distribution network is available hardened (for a price, which is becoming more competitive by the week).

Since IEC 61158 approved the "standard" fieldbus (there are eight, including PROFIBUS, FOUNDATION Fieldbus, P-Bus, FOUNDATION Fieldbus HSE, ControlNet, and others, most of whose major market share is in the European Union), and that should finalize the standard. Unfortunately, as we have indicated previously, these fieldbuses do not interoperate. Most are fundamentally different, even to the Physical layers, as you may have ascertained from some of the networks described in this chapter.

Supervisory Control and Data Acquisition Systems

Just as DCSs and PLC-based systems have become the predominant automation technologies for plant automation since the 1980s, SCADA systems have been the predominant automation technology for geographically dispersed processes since the 1970s. SCADA systems generally consist of a central computer system (the "host" or master terminal unit [MTU]) and geographically-dispersed remote terminal units (RTUs) that are connected via some form of wide-area telecommunications technology. Figure 6-27 gives a highly simplified block diagram of a generic SCADA system. SCADA systems are used to monitor and control processes, such as liquid and gas pipelines, electric power transmission and distribution, water and sewage systems, and transportation systems. Any process that is spread out over a large geographic area (such that LAN technology isn't viable) is a potential candidate for SCADA technology.

SCADA systems replaced the older telemetry technologies that were used for these applications prior to the invention of reasonably priced computer systems. (Many of the legacy problems and features in older SCADA systems are a result of having to interface them with existing telemetry infrastructure.) The invention of 16-bit minicomputers in the 1970s spurred the development of SCADA, and the continued evolution of computers and increases in computing power allowed SCADA vendors to continue to add more features, functions, and applications to their products.

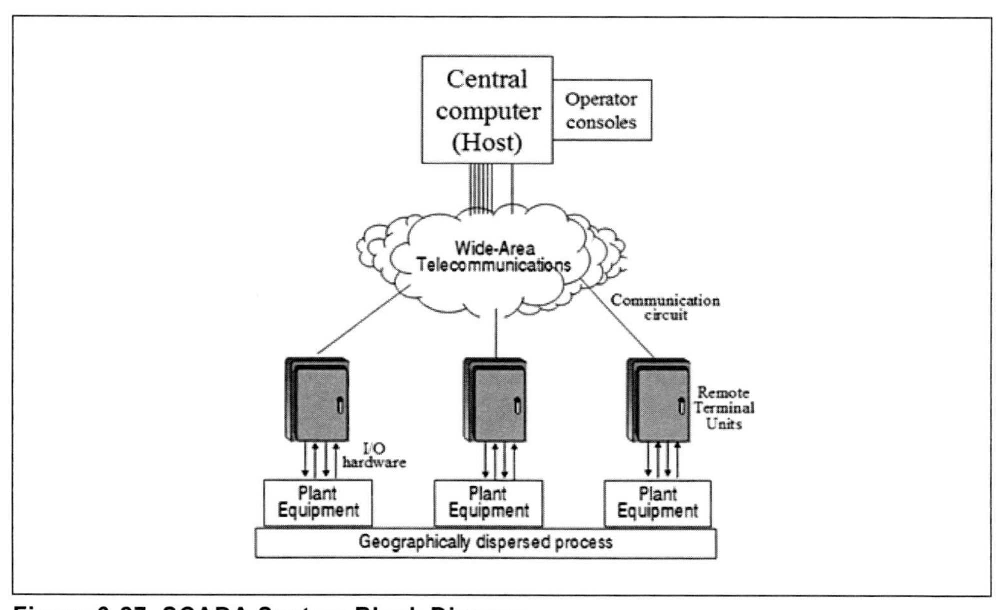

Figure 6-27. SCADA System Block Diagram

As mentioned above, the earliest SCADA systems often had to interface with the existing field equipment and even with the communications infrastructure of telemetry. The RTUs in these systems were essentially I/O multiplexers with just enough digital circuitry (but no computer) to perform analog-to-digital conversion and pulse-train capture. They communicated by sending short messages, consisting of some fixed number of bits that represented the current values of their inputs, or receiving short messages commanding them to change the state of their outputs. These messages were *not* in the form of conventional asynchronous serial communications as we know them; that is, messages broken into multiple 8-bits of data with start/stop bits appended using UART circuitry. These messages often consisted of a long, continuous series of bits, almost (but not exactly) like a synchronous message. Special hardware generally had to be developed to allow computers to send and receive these bit-oriented protocols.

These early RTUs only enabled the remote reading of analog, pulse, and contact inputs and the writing (remote control) of analog, pulse, and contact outputs. There was no additional functionality. All analog I/O was done in terms of counts, the binary value output from the A2D (analog to digital) converter or written to the D2A (digital to analog) converter. Conversion to engineering units was done at the host computer level. These RTUs had few, if any, configuration settings. Each vendor tended to develop their own protocols. Legacy protocols continued to be used, particularly in the electric power industry, even into this decade (2014).

With the development of low-cost microprocessors, these dumb RTUs were supplanted by intelligent RTUs that were capable of performing additional local functions and using conventional asynchronous serial communications based on UARTs. This led to another round of protocols being developed by the various SCADA and RTU vendors, some of which, like DN03.0 and Modbus RTU, became accepted as standards. Figure 6-28 shows how RTU functionality has evolved along with microprocessor technology.

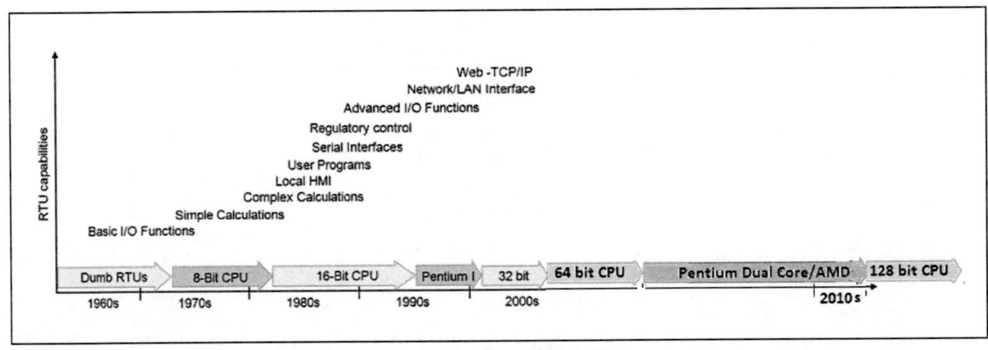

Figure 6-28. RTU Technology/Functionality Evolution

SCADA vendors tended to have certain industry segments where they focused their attentions due to the need to have industry-specific capabilities (e.g., leak detection for liquid pipelines and state estimation for electric power transmission.) Likewise, the RTUs they developed and the protocols used to communicate with the RTUs had industry leanings as well (e.g., American Gas Association [AGA] volumetric calculations for gas pipelines and VAR/Watt calculations for electric power). The complexity and sophistication of the RTU protocols depended heavily on whether the vendor used local configuration of the RTU's programming and logic or they provided the capability of remote configuration and programming via the polling/communication channel. Every SCADA system manufacturer that built their own RTUs preferred to invent their own protocols and, thus, many of them are named for the vendor, for example: Valmet/Tejas series 3 or 5, TRW 9550, Telegyr 800, Harris 5000, and EMC Systems type 1.

Most of those vendors no longer exist but their protocols live on. These protocols were generally designed around standard serial, asynchronous communication technology and, therefore, could be implemented on standard computer platforms. One of the reasons that these protocols are still in use is that it is reasonably easy to replace a central host. It is much more difficult to go to all the field sites and replace the RTUs, especially when multiple RTUs share a common channel. Replacement SCADA systems are usually required to support the protocols used by the existing installed RTUs. If new RTUs are added to an existing polling circuit, they are required to support the protocol used by the other RTUs on that shared channel, hence legacy protocols continue to survive long after their namesake inventors have disappeared.

Depending on the industry and application, a SCADA system may have one polling channel for all of its RTUs or a separate channel for each RTU. The channel configuration selected usually depends on the required data update time. If the application requires fresh data from the field every second or two, dedicated polling channels may be needed. If receiving fresh field data every minute or two is adequate, then a single, shared polling channel may suffice.

In many industry applications, there are multiple RTUs on each polling channel. Figure 6-29 illustrates a shared polling channel. Depending on the bit rate of the channel, the "size" (here, the quantity of I/O) of the RTU, and the efficiency of the protocol, it may take a fraction of a second or up to several seconds to poll an individual RTU and receive an updated set of field measurements and equipment status. If there are multiple RTUs on a common channel, then they must each be polled in turn.

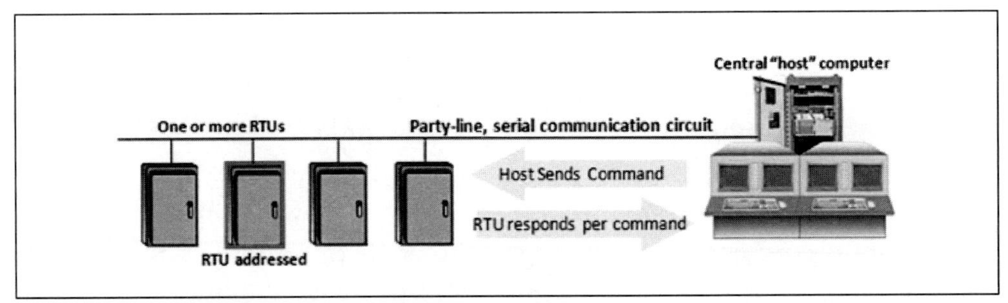

Figure 6-29. RTU Polling via a Shared Channel

All SCADA protocols perform the basic function of retrieving field measurements (RTU inputs) and enabling remote control of RTU outputs. Beyond those basic functions, the protocols vary in the additional features they support. Figure 6-30 shows a breakdown of the sorts of capabilities that are found in different RTU protocols.

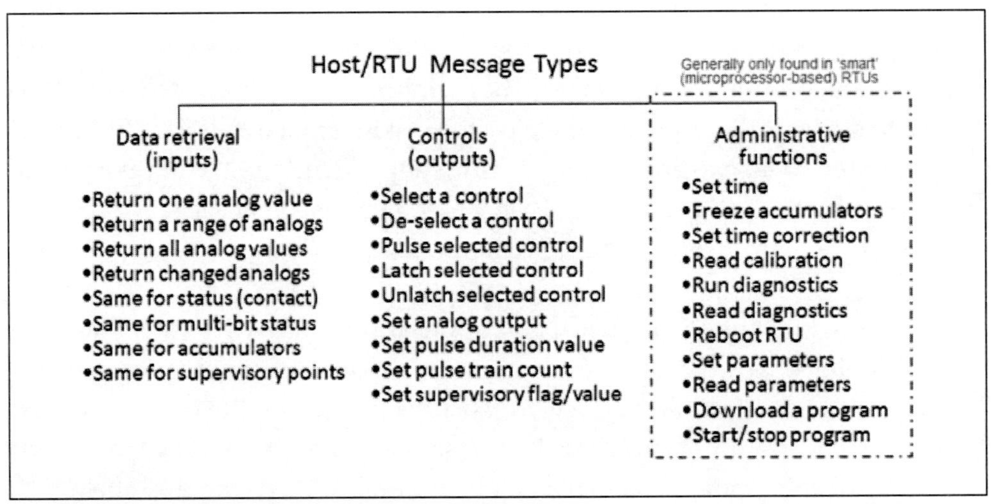

Figure 6-30. RTU Protocol Functions and Features

The extent of the functions supported by an RTU protocol generally corresponds with the capabilities of the RTUs. With the advent of microprocessor-based RTUs, the range of supported functions increased dramatically and the RTU protocols had to be extended correspondingly to support those added features. (Note that it is common to see RTUs that only support a subset of the full function/feature set of a given protocol, especially in the case of new RTUs that were not manufactured by the SCADA system vendor and were added later to an existing SCADA system.)

Early (1960–1980) SCADA vendors devised several methods for minimizing the time required to poll all RTUs and obtain fresh field measurements, this being the same methodology as master slave polling described previously. Most RTU protocols deliver commands, possibly with associated data, to the designated RTUs and expect some response from the RTUs (possibly with associated/requested data)—a process called *polling*, which was discussed earlier. By using shared polling channels, it was possible to designate some RTUs as being more important than others and, consequently, poll those RTUs more often (refer to figure 6-31). It was also possible to request a subset of measurements and inputs (possibly by type—e.g., read-only analog inputs), rather than always requesting a full report of all inputs. As RTUs became intelligent, it was possible to devise schemes whereby the RTU kept track of input changes and only reported input changes since the last poll, a capability called *exception reporting*, as opposed to sending the same unchanged values over and over again (particularly in the case of status inputs). Some RTUs could also locally buffer locally time-tagged contact input status changes observed since the prior poll. All of these techniques and features were aimed at optimizing the use of the available polling bandwidth.

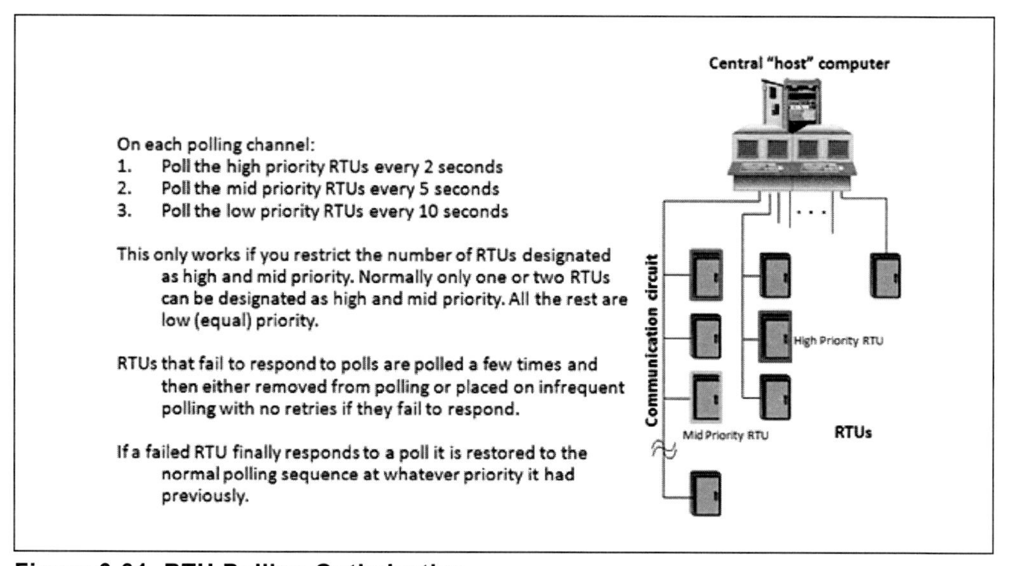

Figure 6-31. RTU Polling Optimization

Wide-Area Communications

SCADA systems need to be able to communicate with their numerous RTUs. In the early decades of SCADA technology, there were limited ways to communicate over long distances: UHF/VHF radio (including microwave) and the public telephone system, as communication satellites were not yet developed. Of course a company could always purchase telephone equipment (land lines and central offices including microwave equipment) and build their own communication system. That turned out to be the only option for many long-haul pipeline and power transmission companies. A common factor with all these technologies was that they were analog (designed to carry voice, not data), so SCADA systems, including the RTUs, had to use modems to encode digital information for analog transmission and live within the bandwidth capabilities of these communication technologies.

In many instances, a SCADA system might need to employ a mixture of communication technologies, and even call on various RTUs to act as local data concentrators, in order to reach all the necessary field locations. Figure 6-32 shows a simplified scheme that uses a mix of leased-analog telephone lines and radio to reach all the RTU locations. This design includes the use of one RTU that will poll a set of geographically adjacent sub-RTUs and incorporate their data into its database for transmission to the central host computer. That same RTU will relay commands from the host to the applicable sub-RTU when needed. A redundant central host is required because the host is common to all the RTUs and if it is inoperable, then the whole system is inoperable.

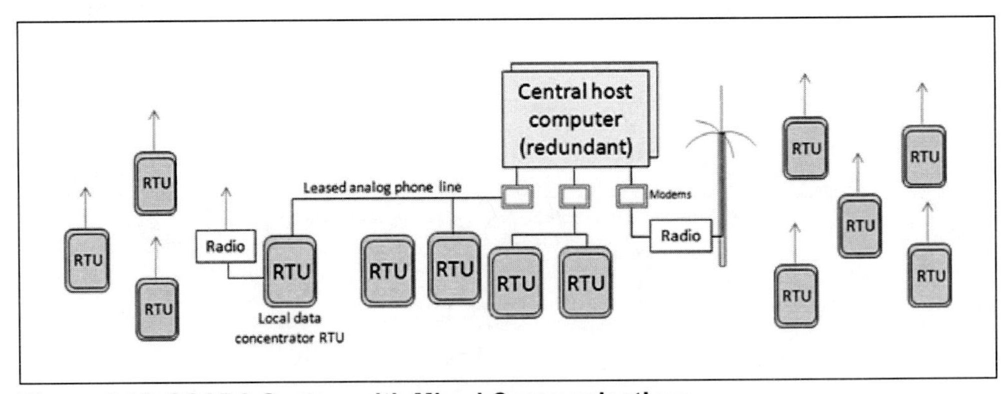

Figure 6-32. SCADA System with Mixed Communications

Up until even recently, it has been typical to use bit rates in the 1,200 bps to 2,400 bps range for RTU polling. Due to these speed limitations, the protocols used for RTU communications tend to be feature-limited and designed with minimal overhead (i.e., no security or authentication features). Figure 6-33 shows a simplified message frame structure for a generic RTU protocol with a polling (command) and response (possibly error) message. Note that only the data/parameter field tends to be variable-length or optional. Most other fields in the frame are fixed in size and location.

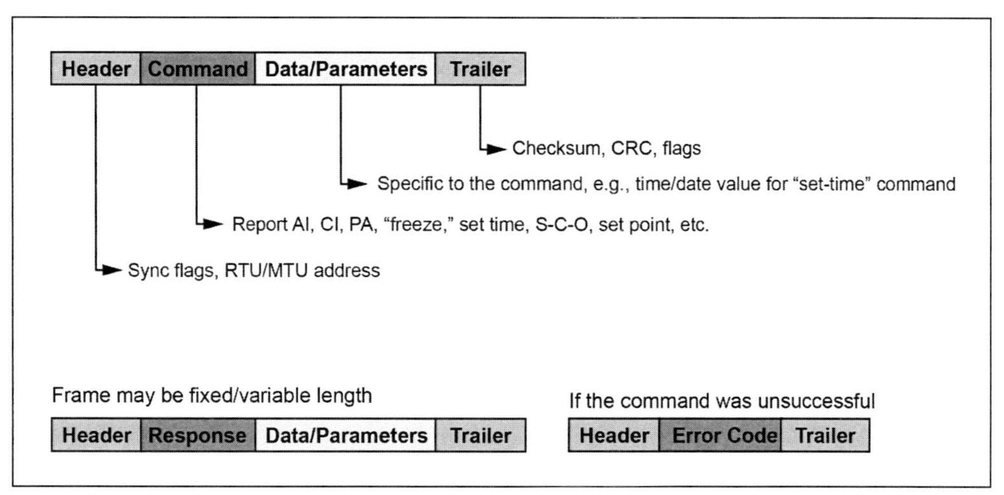

Figure 6-33. RTU Protocol General Frame Structure

Unlike modern wide area network (WAN) protocols, such as TCP/IP, that are designed to transport any type of information, RTU protocols are generally designed with a fixed set of predefined commands and responses, with no ability to create customized variations. The protocol specification lists the supported message/command types and that is all the user has to work with.

Other than when a human operator or supervisory software application wants to send a command to change an RTU's control output or some setting or parameter, the main function of RTU protocols is to fetch field measurements and status indicators and use them to update/refresh a "latest received" value table in the SCADA host. Data from the table is selected for updating displays, generating trends, identifying alarm conditions, computing values for reports, and feeding supervisory application programs. This is illustrated in figure 6-34. When observed with one of several protocol analyzers, the typical polling sequence on a given polling channel will consist of a series of data requests by the SCADA host addressed to various RTUs, with each immediately followed by the corresponding response from the RTU containing the requested current RTU measurement data.

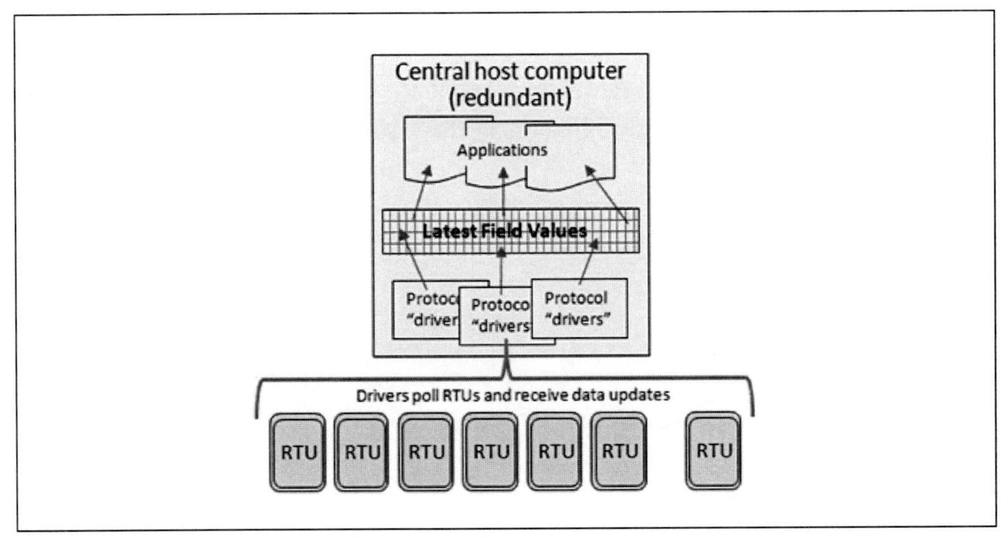

Figure 6-34. Updating Field Data via Polling

Modbus RTU Protocol

The most basic form of RTU protocol is one that performs master-slave polling, wherein the SCADA host sends data requests to RTUs (one at a time on each polling channel) and the RTUs respond with the requested data. A protocol that exemplifies this concept is Modbus RTU, which was originally developed by Modicon, a major PLC manufacturer, for communicating with their PLCs, yet it has been extensively used for other purposes and with other devices. It was described in some detail earlier in this chapter.

The basic frame structure of Modbus RTU protocol was shown in figure 6-6. In the Modbus protocol design, all data and I/O signals are accessed by assigning them to some 16-bit register with a unique address, much like RAM memory. Data is exchanged by reading from the particular register(s) assigned to the desired data. Commands are delivered by writing to the particular register(s) assigned to the desired output or parameter/setting. As a result of this simple scheme, there are only a few commands (*function codes*) needed, as shown in Table 6-4. It is possible to develop custom messages using the available unassigned function codes; however, problems can arise with device incompatibility on a shared communication channel. The specific details of the Modbus protocol can be found online at www.modbus.org (the organization that controls and oversees the Modbus standard).

Table 6-4. Common/Public Modbus Function Codes

				Function Codes		
				code	Sub code	(hex)
Data Access	Bit Access	Physical Discrete Inputs	Read Input Discrete	02		02
		Internal Bits or Physical Coils	Read Coils	01		01
			Write Single Coil	05		05
			Write Multiple Coils	15		0F
	16 bits access	Physical Input Registers	Read Input Register	04		04
		Internal Registers or Physical Output Registers	Read Multiple Registers	03		03
			Write Single Register	06		06
			Write Multiple Registers	16		10
			Read/Write Multiple Registers	23		17
			Mask Write Register	22		16
	File Record Access		Read File Record	20	6	14
			Write File Record	21	6	15

The ability to directly control an individual contact output of the PLC/RTU is managed using a construct called a *coil*. Coils are discrete outputs and they can be individually controlled using code 05; groups of them can be controlled simultaneously using code 15. Most ModbusRT messages (particularly commands) tend to be reasonably short, but a massive data request could generate a long response message—due to the 2-byte (16-bit) message length field, a response could be as big as 65 kb. It is recommended that large blocks of data are requested from the RTU using a series of small data requests, rather than by using one huge request, particularly on a shared channel with other RTUs that need to be polled as well.

With a device that communicates using Modbus protocol, it is necessary to obtain vendor documentation that defines the function codes supported by the device, as well as a mapping of data-to-register addresses, so that the host driver software can be configured appropriately. Otherwise, there is no way to know how to successfully interact with the device. The protocol specification includes some more advanced functionality, such as file transfers and device and vendor identification messages, but many devices only support the minimal function codes needed for their specific capabilities. The Modbus protocol also defines a range of error message codes that can be returned by a device (e.g., invalid function code or register address). Not all vendors make use of these options and the device may merely ignore such messages without returning any error message/response.

Because so many SCADA and RTU vendors have come and gone, some users of SCADA technology (e.g., the domestic water and wastewater industry) eventually adopted PLCs as their technology of choice in place of traditional RTUs. With SCADA applications in those industry segments, it is common to see Modbus protocol used for communications to the field sites.

DNP3 Protocol

Modbus is probably the best example of a simple, low-overhead RTU protocol. At the other end of the spectrum is the Distributed Network Protocol (DNP), originally developed by Westronic, Inc. for the U.S. electric utility marketplace but which now is an international standard. DNP3 defines the rules for remotely located computers, or RTUs, and master station computers (SCADA hosts) to communicate data and control commands. DNP3 is a non-proprietary protocol standard that is available for use by any manufacturer or vendor. The details of the standard can be found at www.dnp.org.

Unlike Modbus, which only deals with 16-bit integer values and discrete bits, DNP3 supports several data types, such as floating point and integer values, as well as multi-bit status indicators and counters. Since it was initially designed for electric utility operations, the protocol incorporates a large amount of error checking functionality. Data values can also have optional associated time and quality tags. DNP3 supports RTU configuration downloading and can support conventional polling, reporting of changes since prior poll, and unsolicited report-by-exception. With DNP3, suitably configured RTUs on a common channel can inter-communicate on a peer-to-peer basis. Figure 6-35 shows the basic message frame structure and header used by DNP3. The DNP3 message header includes a CRC check code for the header itself.

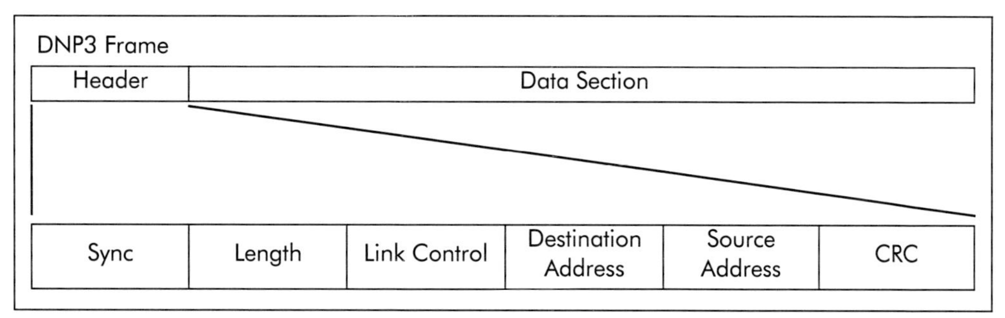

Figure 6-35. DNP3 Protocol Frame Structure

The data payload (data section) in the link-layer frame, as shown in figure 6-36, contains a pair of cyclic redundancy check (CRC) octets for every 16 data octets. This provides a high degree of assurance that communication errors can be detected. The maximum number of octets in the data payload is 250, not including CRC octets. (The maximum length link-layer frame is 292 octets, if all the CRC and header octets are counted.) Block 0 is a header block.

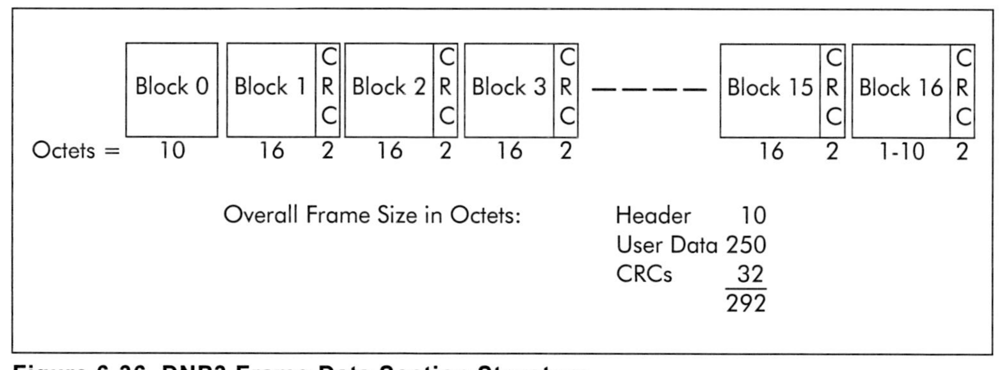

Figure 6-36. DNP3 Frame Data Section Structure

The block(s) of the data section can be delivering a command and associated data to an RTU or returning requested data (objects of specified variation) to the host. DNP3 supports several data types. The individual analog (numeric) type variations defined in DNP3 include:

- A 32-bit integer value with flag
- A 16-bit integer value with flag
- A 32-bit integer value with flag and event time
- A 16-bit integer value with flag and event time

- A 32-bit floating point value with flag
- A 64-bit floating point value with flag
- A 32-bit floating point value with flag and event time
- A 64-bit floating point value with flag and event time

The flag is a single octet with individual bit fields that indicate: whether the source or the RTU is on-line, if the data source restarted, if the communications are lost with a downstream source, if the data is forced, and/or if the value is over range. Common data types can be grouped and fetched as blocks of such values and variations.

In addition to commands for data fetching and controlling outputs, DNP3 includes commands for other functions, for example the host/master can: set the time in the RTU, transmit *freeze accumulator* requests, transmit requests for control operations, and set analog output values using select-check-operate or direct-operate commands.

Unlike with Modbus and other similar protocols, there is a certification process for devices and SCADA systems that speak DNP3. This process ensures compatibility and compliance with the standard and establishes the supported DNP3 compliance level: 1, 2, or 3. As with Modbus, a vendor of a device that uses DNP3 must provide a profile specification for the device. This profile defines the DNP3 compliance level(s) supported, the object types and variations supported, the operating mode (master/slave) of the device, and the function and quality codes supported. This information is needed in order to configure the host/SCADA that will be using DNP3 to communicate with the device.

Select-Check-Operate Protocols

The early RTUs had to contend with poor quality communication channels that could suffer from environmental factors (e.g., mountains, large objects in line-of-sight, temperature extremes) and be subject to noise and interference. For that reason, RTU protocols included check codes to permit the identification of bad messages. Prior to the invention of CRC calculations, the check codes were only moderately reliable. It was possible to send a message and have it damaged in transmission, yet accepted and executed by the RTU. To prevent this occurrence, a special class of protocols called *select-check-operate* (S-C-O) (also called *select-before-operate* protocols) was developed. In this type of protocol **no single message could be sent to cause the state of an RTU control output to be altered**. It was necessary to go through a series of messages and responses to effect an output change:

1. Send a select message to the target RTU stating the output and desired state.
2. Receive a confirmation from the target RTU confirming those details.
3. Send a check message to the target RTU confirming you both agreed.
4. Receive a confirmation of the check message from the target RTU.
5. Send an execute command to the target RTU.
6. Receive a confirmation of command execution from the target RTU.

Once an S-C-O sequence was begun, an RTU would generally not accept another until the ongoing one was completed or a timeout occurred (that is, the full message exchange series was not completed within the allowed timeframe, which cancelled the entire action). This approach was primarily used when RTUs were dumb, in order to put a human operator in the loop as the intelligence that kept a bad or broken message from operating field devices. With the arrival of smart RTUs and much more reliable communication error detection and correction techniques, S-C-O protocols became less of a requirement and more of a legacy/backwards-compatibility issue—but note that DNP3 supports an S-C-O sequence.

Note on SCADA RTU Performance

Because the communication systems available for SCADA use could have momentary disruptions (milliseconds) or even long-duration outages (months), the SCADA system protocol software (the "driver") usually supported timeout and retry logic. When a command was sent to an RTU, the SCADA system driver started a timer. If the expected response did not arrive from the RTU before the timer expired (a timeout condition), the driver would send a repeated command to the RTU, again with a timer. Most systems allowed for some number of retries (e.g., three or four) after which they would skip on to the next RTU in the polling sequence. This often resulted in setting various flags or indicators to both alert personnel of the problem and to allow the driver to either remove the RTU from polling or place the RTU on a separate periodic test-poll list.

Data Concentrator RTUs

We have already mentioned that in some cases an RTU may be a local polling master over other RTUs due to communication requirements. In the electric utility market, RTUs have become one means for extracting and consolidating all the information from the other numerous intelligent electrical devices (IEDs) in a substation. An IED has local control intelligence, performs electrical protection functions, has the ability to monitor processes, and communicates directly to a SCADA system. Many substation devices, such as protective relays, meters, sequence of event recorders, and tap changers (a device that switches to different voltage taps on transformers to change voltage, current, or impedance), contain information that is needed for supervisory applications. Some of this data (e.g., power factors, harmonics) is computed by the devices themselves. It has become more common to see RTUs (even if they are called something else) with no (or minimal) I/O hardware but with numerous serial ports being used to communicate with and to consolidate substation IED information. Figure 6-37 shows the general concept of such an RTU. Many IEDs support the Modbus and/or DNP3 protocol, which makes it possible for them to exchange data via an asynchronous (EIA/TIA-232) serial connection.

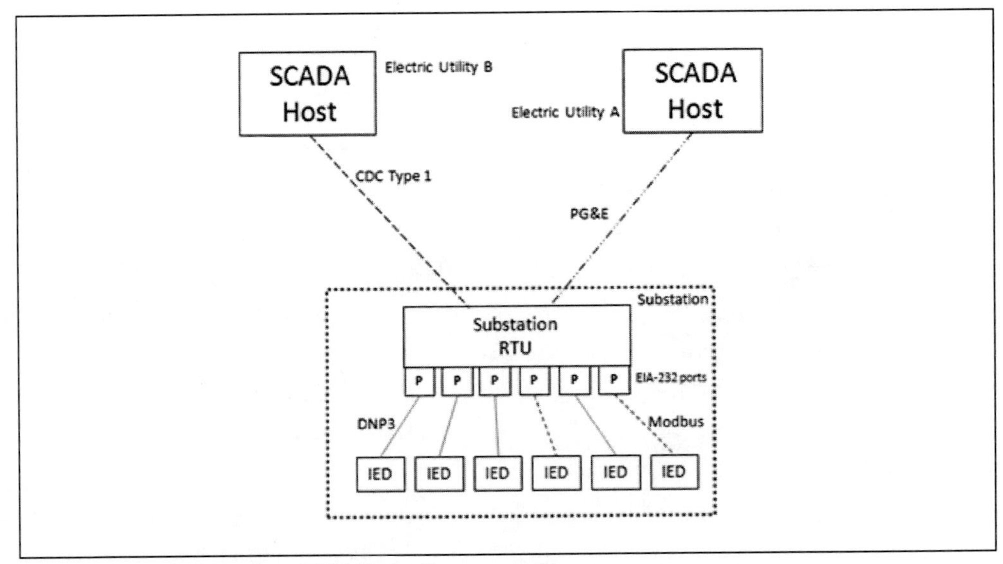

Figure 6-37. Substation RTU Data Concentrator

For various reasons, an RTU used in an electric power application, such as a substation data concentrator, may also have to communicate with more than one SCADA host. It may have to support different protocols for each host and those protocols might be different from the protocols used to extract data from the substation IEDs. It might have to allow supervisory control access to one of the hosts, but only data access to the other host. The RTU in such applications provides a virtual RTU to each host and hides the details of the IED communications.

Communications Security

None of the RTU protocols traditionally developed for SCADA use included any security mechanisms. They did incorporate error detection and recovery capabilities, however, that is not the same thing. In recent years, there has been a growing concern about the possibility of malicious manipulation of SCADA systems. Although it could be possible to add security features to most SCADA protocols, this would require firmware updates to all the installed RTUs and many of those are legacy devices with no vendor support. An alternative is to add "link encryptors," hardware devices added in-line at both ends of the communication channel.

One problem with this approach is that many such devices are only usable on point-to-point circuits and will not work on a multi-dropped channel. Another problem is that many of these devices introduce a time delay as they perform message encryption and decryption that often leads to timeout faults at the host and/or RTU, causing the channel to be declared as failed. There are currently few off-the-shelf solutions for addressing the problem of communication security on low speed, asynchronous RTU polling channels. This is not the case for TCP/IP-based RTU protocols. Communications security will be discussed in more detail in Chapter 9.

Modern WANs

By the early 2000s, the computing power of RTUs reached the point where it was possible to implement both Ethernet and TCP/IP networking hardware and software in an RTU. Some RTU manufacturers even began running embedded versions of commercial operating systems, such as Windows and Linux, in their products. This allowed the use of well-known software applications and IP-based IT protocols, such as: telnet, http, and ftp; IP-based SCADA protocols, such as the Inter-Control Center Protocol (ICCP) and the utility communications architecture (UCA.2); some protocols from the European IEC 61850/IEC 60870 standards for substation automation; IED communications; and SCADA. The IEC (specifically, Technical Committee 57) has been developing standards for electric utility SCADA use including:

- **IEC 60870-5-101 Transmission Protocols:** This is a companion standard especially for basic SCADA tasks (very similar to DNP3).
- **IEC 60870-5-102:** This is a companion standard for transmitting meter data in electric power systems.
- **IEC 60870-5-103 Transmission Protocols**: This is a companion standard for the informative interface of protection relay equipment.
- **IEC 60870-5-104 Transmission Protocols:** This describes network access for IEC 60870-5-101 using standard transport profiles (similar to the TCP/IP version of DNP3).

In the early 2000s, some previously asynchronous serial protocols, such as Modbus RTU and DNP, were updated to TCP/IP-based versions. This was primarily done by separating the command/data portion of the protocol messages (the protocol data unit [PDU]) and transporting them as "application layer" data or messages, using either TCP or UDP plus IP to form an overall application data unit (ADU), and making use of the addressing and routing and end-point delivery mechanisms of those protocols. Both of these protocols have an assigned TCP and/or UDP port number: well-known TCP port 502 for Modbus and registered TCP port 20000 for DNP3.0.

The conversion to IP-based delivery did not add any features or functionality to ModbusRT, except that with Modbus/TCP there is no concept of a master or slave. All devices are peers and any device can initiate communications using standard Ethernet media access mechanisms. Figure 6-38 shows a Wireshark (an open source network packet monitor in wide use throughout the IT industry) capture of a Modbus/TCP "read coils" message. The "message" in a Modbus/IP datagram is exactly the same set of bytes found in an asynchronous serial Modbus message: transaction number (2 bytes=00 00), protocol (2 bytes=00 00), length (1 byte=06), RTU address (1 byte=01), and function code (1 byte=01) followed by any associated data (starting coil number=00, number of coils=01).

In the TCP portion of the capture, the destination TCP port number is 502. The IP portion of the message shows the IP address of the sending node and destination node. The Ethernet frame shows the Ethernet MAC addresses for the

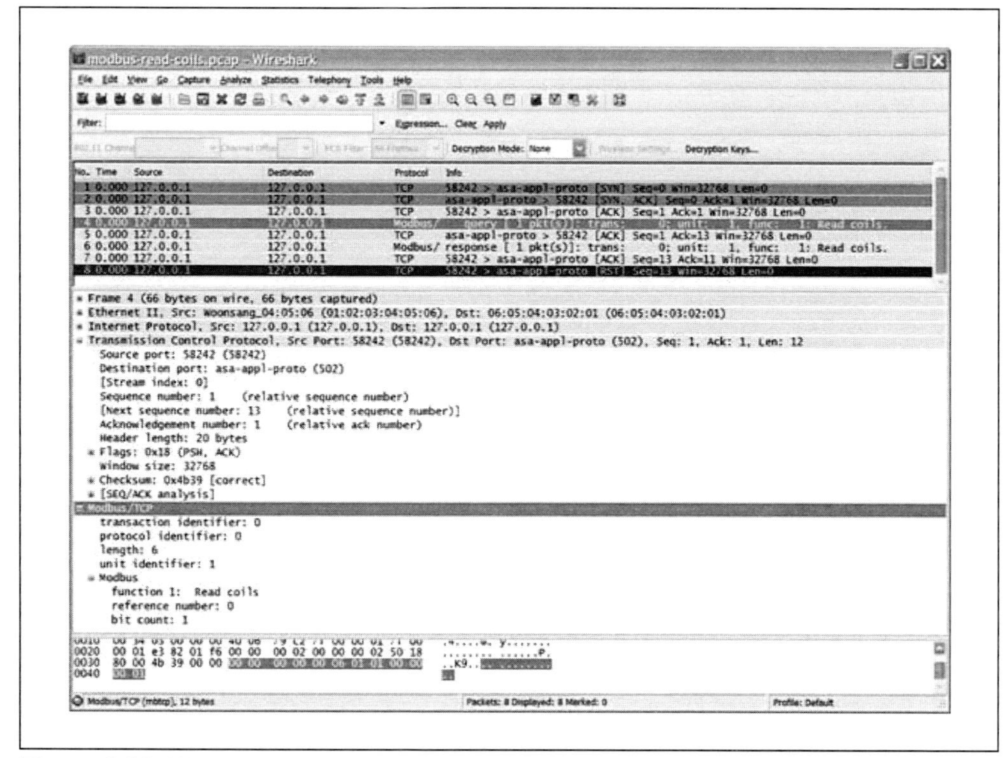

Figure 6-38. Wireshark Capture of a Modbus/TCP Read Coils Message

two nodes. Modbus/TCP relies on the delivery mechanisms of Ethernet and TCP/IP to achieve source-to-destination message delivery and error detection and recovery.

As with Modbus, DNP3 was also turned into an Ethernet-TCP/IP Application layer protocol. The Wireshark message capture shown in figure 6-39 gives the details of a “write” (code 02) command message from station 4 to station 3 on the common network. As with Modbus, this version of DNP3 relies on Ethernet and TCP/IP to provide source to destination delivery. The destination port is 20000.

The message capture indicates that this is the first or last (only) portion of a DNP3 message and that the data is being written to an object of type 50 with variation 01 (time and date); in other words, this message is setting the time and date in the destination device to Aug. 25, 2006 at 11:56:00.89.

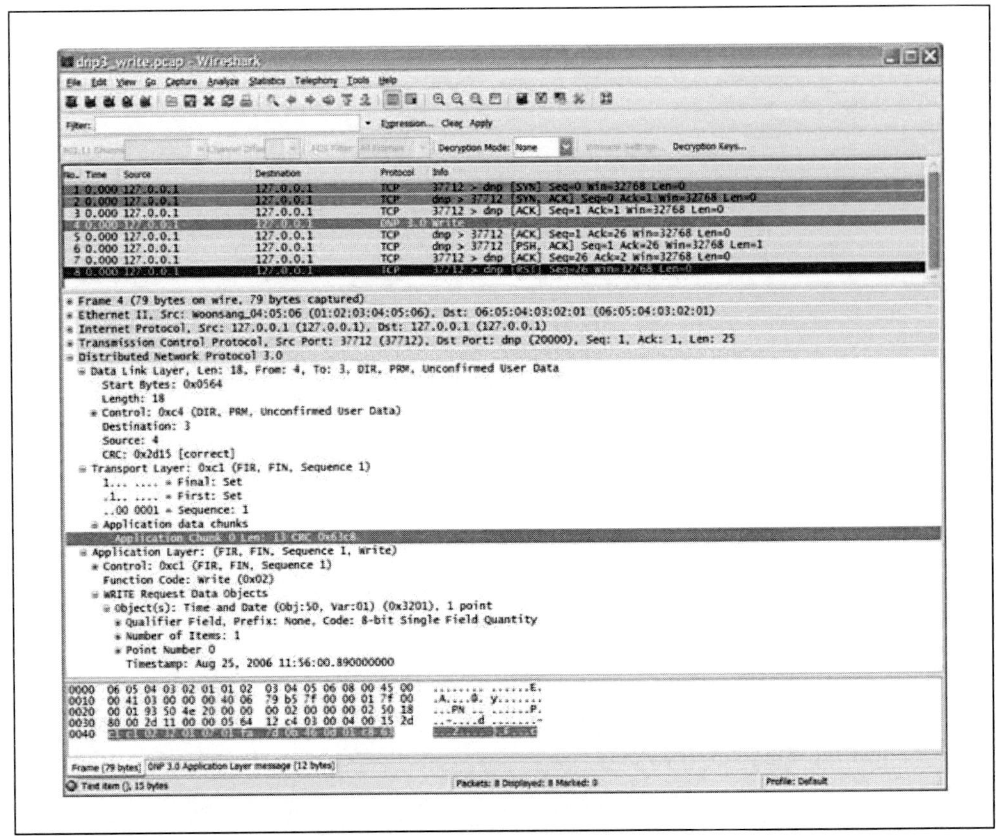

Figure 6-39. Wireshark Capture of a DNP3 TCP Message

Taking IP to the Field

As was mentioned earlier, many users of SCADA technology had to build their own wide-area communications infrastructure. Long-haul pipeline operators and electric power transmission companies, as well as others, often built their own networks. In the distant past, those would have been based on technologies such as microwave repeaters and telephone multiplexers, but in the last few decades that infrastructure is far more likely to be some form of digital network based on Synchronous Optical Networking (SONET) or Asynchronous Transfer Mode (ATM) technology linked together via fiber-optic cables. It is easy to run TCP/IP networking over those WAN technologies and to connect them into a conventional corporate WAN. For this reason, it has become more and more common to see corporations extending their corporate networks out to field sites. If the SCADA system was capable of supporting any of the IP-based protocols we have just discussed and the RTUs also supported them (sometimes called being *IP-ready*), then leased phone lines and radio could be eliminated (except possibly as a backup) and the polling of RTUs could be performed across the WAN.

In some cases, SCADA system owners need to continue to use serial, asynchronous, low-speed communications to poll legacy RTUs, even if they have brought TCP/IP networking to the sites where the RTUs are located. In such instances, it is common to see "terminal servers" used to provide a local asynchronous, serial interface at the field site and possibly at the host end as well. Figure 6-40 illustrates how terminal servers can be used to create a virtual serial channel through an Ethernet-TCP/IP WAN/LAN, so that legacy RTUs and protocols can coexist with modern networking technology. Depending on the SCADA host operating system, a virtual COM port can be created, thus eliminating the terminal server at that end of the connection.

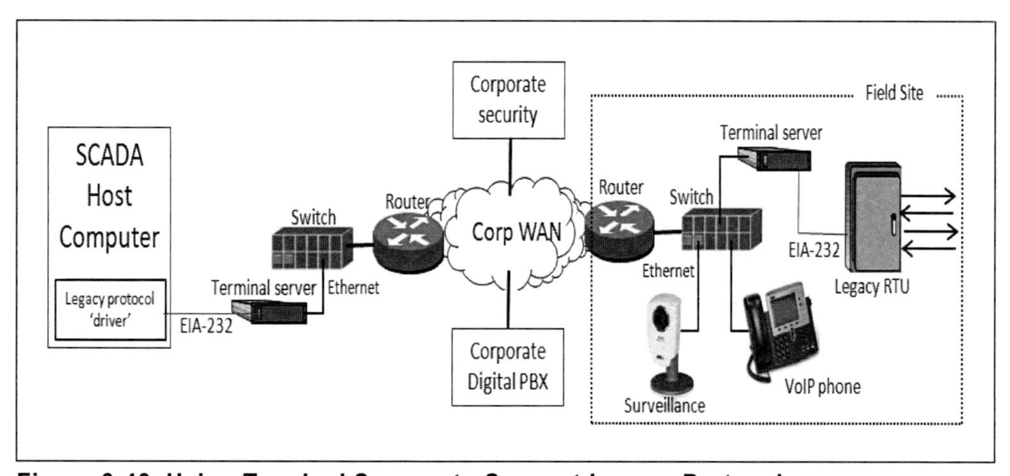

Figure 6-40. Using Terminal Servers to Support Legacy Protocols

By extending a corporate WAN to the field, it is possible to provide added capabilities at the field sites, such as remote video surveillance, and tying the sites into the corporate telephone system. This also potentially allows personnel who are in the field to plug their laptop PC into the local switch and access corporate email, the web, and corporate servers. Of course, all of this has to be done in a cyber-secure manner.

The Death of Analog Telephone Lines

Especially in urban areas, for many users of SCADA systems (e.g., water/wastewater, gas, and electric utilities), the most common means for establishing communications to RTU field locations was originally to pay the local telephone company to do it for them. Telephone companies would set up leased analog phone lines that would originate at the SCADA facility and terminate at the various field sites. Using conventional modems, the SCADA system and RTUs would interface with these lines and communicate. However, beginning at the end of the 1990s, many telephone companies were trying to eliminate all analog equipment and were unwilling to provide or continue to support those analog leased lines.

The typical alternative offered to replace those lines was a frame relay circuit, which is a digital networking technology. Unfortunately, such a circuit is totally incompatible with the hardware and software used for analog phone line communications. The solution was to place a gateway/router device called a *frame-relay access device* or *frame-relay assembler/disassembler* (FRAD) between the RTU and the circuit at the receiving end and between the SCADA system and the circuit at the originating end. The FRAD essentially did the same thing as the terminal server in the prior example: it provided a virtual serial, asynchronous channel that the SCADA system and RTU could use to communicate through a frame relay WAN. Different manufacturers put different capabilities into their FRAD products but several targeted the SCADA "upgrade" market to provide solutions to the disappearing analog line problem.

Summary

This has been a long chapter. The main requirements for an industrial network were defined in the beginning of the chapter, but the point should be made that all of these industrial networks (including SCADA and assorted fieldbuses) attempted or are attempting to meet the same goals. Some industrial networks try to be overly broad, and some are dedicated only to process control. Today, it takes a network guy and a process guy just to determine the design of a new industrial network. Vendors will provide a turnkey proprietary package (the margins are much higher for the vendors with this option), but then the user is locked into a proprietary solution that may work correctly for some, or even all, of the combined processes in a facility; however, that leaves the user at the mercy of the marketplace, vendors' whims, and changing process requirements.

If an industrial company chooses to go it alone and build a network out of components, it will be able to control not only the final design, but the project process, checkpoints, component selection, and installation; however, it will cost more in time and personnel. A standard industrial bus (those described in this chapter and those to come) would certainly reduce the time spent determining compatibility and interoperability. While there are many to choose from, a network that fits into your automation niche (and has a track record in that niche) will probably be the best choice.

Bibliography

Note that Internet links were correct at the time of writing but may change.

Boggs, D.R., Mogul, J.C., and Kent, C.A. "Measured Capacity of an Ethernet: Myths and Reality." *ACM SIGCOMM Computer Communication Review*, 25, no. 1 (January 1995).

Byrnes, E. *Instructors Notebook ISA Course FG21C*. Research Triangle Park: ISA, 2000.

———. *Instructors Notebook ISA Course FG30C*. Research Triangle Park: ISA, 2003.

Edison, J. and Cole, W. "Ethernet Rules Closed Loop System." *InTech* 45, no. 6 (June 1998): 39–42.

Fieldbus Foundation (organization website) http://www.fieldbus.org

Fisher-Rosemount, Inc. *Fieldbus Technical Review*. Austin: Fisher-Rosemount, 1998.

GGH Marketing Communications, Ltd. *The Industrial Ethernet Book*. http://ethernet.industrial-networking.com

International P-NET User Organization. "The P-Net Fieldbus" in *Principles of P-Net*. http://www.p-net.dk/booklet/bookpg04.html

ISA100Wireless Compliance Institute. *The Technology Behind the ISA100.11a Standard – An Exploration*. Webinar, no date given.

OPC Foundation. http://www.opcfoundation.org

Pinto, J. "The Great Fieldbus Debate – Is Over!" http://www.jimpinto.com/writings/debate.html

Rockwell Automation, Inc. Product Catalogs. http://www.ab.com/catalogs

Schneider Electric SE. Telemechanique PLCs. http://www.schneider-electric.us/sites/us/en/company/profile/history/telemecanique-transition.page

Schneider Electric SE. Modicon PLCs. http://www2.schneider-electric.com/sites/corporate/en/products-services/automation-control/automation-control.page

Shaw, W.T. *Cybersecurity for SCADA Systems*. PennWell Corporation, 2006.

Smar International Corp. http://www.smar.com

"Wireless Systems for Industrial Automation: Process Control and Related Applications." ANSI/ISA-100.11a-2011. Research Triangle Park, NC: International Society of Automation, 2011.

7 Wide Area Networks

This chapter discusses communications over a large geographical area (in contrast to a local area network [LAN]) where the media and/or infrastructure is not owned by the entity doing the communicating. That is, if you are more than just campus–wide and if you rent, lease, or otherwise do not own the media carrying your data, you are using a *wide area network* (WAN). This network may include the public switched network (including satellite and cell phone technologies) or leased portions of the network for private lines. Transmission over a WAN is almost always performed in bit serial fashion (one bit after another).

Why should someone who is primarily interested in instrument loops care about WAN use or technologies? Because business drivers and purposes have changed but media characteristics haven't—a twisted copper line pair has the same impedance regardless of the network type, although the impedance may have a greater or lesser tolerance. Although the modems and digital lines described here may be part of your immediate network, when a LAN wants to communicate with the outside world (and in the enterprise scheme of things, chances are it will), it will do so using one of the wide area technologies described in this chapter—or, at a minimum, it will employ a means that can trace its parentage back to one of these technologies.

In discussing WANs, this chapter proceeds in an almost chronological fashion, beginning with wireline and wireline modems (including a brief explanation of modulation—and we do mean brief) and working up to the digital line offerings. A careful read will give you a good idea of where many of the serial standards (EIA-232 specifically) had their start and why.

New students of data communications may wonder why wireline modems were first used and why they were so slow, particularly if they remember their own experience with the Internet and 56 Kbps modems. First, we need to look back to the early 1960s. What network could then be found serving almost all businesses? The public switched telephone network, of course. As you read this chapter, you may notice that a telephone line is not necessarily an ideal data path. Still, the telephone system served most businesses, it was relatively inexpensive compared to the alternatives of the time, and it could be used for data transmission.

Author Thompson remembers quite well listening to a Bell Systems engineer in the mid-1960s who stated that data could not and would never be transmitted down a voice-grade line (your standard telephone line) any faster than 2,400 bits per second. In 2014, 56 Kbps (kilobits per second) download and 33.6 Kbps (kilobits per second) upload was a standard (V.92), and DSL lines using that same copper pair exceeded 3 Mbps (megabits per second). How was this increase accomplished? What data transmission method should you choose now? Those questions are what this chapter is all about.

Keep in mind that industrial control systems use LANs and are not normally run over a WAN (the Internet is also a WAN). However, many supervisory control and data acquisition systems (SCADA) do use remote inputs; many of which are obtained over a WAN using WAN techniques. As the enterprise-wide communications effort broadens, the need to interface with users who are geographically remote will become essential for any complete control system. Most companies cannot afford a WAN infrastructure (which would be a wide area LAN; the U.S. government employs these), so they lease the service needed. This is a recurring cost but it has become less expensive because of the Internet.

One last thought before you start this chapter—many of the LAN developments used routinely in industrial data communications (such as TCP/IP) had their start in the WAN, where limited bandwidth, noisy media, and the constant need for improved data speeds are the norm.

Wireline Transmission

This section describes WAN devices that fit into the first two layers of the ISO OSI (Open Systems Interconnection) model, as do many of the WAN technologies: Layer 1, the Physical layer, and (above it) Layer 2, the Data Link layer. These are sometimes referred to as *levels* in texts written before the creation of the OSI model. Layer 1, the Physical layer, provides the physical connection—the electrical and mechanical means to establish, maintain, and end physical connections between Data Link points. Layer 1 also provides the functional and procedural means for ensuring the connections, including handshakes. Layer 2, the Data Link layer, provides the functional means (software and hardware) to establish, maintain, and disconnect data link connections between data terminals, modems, gateways, and the like. Please understand that there may be little demarcation between Layer 1 and Layer 2, particularly if both sets of functions are performed in a single piece of equipment, as may be the case with modems that are integrated into servers.

Most long-distance data transmission is serial; in fact, almost all high-speed data transmission is serial, except for very local areas like a motherboard. Serial transmission makes possible two major types of links: *switched* and *permanent*. A *switched* connection is much like the landline telephone system: a separate connection is made for each communication. The connections are originated at the beginning of a communication (off-hook and dial) and taken down at the end of communication (hang-up). A *permanent* connection allows continuous communication and the lines are normally leased 24 hours a day, seven days a week. It uses the same media but doesn't go through the switching equipment.

Since the length of the line introduces distortions and line losses to electrical signals that have a large direct current (DC) component (known as baseband signals), an alternating current (AC) is used as a carrier for the data. This is accomplished through the process of modulation and demodulation, which is a method for translating frequencies by taking the DC signaling rate and transforming it into a higher frequency range, thus eliminating the DC component.

Modulation is a term used to describe the *impression of intelligence on a carrier*. In other words, modulation consists of modifying some sort of carrier (in this case, an electrical waveform) with some sort of data. This data is the *intelligence*. Demodulation is the reverse, removing the intelligence from the carrier.

It is actually the more difficult of the two processes. (Note that *intelligence* as used here does not refer to *intelligent devices*.)

Many, many books have been written about modulation and demodulation. The following information is only a concise simplification of modulation and its effects on circuit behaviors. We will detail only the effects; the mathematical models are not shown. Since almost all electrical communications over any great distance or at a high rate of speed use one or more modulation techniques, grasping them is essential to understanding industrial applications in which data is normally transferred some distance.

Carrier Concepts

Why is modulation needed? Although there are many reasons, for our purposes here there is only one: to translate (move) a given signal's data (intelligence) to a different frequency. The reason that this is necessary will become evident in the discussion that follows.

Wireline Effects on a DC Signal

People are normally inclined to think of data communications through a wire as being instantaneous. This is because their observations are usually based on short lengths of wire and are focused on such effects as turning on a car's headlights, which seem to come on immediately. However, the transmission time from the moment the switch contacts are made (closed) until current is flowing through the lamp filaments is definable. This is particularly true of signals that have a fast voltage rise (or fall) time such as data signals. Figure 7-1 illustrates the waveforms, both ideal and real, for a simple series circuit that consists of a load (resistor), a source (battery), a switch, and a relatively long length of connecting wire. The switch's ideal output is an abrupt voltage change, called a *step voltage*. At the end of a long length of wire the end result is something less than abrupt.

NOTE The following discussion of wireline waveform changes calls for basic electrical knowledge. Skipping it will dilute your understanding of why modems are necessary but not of what they do, so if you have a limited electrical background (or none at all) you may want to skip to the next section, "Sine Wave as a Carrier."

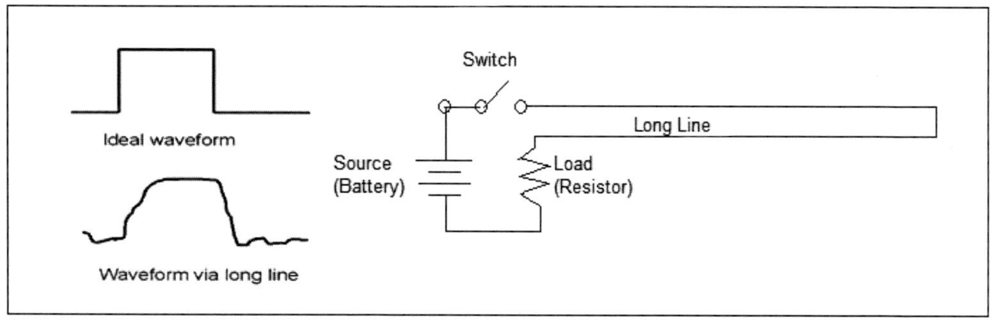

Figure 7-1. Wireline Waveform Changes

To see the reason for this output change, we must analyze the step voltage and understand that any conductor, such as a wire, has a series inductance (to itself) and a parallel capacitance to some other conductive body (such as ground). The longer the wire, the greater the inductive and capacitive effects will be. Figure 7-2 illustrates the lumped–constant schematic. *Lumped* merely means that the inductance represented by the inductor symbol is not localized but is distributed evenly throughout the conductor. The state of being lumped is represented by the schematic symbol for an inductor whose value represents the total inductance throughout the wire; the same is true for the capacitor, a single symbol representing the lumped capacitance of a segment of line. The resistor represents the ohmic (copper) loss that contributes to the attenuation of the signal strength over the length of the wire.

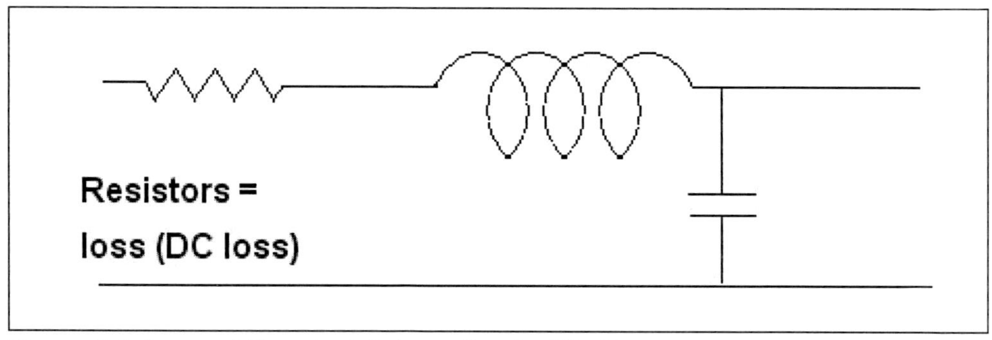

Figure 7-2. Lumped Constant Schematic of a Wireline

Observing the schematic representation of the lumped constant circuit shows it to be filtered (a long line will have multiple sections, hence multiple filter sections). Arranged as shown in figure 7-2, it is a *low-pass* filter, meaning that the high frequencies are attenuated (possibly totally), while the lower frequencies *pass through*, far less attenuated.

Analyzing the signal of a step waveform (where a signal goes from one value to another value with a sharp—almost vertical on a graph—rise time) is complex and involves the use of higher mathematics. However, it should be easy to see that the step waveform signal (with the abrupt rise or fall) can represent any on-off signal that periodically (on a scheduled rate, perhaps 60 times a second) changes state, such as a square wave. A square wave itself can be thought of as a sine wave whose period is the sum of one *on* and one *off* state, or two element times. The square wave's leading and falling edges consist, for the most part, of the higher-frequency components. The level state is made up of the lower-frequency components—primarily the fundamental sine wave. Developing a square wave from the fundamental sine wave and its odd harmonics is performed by Fourier analysis techniques (quite beyond the scope of this text); all that is necessary to know is that the resultant square wave shape can be modeled by its fundamental frequency and odd harmonics. Figure 7-3 is an example of modeling a square wave; the figure on the left is using the fundamental and the 3rd harmonic, the one on the right, the 3rd, 5th, 7th, 9th, 11th, 13th, and 15th harmonic. (These figures were developed using MatLab software.) Note that the fundamental frequency sets the level, while the odd harmonics contribute to the rise time. This is the reason a long wireline, which is a low pass filter, rounds off the transitions.

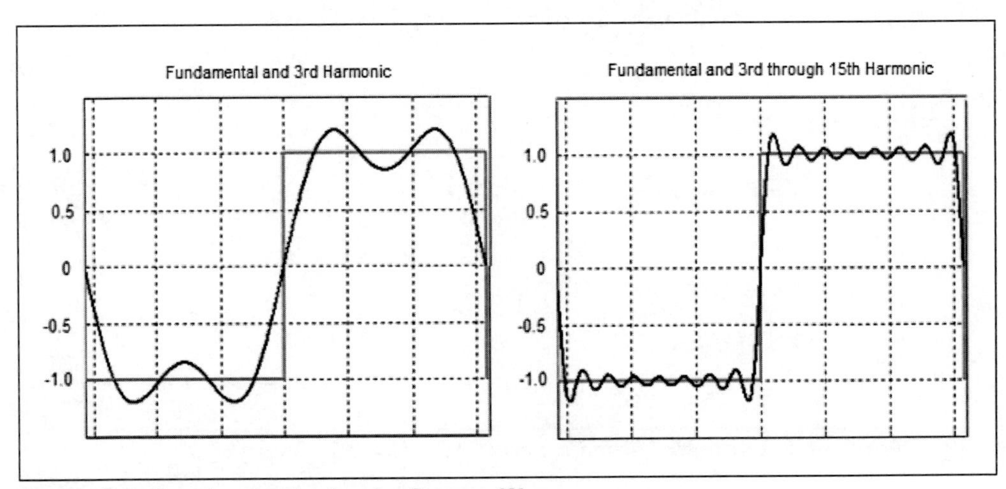

Figure 7-3. Fourier Analysis of a Square Wave

When a step voltage change, or any other large swing in voltage, occurs and a device attempts to transmit this change down a long metallic wireline, the line's lumped capacitance and inductance and the line's copper losses (the DC resistance) combine to reduce the output waveform's amplitude and attenuate its higher-frequency components. The faster the rise time (or fall time), the greater the attenuation. In addition, the metallic line acts as a delay line. That is, the reactive time constants are different for different frequency components of the square wave. This results in different parts of the waveform arriving at different times. The net result is a distorted waveform. The attenuation for low- or high-frequency components is called *amplitude distortion* and the different transmission times result in an effect called *phase distortion*.

The number of decisions (signal changes) that a medium can support in one second is called the *line modulation rate* or *baud rate.* A medium's required baud rate may be determined by taking the smallest element (in the case of digital data, a binary digit or bit) that you wish to transmit (the shortest decision time) and dividing this time into 1. A standard telephone wireline of 300 to 3,300 Hz is generally capable of 1,200 baud. That is 1,200 decisions (signal changes) a second. For duplex operation, this is 600 decisions in both directions (adding up to 1,200). This is the maximum that typical analog telephone line media can support.

Sine Wave as a Carrier

The sine wave is used as a carrier because a sine wave cannot be integrated or differentiated mathematically. The amplitude or phase angle of a sine wave may change, but the shape is still that of a sine wave. Long wireline processes affect a square wave exactly as they affect the ID functions of a proportional-integral-derivative (PID) controller (widely used in industrial control systems). Integration (the averaging of change over time) causes a square wave to become a triangular waveform, while differentiation (the rate of change over time) on a square wave causes it to become a peaked waveform. These processes do not affect a sine waveform. The amplitude can be reduced and the entire waveform can be shifted in phase, however, the sine wave's shape cannot be changed through a linear or passive device. That a sine wave cannot be integrated or differentiated by linear or passive devices, that a square wave is made up of a fundamental sine wave and its odd harmonics, and that an electrical line acts as a low–pass filter are physical phenomena; we may not necessarily know why, but we can use the rules. Because a sine wave will retain its waveform through linear processes, it is used to "carry" digital data.

Modulation

Modulation is the process whereby a carrier's characteristic waveform is modified to contain the data that is to be transmitted. The sine wave has three characteristics that may be modified:

1. Amplitude
2. Phase
3. Frequency

Amplitude is the magnitude of a signal (could be stated as the rms or peak voltage or current). Frequency and phase are collectively considered angular momentum; a change in frequency comes about from a continuing change in phase.

There are many ways to impress intelligence on a waveform. Even the lack of change is data. The first modulation schemes consisted of merely turning the carrier on and off. The on and off states are comparable to the dots and dashes of the Morse code scheme. (It is important not to confuse coding and modulation. Modulation is the altering of a carrier's characteristics, while coding is a scheme to determine what an arrangement of bits represents.) Over time, the on-off modulation of the carrier evolved into another form of modulation called *frequency shift keying*, a form of amplitude modulation (not *frequency modulation*, as it is sometimes mistakenly called). In frequency shift keying, two tones are employed: one tone to represent a 1 and another tone to represent a 0. An example is the Highway Addressable Remote Transducers (HART) protocol, which uses 1,200 Hz for a 1 and 2,200 Hz for a 0.

Amplitude modulation and frequency modulation, known as AM and FM, respectively, are familiar to most people through their radios and home stereo system receivers. AM is generally thought of as rather noisy and limited in frequency response but able to be transmitted over great distances. FM is considered quiet and capable of providing good frequency response but unable to travel well, even from one city to another. These generalizations, while accurate, have less to do with the type of modulation than with the frequency range in which the two different modulations operate.

Probably less familiar is the nature of the analog TV signal, now a legacy technology in the United States. An American analog broadcast television signal has a 6 MHz baseband whose range covers 0 to 6 MHz. The sound is a subcarrier of 4.5 MHz, which is frequency modulated. The color information is quadrature modulated (independent sideband AM [ISBAM]) with a 3.58 MHz reference carrier. This entire signal (sound plus color) is impressed upon the video carrier by vestigial sideband amplitude modulation (VSBAM). At the height of analog TV transmission, even more features were put into the baseband signal, such as text and stereo sound.

The same information is now transmitted digitally and four channels (standard) will fit into one analog channel. Along with the DVD/Blu-Ray component, the modern digital TV is quite probably one of the most technically sophisticated devices in the average household and we haven't even approached high definition TV (HDTV, which is also a digital technique), which brings yet another layer of complexity. All the broadcast modulation techniques just mentioned have been used to transmit data at one time or another. In contemporary use, they are normally combined, as will be seen shortly.

Another reason for using modulation techniques involves the resistive losses in a long wireline. A DC current (or an AC voltage superimposed on a DC current) traveling through the line resistance will dissipate energy in the form of heat; the higher the current, the greater the heat. Although the DC input voltage may be raised to compensate so that the output current will be adequate, the higher voltage causes even greater energy loss. Alternating current can use transformers to match impedances and to select the correct voltage/amperage ratio for transmission. Since watts (power) = voltage x current, raising the voltage to a high rate causes a lesser amount of current to be required for the same power to travel through the line. Less current equals less heat dissipated. By using a transformer, it easier to raise an AC voltage—far easier than trying to raise a DC voltage. Seldom is a DC signal transmitted long distances over a wireline.

Before it is modulated or encoded, a digital data signal has a maximum frequency (alternating 1s and 0s) that is one-half the bit-per-second (bps) rate and a minimum frequency that is zero when transmitting steady 1s or 0s, which at 0 Hz is a direct current (DC). DC is not coupled across a transformer (no lines of flux are being changed) and a voice-grade wireline has at least two transformers, one at each end. It should be obvious that we will have to change the DC component of our data signal if we want to transmit this signal at a fast data rate any distance at all (usually more than 100 m; 330 ft.) over a voice-grade wireline.

Amplitude Modulation

Amplitude modulation (AM) is the process whereby the amplitude of a carrier wave (usually a sine wave) is varied according to the data that is to be transmitted. AM involves more than simply turning the carrier on or off, although that in itself is a legitimate form of modulation. To better explain amplitude modulation, we will use the classic case in which the carrier is a sine wave at some frequency many times higher than the frequency of the modulating signal, which is also a sine wave. Figure 7-4 illustrates the frequency spectrum occupied as a result of modulation. The result may be graphically plotted, or a trigonometric identity may be used. In either case, the outcome is the same. The output actually consists of four different frequencies after modulation: the carrier frequency, the modulating frequency, and two new frequencies that are the result of modulation:

1. The instantaneous sum of the carrier and the modulating frequency
2. The instantaneous difference of the carrier and the modulating frequency

Figure 7-4 shows the results of a carrier at 1 MHz and a modulating frequency of 1 KHz. The sum and difference frequencies are then 1.001 MHz and 0.999 MHz and are called *side frequencies*. It is these new frequencies that actually contain the intelligence. The change in the carrier's amplitude at the modulating frequency rate causes the side frequencies which are referred to as *sidebands*.

It is important to note that total signal (carrier ±side frequencies) occupies 0.999 MHz to 1.001 MHz. If the receiving bandwidth is less than this, the data cannot be received (demodulated) correctly. In this context, *bandwidth* means the span of frequency, measured in hertz; a signal occupies and is determined by the difference between the highest- and lowest-frequency components that are significant. In this case, subtracting the lowest frequency component (0.999 MHz) from the highest frequency component (1.001 MHz) (determined the same way as range is determined in instrumentation) equals 0.002 MHz, which is 2 KHz (determined the same way as span is determined in instrumentation). If the modulating frequency is a band of frequencies, say 20 Hz to 5 KHz, then the resulting required output bandwidth is 10 KHz, with each

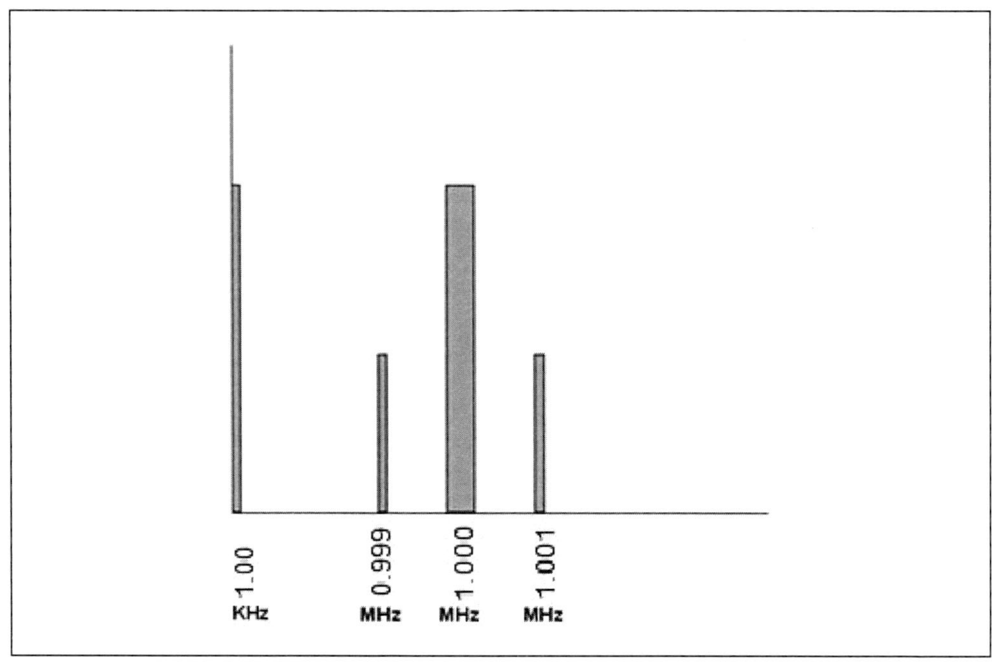

Figure 7-4. Carrier and Sidebands

band occupying a 0.02–5 KHz range on each side of the carrier. This is known as *double-sideband amplitude modulation* (DSBAM). The 1.0 MHz carrier used in this example corresponds to the 1.0 or 100 position on the AM radio dial and the sideband frequencies roughly correspond to those that the radio receives. In data transmission applications, the carrier frequency is not very high because the losses in a twisted-wire pair, as used in the bulk of installed telephone systems today, would not be economical. Wirelines (normal telephone lines) have a nominal frequency response of 0.3 to 3.3 KHz. If the signal has a frequency range of 20 Hz to 5 KHz, the wireline passes only 300 Hz to 3.3 KHz. This is (usually) due to the matching transformers (matching line impedances), which results in limiting response to just the minimal usable voice components. DC cannot pass through a transformer and the transformer's ability to transfer power below a certain frequency (in this case 300 Hz) falls off. At the upper frequency limit (3.3 KHz), the inter-wire capacitances of the transformer limits the useful upper frequency. This results in portions of the signal bandwidth being lost due to attenuation of the frequencies not passed by the wireline.

Double-Sideband AM

The sidebands of a double-sideband amplitude modulation (DSBAM) signal are redundant (contain the same data). When AM is used for data transmission over a wireline, the carrier frequency is a lot lower than the typical DSBAM radio transmission. In the data case, the carrier frequency is usually just above the signaling rate, rather than a number of magnitudes greater, as is the AM radio example. DSBAM can be used for data transmission. However, DSBAM requires two cycles of bandwidth for every one cycle of the modulation frequency. This waste of bandwidth may be offset by the fact that it is far simpler to detect and demodulate DSBAM than most other modulation schemes.

Vestigial-Sideband AM

The scheme most often used for AM data transmission is vestigial-sideband amplitude modulation (VSBAM). This form is used because the bandwidth it requires is a little more than one-half that required for a DSBAM signal. This is the reason VSBAM was used in analog commercial broadcast television. The upper sideband is selected for television; however, either sideband could be used. In VSBAM, the carrier is not fully suppressed and a portion (a *vestige*) of the upper sideband (for data transmission) is transmitted. It is important that some carrier frequency be received at detection so that the signal will be demodulated correctly and in phase. (This is necessary because of filter characteristics; filters do not abruptly cut off at a certain frequency but have certain *roll-off* characteristics starting at 0.707 of 100% amplitude on both the upper and lower frequency sides—for a bandpass filter—as to how quickly the signal past these points is attenuated.) Figure 7-5 illustrates the spectrum of a vestigial sideband signal using the lower sideband.

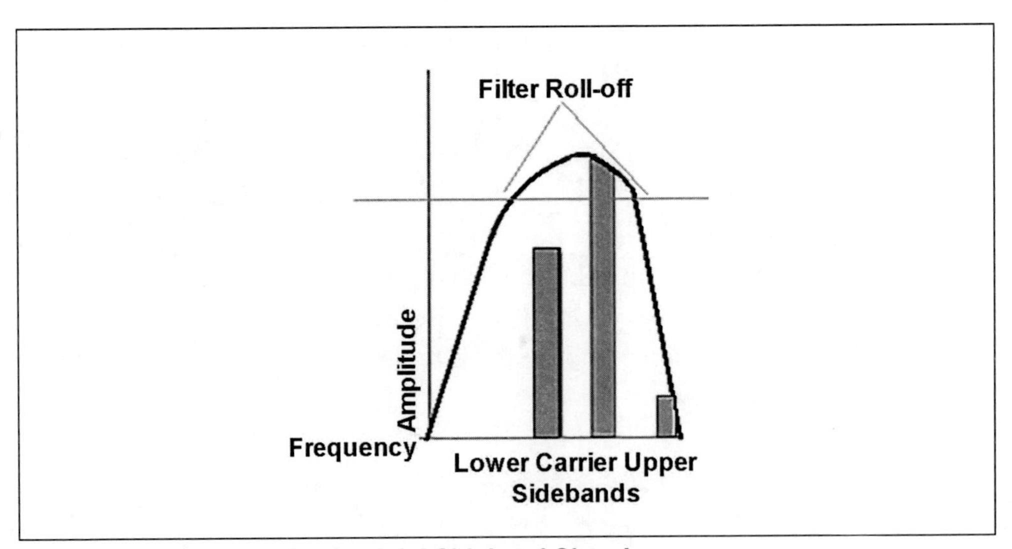

Figure 7-5. Spectrum of a Vestigial Sideband Signal

Phase is not terribly important to analog voice users; however, it is very important for digital signals. Phase distortion is caused by the inductive and capacitive reactive components of the medium and gives different parts of the signal different delays. Parts of the signal are not supposed to arrive later than other parts, particularly as all parts started at the same time. This distortion makes detection of signal changes difficult.

Single-Sideband AM

Many people are familiar with single-sideband AM (SSBAM) from its use in amateur and Citizens Band radios. In a single-sideband signal system, one sideband is used to transmit one data stream. During modulation, the carrier signal is totally removed (the intelligence is in the sidebands) and one side-band is eliminated along with the carrier, then all of the transmit power is employed in only one sideband (either the upper or lower sideband) to send a single data stream. Single-sideband transmission conserves bandwidth, but the receiver must be more complex than with a DSBAM signal. For demodulation to take place, the receiver must generate a signal that is the same frequency as the carrier—the substitute carrier frequency— which is then used to demodulate the sideband. Although this modulation/demodulation is quite readily done today, the phase of the receiver-generated carrier is not always the same as that of the carrier that performed the modulation. This is what gives voice SSBAM that peculiar audio quality: the difficulty in correctly reproducing the carrier in the correct phase at the receiver.

SSBAM is seldom used as the sole modulation method for digital signals. It is used in analog multiplexing schemes and in the radio transmission of data, after the data has been converted from its baseband frequency by some other modulation method. Figure 7-6 illustrates the frequency spectrum of single sideband.

Independent-Sideband AM

Independent-sideband AM (ISBAM) is a single-sideband type of signal used in telecommunications. It is similar to the SSBAM signal except that both the upper and lower sidebands are transmitted with the carrier suppressed. Each sideband carries independent (from the other sideband) data. In fact, the standard analog ISBAM transmission has four 3 KHz voice channels. This multichannel scheme is not directly used in data transmission.

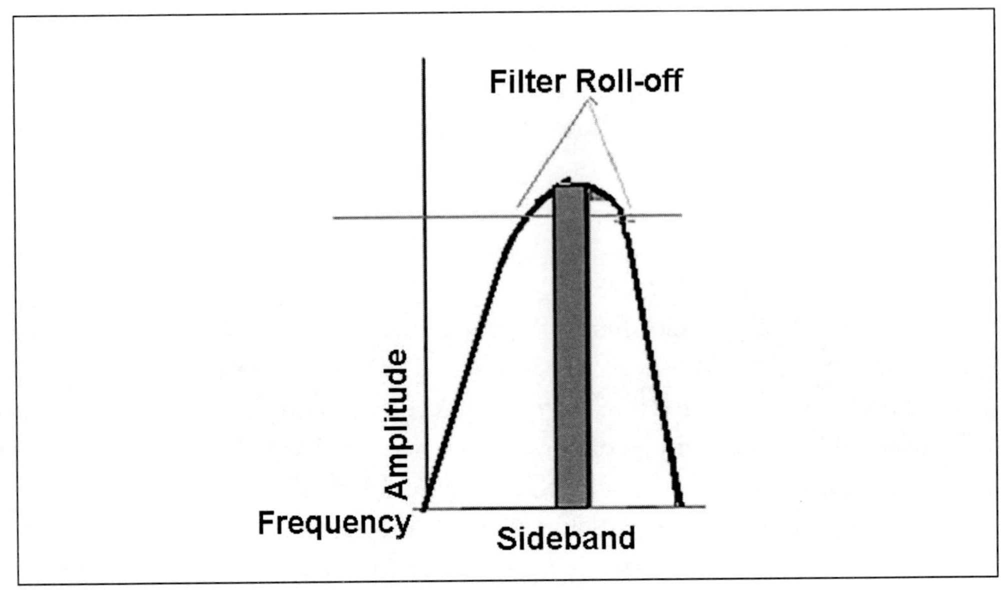

Figure 7-6. Single Sideband

An important characteristic of ISBAM is that when the upper sideband is used for one data channel and a lower sideband is used for another channel, they cannot demodulate each other. If they could, it would result in distortion. You will see this technique used in carrierless amplitude and phase modulation (CAP), which is used in digital subscriber line (DSL) transport. Figure 7-7 illustrates the frequency spectrum of an independent sideband signal.

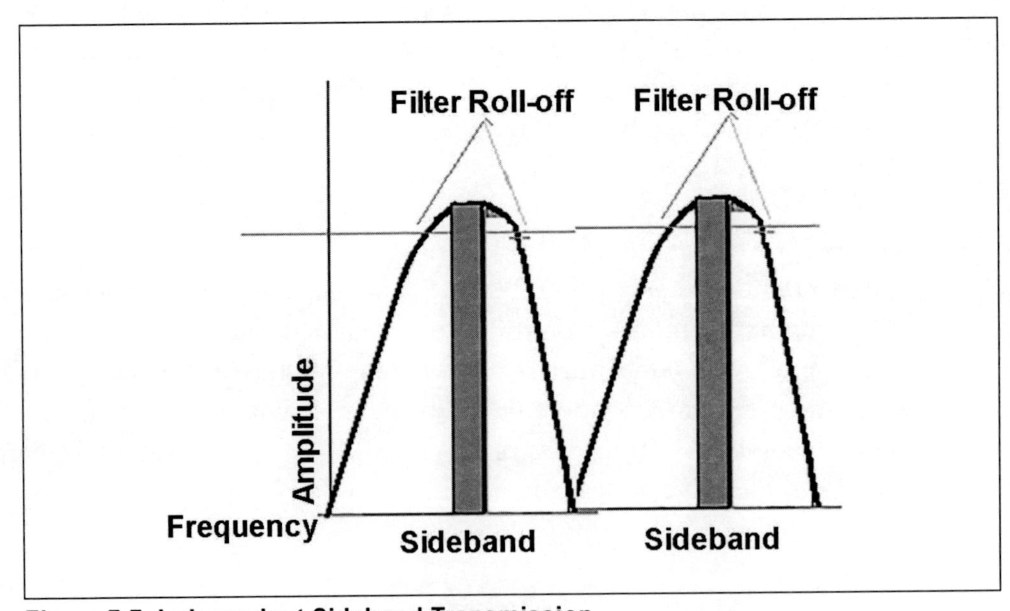

Figure 7-7. Independent Sideband Transmission

Frequency Shift Keying

One of the most popular and oldest methods (since 1910) of frequency modulation for digital data is frequency shift keying (FSK). This method has two (or more) tones or carrier frequencies in the audio range: (1) a tone (A) is keyed ON to represent a 1 state, the other tone (B) is OFF; and (2) the B tone is keyed ON to represent a 0 state, the A tone is OFF. A rule-of-thumb requirement for this method is that the tones' frequencies should be separated by approximately the same number of hertz as the signal's bps rate. As an example, if data is to be transmitted at 1,200 bps, then two tones, 1,200 and 2,400 Hz, would satisfy this rule. Actual practice allows less separation, but inter-symbol noise (the inability of the demodulator to correctly determine the signal state as a 1 or a 0 tone) will increase. The bandwidth of a typical FSK signal is illustrated in figure 7-8.

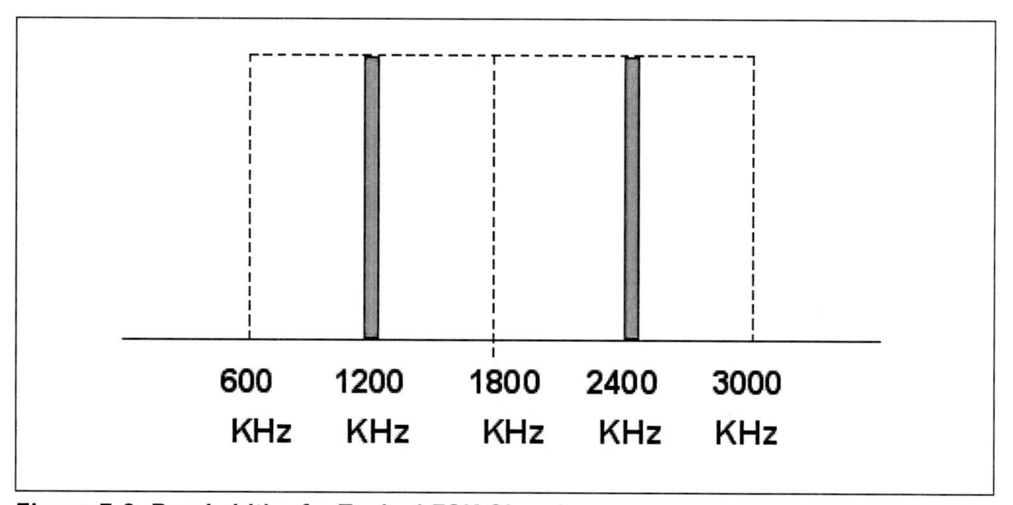

Figure 7-8. Bandwidth of a Typical FSK Signal

Why the rule of thumb? Why couldn't a system have two tones—for example, 200 Hz apart, keyed on or off at 1,200 bps? It could, but the signal could not be detected correctly. The data signal's maximum base frequency is 600 Hz. (The fastest rate of data change is alternate 1s and 0s: if each bit occupies 1/1,200 of a second and a 0 follows a 1 bit, then you have a cycle [two alternations] occurring in 1/600 of a second, or 600 Hz.) A 1,200 Hz signal modulated at 600 Hz has a bandwidth of 600 Hz to 1,800 Hz. A 2,400 Hz signal modulated at 600 Hz has a bandwidth of 1,800 Hz to 3,000 Hz. If the sidebands overlap they will demodulate each other, producing a signal that bears little resemblance to the original.

Frequency shift keying is employed because it is an easy process and it fits in a binary digital transmission scheme. Even the frequency shift control is just a matter of a few logic gates. However, FSK occupies two cycles of bandwidth for every bit per second, which requires double the bandwidth required for other methods; in this method we are trading bandwidth for simplicity. A 1,200 bps signal requires 2,400 Hz of bandwidth. Nonetheless, this is the method used in all low-speed modems (300 bps and less). It is also the method used for HART, in which a single 1,200 bps signal occupies 2,400 Hz of bandwidth.

Duo-binary encoding reduces the bandwidth required for FSK by causing the output to transmit a full cycle of the upper FSK tone and a half cycle of the lower FSK tone whenever a data transition occurs. To the line, this appears to be a 600 Hz signal that uses tones at 1,200 Hz and 2,400 Hz. Using duo-binary encoding causes three tones to be output: 1,200 Hz, 2,400 Hz, and 1,800 Hz (if an alternating digital signal is used—the fastest rate of change for a binary signal). This last output, 1,800 Hz, is the average frequency output because duo-binary encoding does not allow a direct transition from one frequency to the other. This technique gives one cycle of required bandwidth for one bit time.

Another method for reducing the media bandwidth required, known as *bi-phase* but more popularly as Manchester encoding, places the state of the binary signal into a transition. If the signal transitions from a 0 to a 1, it represents a 0 state. Transitioning from a 1 to a 0 represents a 1 state; a return to the opposite state will be at clock (twice the data rate) and is ignored. The net effect for a data signal that is alternating 1s and 0s is the same output as duo-binary encoding, that is, three frequency components. Manchester encoding is used quite often at speeds of up to 10 Gbps (gigabits per second); it is illustrated in figure 7-9.

Frequency Modulation

In frequency modulation (FM), the modulating signal varies the carrier's frequency change from rest according to the modulating signal's amplitude. Moreover, the carrier's frequency changes vary at the modulating frequency. An important fact to remember about FM is that the output carrier's amplitude remains constant; only the frequency changes. Most noise involves amplitude; that is, it tends to ride on signals above and below those of the average carrier. By slicing off the top and bottom amplitude of the received

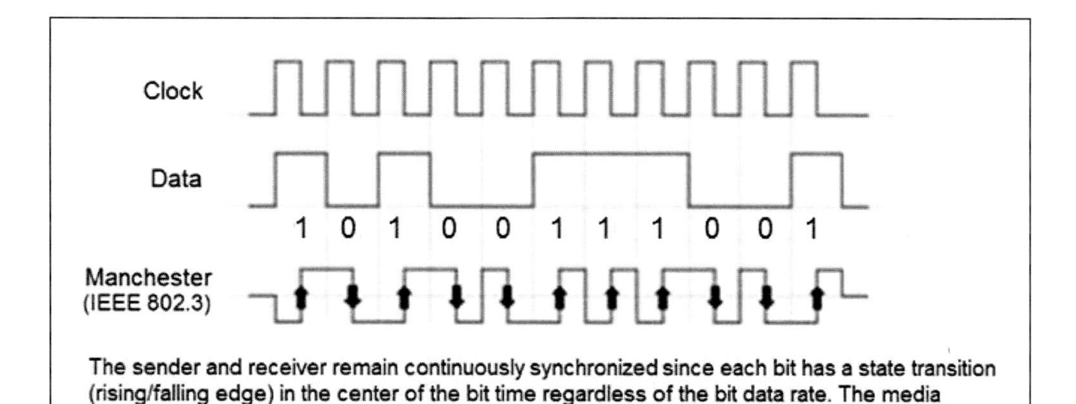

Figure 7-9. Manchester Encoding

carrier (a process known as limiting), most of the amplitude-type noise can be removed from the FM signal.

All this is not without cost, however. Bandwidth is needed to properly demodulate an FM signal. There are two kinds of FM: wideband and narrowband. Most of FM's appealing characteristics are inherent only in the wideband type. An important fact to remember about FM is that the *deviation*—the amount by which the carrier frequency is varied—depends only on the modulating signal's amplitude. The frequency of the modulating signal determines at what rate the deviation takes effect, not the amount of deviation.

An FM signal has an infinite number of sidebands. They are spread out over the frequency spectrum, above and below the carrier frequency at differing amplitudes. Their relationship to the carrier is such that each sideband is an integral (whole number) multiple of the modulating frequency away from the carrier. That is, if the modulating frequency is represented by "fm," then the sidebands are separated from the carrier frequency by $\pm$1fm, $\pm$2fm, and so on. The amount of energy in each sideband decreases as its distance from the carrier increases, eventually becoming insignificant. The decrease is not characterized by a simple linear relationship but rather is based on a mathematical treatment using Bissell functions. This treatment is beyond the scope of this discussion and may be found in any standard text on frequency modulation. What is an FM signal's bandwidth requirement? Without encoding, three cycles of bandwidth are typically required for each bit per second of data.

We stated earlier that there were two kinds of FM: wideband and narrowband. The dividing point between them is their deviation ratio, or the ratio of the maximum carrier deviation divided by the maximum modulating frequency. As an example, if the maximum deviation is 75 KHz and the maximum modulating frequency is 15 KHz, then the deviation ratio is 5 (75 ÷ 15 = 5). Signals that have a deviation ratio greater than 3 are considered wideband; signals with a deviation ratio that is less than 3 are considered narrowband. Because the FM signal requires a large bandwidth in uncoded form, FM is seldom used in digital transmission as the sole means of modulation.

Phase Modulation

Phase modulation (PM) is essentially like frequency modulation, except that the amplitude of the modulating signal causes a shift in the reference (or center) carrier phase. The basic difference between FM and PM is this: for a particular modulating signal value of amplitude, the phase modulator's output amplitude is a constant and the phase modulator's output phase (as measured from the reference phase) varies only with the modulating signal's amplitude and not the modulating signal's frequency. The output of the FM modulator, on the other hand, varies directly with the modulating signal's amplitude but inversely with the modulating signal's frequency. In other words, when using FM: the lower the frequency, the greater the deviation for a given amplitude. From a practical point of view, that is all there is in the way of theoretical difference. It is possible to obtain FM from a PM generation or PM from an FM generation; it is all in how the modulating signal is presented. (Note that a continuous change in phase is a change in frequency.)

Encoding Data

In our prior discussions, we determined that the baud rate was the maximum signal change rate the medium could support. The rate is specified in baud (*per second* is understood in the definition) and is called the *baud rate*, line rate, or line signaling rate. Bits per second (bps) is the rate at which data can be passed. The baud rate and the data rate can be quite different. We also discussed the attributes of transmission line *lumped characteristics* and determined that any given line will suffer from signal degradation as the signaling frequency (baud rate) increases. At some frequency the signal won't get through the line, that is, you can transmit at that rate but it cannot be detected at the receiver at that rate. Thus, there is a speed limit of sorts on any transmission line. The technical term for this is *bandwidth*. In 1948, Claude Shannon stated that the maximum data rate possible on a communication channel is proportional to the bandwidth of that channel (if you want the full technical details, search *Shannon's Law*). Harry Nyquist (who preceded Claude Shannon at Bell Labs) worked in this same area and also determined that the fastest signaling rate is proportional to the bandwidth.-

To increase the data rate for a particular modulation scheme without increasing the line's baud rate requirements, the digital data may be encoded. One of the earliest examples of encoding is a process called *di-bit encoding*. In it, a buffer holds the input data to be transmitted and signal changes on the medium are made on every 2 bits. As an example, let's look at a data rate of 2,400 bps. Logic is constructed so a decision is made every 1/1,200th of a second, rather than at the bit rate of 1/2,400th of a second. Examining the bits in pairs is the reason for this decrease in decision-making frequency. Four combinations are possible from the 2 bits but a decision about which one of the four combinations is needed only has to be made 1,200 times a second, rather than 2,400 times a second. This is now a four-state signal (00, 01, 10, and 11) as illustrated in figure 7-10.

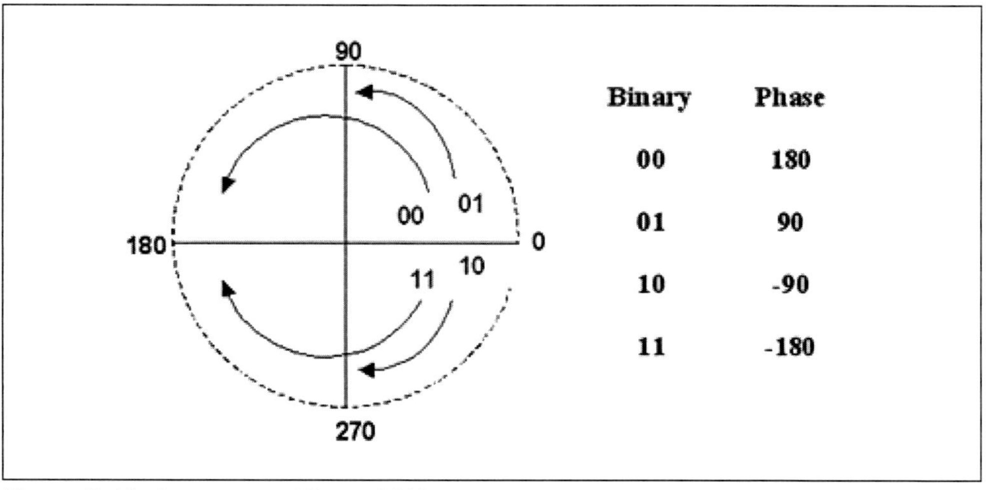

Binary	Phase
00	180
01	90
10	-90
11	-180

Figure 7-10. Quaternary Phase Shift

This type of phase modulation is called *continuous phase modulation* or *continuous phase shift keying* (PSK) and the resulting four-state modulation is referred to as *quaternary phase shift keying* (QPSK). Figure 7-10 illustrates the phase modulation technique that is used to transmit the four states. The detection process is differentially coherent detection, which means that the last transmitted phase becomes the reference for the next transmitted phase. A long string of 1s or 0s could cause a continuous change in phase resulting in a change in the transmit frequency. If the change is significant enough, the receiver will lose synchronization. This can be remedied by adding circuitry and logic that cause predictable state changes at the modulator and removing the changes upon demodulation; this function is known as a *scrambler*.

Of course, the laws of physics won't give something for nothing. In this case, the something is signal-to-noise ratio. This is a hard and fast rule: upward changes in the data rate will require a better (than before the higher data rate) signal-to-noise ratio, plus tighter tolerance for other unmentioned factors (cross talk, impedance, etc.). Multiple-bit systems form the basis for modern data transmission technology as they increase the data rate, leaving the baud rate intact (you can also use multiple bit techniques with slower media to achieve a higher data rate). However, even in this case, all the media's attributes (particularly signal-to-noise ratio) must be controlled to a higher degree of conformity than with the lower data rate signals.

Tri-bit systems are those that use eight different phase states (to represent three data bits) to encode the data signal. Newer techniques have six (or more) data bits encoded by one decision (selection of a phase/amplitude state known as a symbol). A selection of a symbol is then made from the available data states per quadrant. One can double the available data states by choosing half power or full power amplitude modulation. In addition, choosing the amount of phase change (along with the amplitude modulation, if used) that makes the largest change of phase (and signal amplitude if used) for the bit combination being encoded is part of a technique called *trellis-coded modulation*, so called because when the possible states are graphically plotted on a grid, the resulting image is that of a trellis. Trellis-coded modulation requires using a selection algorithm, which is usually executed by a microprocessor. In most standard modems, phase shifts and the amplitude modulation of the particular phase are selected, which results in a particularly large trellis (a large number of symbols). Trellis-coded modulation is how 33.6 Kbps duplex (a total of 67.2 Kbps for both directions) makes it down a 1,200-baud wireline. With this kind of encoding, only 600 symbols are transmitted in each direction per second, which does not exceed the bandwidth of the line but each symbol is associated with a 6-bit binary value.

Figure 7-11 is a partial representation of trellis coding, showing the first 90° of phase/amplitude differences. Twelve symbols are possible: six for each 15°

change in phase shifts at half amplitude and another six for each 15° change in phase shifts at full amplitude. For four quadrants, this means 48 possible states. Five bits are typically used; this gives 32 states required and 32 times 600 (the allowable number of symbols transmitted in one direction) gives a data rate of 19,200 bps.

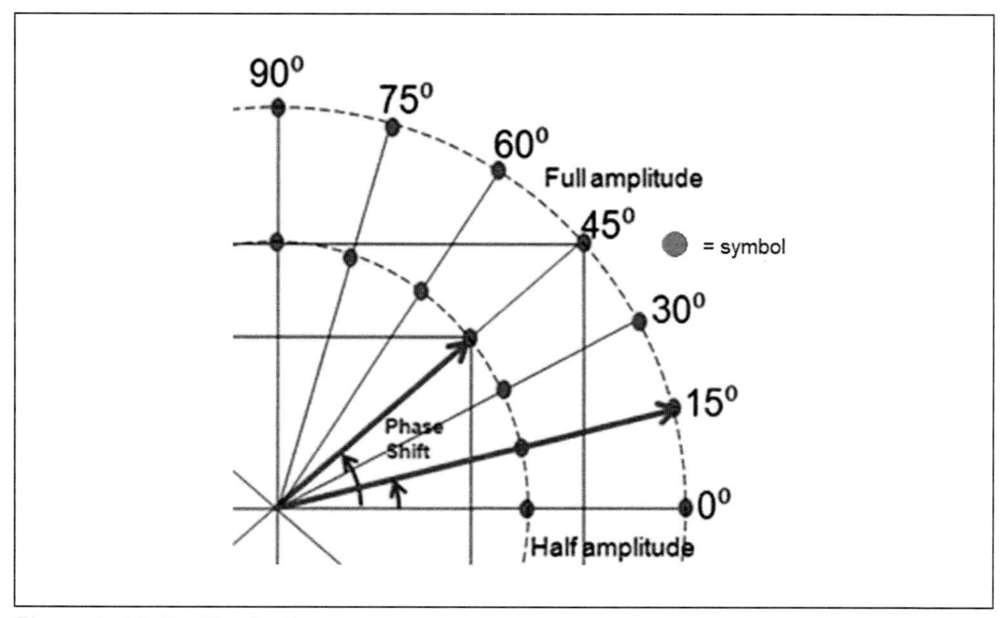

Figure 7-11. Trellis Coding

There are more symbols than decisions because the algorithm always chooses the maximum deviation for each selected symbol. The receive demodulator must employ the same algorithm. The point of this discussion is to emphasize the importance of having identical selection algorithms at each end. As demodulation is the reverse of modulation, both the receiver and the transmitter must possess the same algorithms, without them—detection is impossible.

Various modern communication schemes use trellis encoding and employ up to 1024 symbols, which allows a 10-bit binary number to be delivered for each symbol transmitted (or a data rate 10 times the actual signaling/baud rate).

Modulation Summary

Only three components of a sine wave may be modulated: amplitude, frequency, and phase. Frequency modulation (FM) and phase modulation (PM) are usually lumped together as *angle modulation*. Frequency (phase) modulation changes the carrier's frequency (phase) based on the modulating frequency's amplitude. These carrier changes take place at the modulating signal's frequency. Amplitude modulation (AM) changes the carrier's amplitude according to the modulating signal's amplitude and the changes in the carrier are at the modulating signal's frequency. Regardless of which modulation scheme is selected (AM, FM, PM, or a combination of techniques), the following rule applies: encoding the data effectively makes it possible for transmission lines to support higher data rates while still staying within the available bandwidth.

Wireline Modems

Serial modems originated in the 1960s with point-to-point serial data transmission, which, at the time, was the most economical way to transmit data over long distances. Networks were composed of interconnected, point-to-point transmission media. At that time, most loops were local (meaning they were voice-grade lines operated and maintained through a telephone central office), though several could operate through modems for a dedicated point-to-point line (which had a different tariff then the regular switched telephone lines).

Over the years (up to 1974 or so), some things in the modem world became more or less standardized. The modem's parameters were expressed in terms of the (pre-divestiture) AT&T (Bell System) specifications. These specifications were both historical and contemporary. Long-distance data communication, although more rapid than four decades ago, still uses some of the same specifications, as well as all the old technical jargon.

Definitions

Let's start our discussion of modems by defining key terms and the technical rationale for them.

- **Asynchronous** – In general, asynchronous means that something may occur at any time and is not tied to a clock. The old *start-stop* teletypewriter signal, with its 5 bits representing the text (see chapter 1) and one start bit and one (1.45 or 2) stop bit(s), is a good example. The reason for the partial stop bit time (it was generated mechanically—not electronically—and converted to an electrical signal) was to ensure that the receiver had enough time to mechanically impress the typeface on the paper medium before the next start bit. This signal used motor speed as the main synchronizing element and the start-stop bits synchronized each character. In today's vernacular, any start-stop signal is assumed to be asynchronous.

- **Synchronous** – Synchronous generally means tied to a common clock, the clock signal is transmitted along with the data. Originally meaning a signal that had no need for start-stop synchronization, synchronous used bit timing so each *bit of data was accounted for.* This is the basis of all modern data communications.
- **Baud Rate and Bits per Second** – Baud rate and bits per second are often used interchangeably, however, this is incorrect. *Baud rate* is the line modulation rate (line rate), that is, what the transmission medium must support in order to pass the data. Data rate, in bits per second, is the transmission speed the device is capable of transmitting or receiving. The bit rate (throughput—generally in bits per second [bps]) of a modem depends on the baud rate, the condition of the line, the packetizing method used, the amount of data compression, and the encoding scheme used (see the "Encoding Data" section in this chapter). *Baud per second* describes a change in transmission bandwidth, not a data rate. A V.90 56 Kbps modem (56 Kbps is the data rate) requires a 1200-baud line, that is, a line that has a bandwidth great enough for 0.0008333-second rectangular pulses.
- **Preamble** – A preamble is a Physical layer signal that is used to synchronize connected devices. In many systems, a burst of carrier signal (or data clock for baseband) is sent to synchronize receivers. This is followed by bit patterns to synchronize the bit timing, then control patterns (characters or a unique bit arrangement) to synchronize messages.
- **Synchrony** – In the context of modems, synchrony relates to the way two modems get their bit-rate clocks in phase—or how they know when they are in time with each other. Many modems supply a *receive* or *recovered* clock that may be used to synchronize the data terminal equipment (DTE). In most modern communications devices, the transmit signal's transitions are performed at the transmitter's clock time so the transmitter's clock timing is inherent in the data signal's transitions. These may be recovered on the receive end and the receive DTE's clock will be adjusted until it is in phase with the transmitter's clock.

As one works through the various modems and networks available, a pattern emerges: first, frequency synchrony is achieved, then bit synchrony. Although methods differ, this is the pattern in all synchronous systems. In general, an asynchronous modem will not supply a recovered (receive) clock and a synchronous modem *must* supply it, in order to account for each bit.

Modem Types

Wireline modems are for use over dial-up or leased telephone-type lines and are divided into two classes: Bell[1] and International Telecommunication Union[2] (ITU), which issues the V and X standards. Table 7-1 lists various standard modems governed by these specifications with their speed and modulation types. Most modern modems are synchronous between the modems, using Link Access Protocol – Modem (LAP-M) packets and may have a synchronous or an asynchronous interface to the DTE. On PC modems the interface is asynchronous, as the PC bus for I/O is asynchronous. Depending on price and application, stand-alone modems may be configurable for an interface through either programming or the use of a dip switch.

In table 7.1, the word *similar*, next to an ITU modem indicates that the modem performance is similar to the Bell model but it does not meet the Bell standard in many performance aspects. The term *bis* indicates that the standard is the second iteration and *terbo* indicates that the standard is in the third iteration.

Worth noting: Typical of many of the low-speed legacy wireline modems is the Bell 212, which contains a Bell 103 set of carrier frequencies. A plug-in card for many PCs, this once was the most widely used modem on dial-up lines. The Bell 212 is capable of identifying what type of modem it is talking to and adjusting itself to become that type. Considering the great differences in modulation schemes, and so on, this is no small feat.

1. The name *Bell* is generally no longer used, just the specification number.
2. ITU is formerly the International Telegraph and Telephone Consultative Committee (CCITT), which was originally the International Telegraph Union.

Table 7-1. Wireline Modem Characteristics

Asynchronous Modems

Speed	Bell Specification	ITU Standard	Modulation Method
300	103/113	V.21 (similar)	Frequency shift keying
1200 (half-duplex)	202	V.23 (similar)	Frequency shift keying
1200 (duplex)	212A	V.22 (similar)	QPSK (di-bit)

Synchronous Modems
(PC internal modems are asynchronous input, but synchronous between modems)

Speed	Bell Specification	ITU Standard	Modulation Method
1200	212A	V.22 (similar)	QPSK (di-bit)
2400	201	V.26 (similar)	QPSK (di-bit)
4800	208A	V.27 (similar)	QPSK (tri-bit)
9600	209A	V.29 (similar)	AM/PSK
9600		V.32	Trellis coded
14400		V.32bis	Trellis coded
19200		V.32terbo	Trellis coded
28800		V.34	Trellis coded
33600		V.34	Trellis coded
56K (download) 33.6K (upload)		V.90	Broadband on download, trellis coded on upload
56K (download) 48K (upload)		V.92	Broadband on download, trellis coded on upload

Faster Modems

Prior to the Bell 212 modem, the Bell 202 modem operated at 1,200 bps using FSK, with one carrier at 1,200 Hz and the other at 2,200 Hz. This transmission scheme used almost all the bandwidth available to the typical two-wire wireline and the modem had to operate in half-duplex mode. This caused throughput problems because the predominant method of error recovery was automatic retransmission query (ARQ), which required that a message receipt acknowledgment be returned to the transmitter after a specific number of blocks were transmitted. To avoid having to perform handshaking every time the line was turned around to send a message receipt, the 202 used a secondary channel. A small portion of the spectrum, near the 300 Hz lower limit, was used for a 5 bps channel. This is technically an asymmetric duplex scheme – each direction at a different line speed. The primary problem was the use of FSK as the modulation scheme and the limited wireline bandwidth. The Bell 202 could actually be called the originator of the asymmetrical data line, in which data in one direction proceeded at a much higher line rate than data in the other direction.

The Bell 212 modem used data encoding to overcome the bandwidth restrictions of a typical dial-up wireline. The 212 modem used di-bit-encoded, quaternary phase shift keying with coherent detection and synchronous modulation/demodulation because it has noise advantages over a four-level AM signal (discussed previously). The 212 modem had a unique answer tone that allowed for the convenient identification of modem speed when the called end answered. This modem could be used in asynchronous or synchronous systems, offering the choice of using or not using a recovered receive clock. After the 212, modem data speed progressed until the V.92 modem was developed. V.92 is the current modem release for the dial up system (2014).

Modem Data Compression and Error Detection

We have not mentioned error detection and correction schemes. Operating above 2,400 bps over a wireline presents many more opportunities for errors than operating at slower speeds. Since strings of repetitive ones and zeros can be encoded in fewer characters (much like Morse code uses letter frequencies to assign the shortest data elements, as previously described in chapter 1), throughput increases. Modem manufacturers offer different data compression, as well as error detection and correction schemes.

Unless both modems are operating with identical schemes for either data compression or error detection and correction, the error detection and correction process will generate errors because the error detection check characters are different for different schemes (refer to the "Cyclic Redundancy Checks" section in chapter 1). Two sets of standardized schemes are in general use: the V.42 international standard, which in the original and *bis* versions allows data compression at two to one (original) and four to one (*bis*), and the Microcom Networking Protocol (MNP), which is discussed below.

Other Modern Modem Features

The MNP comes in various flavors, the first five have been released to the public domain. MNP-1 and MNP-2 are generally concerned with packetizing data, while MNPs 3 through 5 address a range of parameters, including losing the start and stop bits (as does V.42), for a 20 percent gain in throughput, and by packetizing the data. MNPs 6 through 10 are still in the proprietary domain, meaning that it is necessary to have modems that have the appropriate level of MNP software at both ends of the data transmission. V.42 employs the LAP-M frame and uses code lists to represent strings of characters, replacing the strings with the shorter code lists to perform data compression.

How does a modem know what speed to use, which error detection scheme to use, and which data compression scheme to use? Part of modern modem operation is a process called *training* or *negotiation.* In this process, the modem typically starts up at a low speed (1,200 bps). If transmission is not possible at 1,200 bps, there is no use in trying higher speeds. If transmission is possible, the modems then negotiate the speed, data compression, and error-detection scheme the modem at each end will use, arriving at the highest common denominator of operation. This procedure is required because of all the different types and speeds of modems and the various error-detection and compression schemes available.

On the way to the current modem standards, a variety of techniques were developed to permit high-speed data transmission over a 3 KHz wireline. One of the techniques developed is multi-tone. In multi-tone transmission, the data stream is divided into eight (or more) parallel streams. Each stream is di-bit encoded, after which each di-bit encoded stream is phase shift modulated using one of the eight or more tones separated by about twice the tone bandwidth as the carrier—in this context, bandwidth means the span of frequency in hertz. All the tones are used as a simultaneous baseband to a VSBAM modulator. At the receive end, the demodulation process takes place until the signal is reassembled. This scheme is the basis for the discrete multi-tone (DMT) process used in DSL lines.

There is a continual push to increase transmission speeds—especially in the wireline arena. The highest-speed standard modems for wireline at the time this book was written were the V.90 and V.92 (2014) at 56 Kbps. However, because of the need for increased throughput over the wireline medium, efforts are being made to go past these speeds when economically possible. The V.92 standard improves slightly on the V.90 specifications by adding a method for disconnecting the modem long enough to let you know that someone is trying to call you without losing the connection, a feature referred to as

Internet call waiting. In addition, the maximum upload speed has been increased from 33.6 Kbps to 48Kbps.

Most modern high-speed modems now incorporate a *fallback* system that works as follows: if the error-detection rate rises above a preset level, the speed is dropped back to a lower data rate. Some modems drop the data rate in integrals of standard data rates and some by smaller increments. The data error-detection rate is continually monitored and as it improves, the speed is adjusted back up to the desired rate. None of these newer modems would have been possible without the large advances in modern microcircuitry. What would have been literally tons of equipment is now routinely packaged in one integrated circuit (IC).

Modems Summary

In this section we have discussed modems, from low-speed to high-speed, that operate over the typical telephone wireline. Data speeds increased from the high speed of the early 1970s (2,400 bps) to an order of magnitude (or greater) in just four decades (going back from 2014). Even more significant is the fact that one of the first auto-dialing, auto-answer modems with a data rate of 300 bps cost $500 in 1984; and in 2007, one could purchase a 56 Kbps modem with bells and whistles for around $20. This comparison doesn't take into account data compression on the 2007 model, which can increase the throughput to a continuous average of near 2.8 to 1—nearly 250 times the performance for one tenth the cost of a Bell 212—and these modems are more reliable. The explosion in the use of Internet graphics and the need to transmit objects rather than just character-based text have fueled the insatiable appetite for more bytes-per-buck and wringing the highest possible data rate out of the wireline. Can speeds greater than today's wireline modems be achieved? Remembering the engineer who once told the author, "No more than 2400 bps down a wireline...," the author refuses to speculate. For those who need more bandwidth now, there are present-day solutions (2014) at one-tenth the cost of earlier modems. These solutions are outlined in the next section on digital line offerings.

Modems are used for frequency translation, and wireline modems comprise only a small segment of those devices. Other types of modems are used in LANs, cable television[3], and satellite and ground communications, just to

3. CATV stands for Community Antenna Television. This is where the cable modem has its roots.

name a few uses. A judicious use of a search engine will bring forth most of the background for modem data delivery.

WAN Digital Lines

A WAN is commonly understood to be one that serves geographically separated areas. One of the best and largest examples is the Internet, but there are other wide area data networks (e.g., the DOD military networks). Metropolitan area networks (MANs), governed by IEEE 802.6, can be connected as WANs. Intercity links can be considered as WANs and usually operate at 1.544 Mbps, the T1 carrier data rate; T1 has become the standard telco (telephone company) channel. Subscriber rates are available as fractional rates, such as 56 Kbps, 64 Kbps, and so on.

Telephone Lines as Media

One of the original WANs was the landline telephone network, in particular the direct distance dialing (DDD) network that existed for many years prior to digital networks. It was comprised of trunk and subscriber lines and was originally totally analog. Since the 1990s, the DDD has been a digital network with the exception of the subscriber lines; these are still analog. Subscriber lines still make up much of the telephone system and are primarily twisted pair copper lines.

Even at its best, there are two problems with the bandwidth (line rate capacity) of the analog telephone line: (1) frequency-selective amplitude distortion and (2) frequency-selective delay (phase distortion). Since the ear is most sensitive to amplitude distortion, techniques for reducing this distortion have been applied to most phone lines. Although many of the techniques used to reduce amplitude distortion increase phase distortion, it is not as much of an issue because the ear is not very sensitive to phase distortion. In digital transmission, phase distortion problems are corrected as much as possible by applying *conditioning*—using features that may be focused in the line, in the modem, or possibly in both. Higher-speed modems also provide automatic conditioning that is based on pilot tones or bit error rate (BER).

Rather than adding conditioning to analog lines, a subscriber can purchase or lease digital lines. For years, telephone operating companies have been promoting digital lines to customers by leasing a line, a data service unit (DSU), and a channel service unit (CSU). The DSU may be customer owned but the rest of the transmission is the phone company's worry.

Although the public switched telephone network (PSTN) began life as analog, it started becoming digital in the 1960s and today is totally digital, except for the final mile connection (the "last mile") to your home. The phone company digitizes your voice using an 8 kHz sampling rate and 8-bit analog-to-digital circuit. This produces a continuous data rate of 64 Kbps (called a *DS0 channel*), which the phone company has to continuously deliver from your land-line phone to the land-line phone at the other end. (When you use your cell phone, the sampling and digitization are happening in your phone.) Within the phone system, message traffic is routed as data and many concurrent data flows (which could be data or digitized conversations) can be combined, passed through digital switches, and transmitted over digital circuits. When looking at the signaling/data rates of various telephone company standards (such as the T1 and T3), the strange-looking values (e.g., 1.544 Mbps) make sense when you realize that they are an even multiple of 64 Kbps (e.g., a T1 circuit is potentially delivering 24 separate phone conversations, each using 64 Kbps).

Although the telephone company continues to deliver analog subscriber lines to homes that need multiple phone lines and Internet access, businesses are more likely to get one or more T1s from the phone company and hook them to a digital (IP) private branch exchange (PBX) that interfaces via Ethernet LAN with VoIP phones in each office and the company email/web servers. The various telephone/telecommunication service providers offer a range of digital connectivity options into the telephone system and the Internet. (Whether part of the circuit connection is a cable modem, a satellite system, or originated in a cell phone, Internet traffic is run across the same telephone trunk lines and switches as your digitized voice. The Internet and telephone system actually consist of much the same systems, infrastructure, and equipment.)

The source of data-ready lines is typically fractional T1 lines (explained later in this section). Remember, digital data will not transmit on analog telephone lines because it has a DC component that is caused by successive 1s or 0s. Most analog telephone lines have numerous transformers in the terminations and couplings, so a frequency translator, such as a modem, must be used to put digital data onto these lines and recover it at the distant end. As an option, the transformers (or, more correctly, the line-impedance matching devices) may be bypassed and the copper accessed directly. The typical modern voice-grade line by itself is capable (through differing techniques) of transmitting digital data at speeds above 56 Kbps for limited distances. While digital lines will support baseband digital, there should be some provision (either in the hardware or in the DSU) for eliminating long consecutive strings of 1s or 0s so enough transitions will take place to prevent loss of synchronization.

Public Switched Telephone System (Direct Distance Dialing)

The DDD system is still used for calls outside the so-called local access and transport area (LATA). In North America, the ten-digit LATA number includes a three-digit area-code prefix to the normal number, which is a three-digit central office and a four-digit station number. Calls in the LATA or inter-LATA can be routed in many ways: by radio link, satellite, wireline, microwave, or fiber optics. All of these are digital trunks. As mentioned earlier, only the subscriber line, the so-called last mile, is analog; all other transmissions are digital. At the rapid rate that mobile devices (such as *cell phones*) are being used, land line services are becoming a historical novelty; so it really makes little difference anymore what a land line is or is not because DSL, cable modems, satellite links, voice over IP (VoIP), 3G and 4G smart phones, and Ethernet everywhere (through Wi-Fi) generally command the lion's share of technical attention. For land line data transmission, the home computer transmission is digital to the modem, converted to analog to go over the subscriber line to the central office (CO), and converted to digital for long lines. At the destination CO, the transmission is converted back to analog, sent to the destination, and converted back to digital for the computer. That are a lot of wasted conversions where errors can corrupt data; an all-digital solution would be preferable considering the fact that only the subscriber line is analog and the rest of the system is digital. For reference see figure 7-12.

Not all lines in the world are digital, however. Some long lines (outside of North America) are still analog and connect to trunk lines where the circuit is converted from a two-wire (one pair) to a four-wire circuit (one pair for transmit and one pair for receive). Why? Because some lines need amplifiers to make up for loss. These amplifiers only operate in one direction, from the source to the destination, so each pair of a four wire line is unidirectional. If the two- to four-wire conversion devices (for analog lines) are not exactly balanced, then some of one line's energy will slip over into the other line's amplifier. If that happens at the other end also, an oscillator is created that produces, at the least, an echo that is distracting to the person trying to talk or

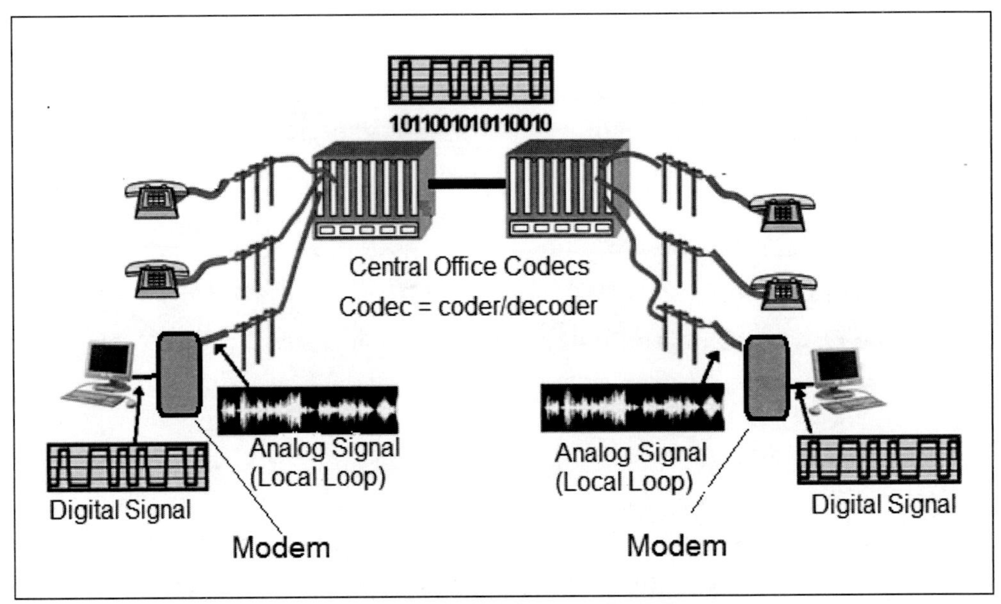

Figure 7-12. The Subscriber Line DDD Telephone System

listen. To deal with this problem, an echo suppressor is placed in both lines. This is an electronic switch that allows only one amplifier to be on at a time. The parties cannot talk simultaneously. One person talks and when he or she falls silent, providing an opening, the other person talks (as is done in normal human communication unless, of course, you are arguing with your significant other). The echo suppressor listens and when there is no energy in one line, it turns the amplifier off. It takes only 100 milliseconds of silence for the echo suppressor to switch the silent party off and switch the talking party on. This action is not noticed in the normal course of human speech. Both *digital* and *analog* lines can have echo suppressors.

Installing duplex modems on analog lines creates problems on lines that have echo suppressors. Communication is not possible in both directions, as required for duplex operation, on lines that have echo suppressors because one of the amplifiers must always be off. This is taken care of in the modem handshake. The answer tone will last in excess of 400 milliseconds. When the echo suppressor hears the answer tone for a sufficient length of time, it turns off the echo-suppressor feature until the next 100 milliseconds of silence occurs. Since the modem transmission is normally bidirectional until the transmission is complete, the echo suppressors are effectively out of the circuit. If the line is dropped, even for only little more than 100 milliseconds, connection must be reestablished.

Packet Switching

Once data is packetized (bundled into packets, frames, or blocks), the packet assembly/disassembly device (PAD) begins the process of breaking up a digital data stream into packets or fragments of packets. Each packet includes its destination and source address, perhaps with packet control and data as well so the packet may be switched (a Layer 2/Data Link layer process) rather than routed. Obviously, these items must be arranged in a specific format. CCITT (now International Telecommunications Union—International Telecommunications Standards [ITU-ITS]) Recommendation X.25 provides a packet-switching scheme to be used over the public network. The scheme provides the technique used by the PAD to create and add the headers to each of the small packets.

The PAD breaks a message up into small packets, each addressed to the destination, and inserts them into the transmission channel when it is available. Each step is connection-oriented: a logical channel number (LCN) is assigned to the packet by the X.25 network cloud management system when the packet is transmitted from the originating station to the network cloud. (Note that there is a limited number of LCNs available and they are reused when the packet switch circuit is finished.) A network cloud is a mechanism or symbology allowing one to see the connection interface but the mechanics in the cloud (topology, protocols, routing) are not observable, only data at the interface. (See figure 7-13 for the X.25 packet switching layout.) At the receive end of the cloud, the packets are then transmitted to the receiver, with another LCN assigned to them.

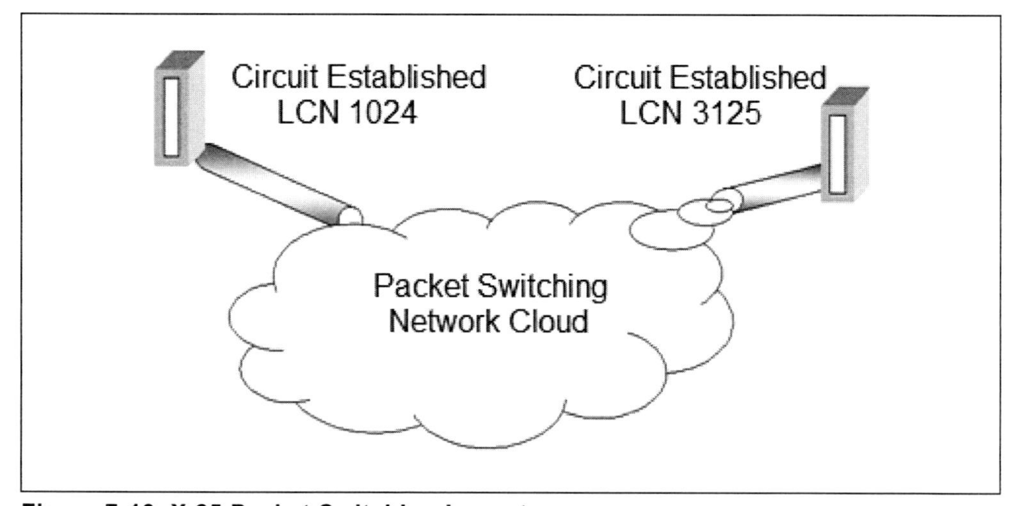

Figure 7-13. X.25 Packet Switching Layout

The computing device at the receive end then reassembles the message from the packets. Various data speeds can be used; however, 56 Kbps is the highest and others are a submultiple of 64 Kbps. Only virtual circuit service (regardless of the topology, methodology, and protocols in the cloud, the interface appears to be a point-to-point connection) is typically used, and that is a connection-oriented service.

To review, in connection-oriented transmissions, a destination is set up and the packets are routed to that source in the same sequence in which they were generated, even if it means the packets must be stored at one point while they wait for an available path. Connectionless service allows the packets to be routed by any available path and they may arrive at the destination out of sequence. It is the responsibility of the receiving device to reassemble the packets in the correct order. (This could be tough on real-time voice or streaming data, which requires predictable timing in order to reassemble a real time signal. Some form of priority for streaming services must be established or a point-to-point connection made.)

X.25 uses an HDLC point-to-point frame in which a protocol data unit (PDU) is enclosed. Packet switching is performed on the public switched network, although it could doubtless be used on private networks. X.25 is available everywhere (in theory, it has been technically superseded by frame relay) as a data network and since these data packets can exist with a digital voice scheme, they become the basis for an integrated network of voice and data. A small amount of time researching X.25 with an Internet search engine will reveal many good texts relating to X.25 and packet switching in general. You are referred to them, if further investigation into X.25 is warranted.

Although the public switched network and X.25 packet switching are viable, proven technologies, they are not now configured to meet the needs of industrial data communications (primarily process control communications), as they are not inherently deterministic. X.25 packet switching is now a legacy protocol. X.25's reliable delivery mechanisms and PADs were needed when the devices communicating were dumb terminals (including point-of-sale terminals) and central mainframe computers, but today every end device is smart and is capable of implementing protocols (e.g., TCP/IP) that provide reliable delivery so that the underlying network does not have to do so (and carry the associated overhead.)

Integrated Services Digital Network

The integrated services digital network (ISDN) service was designed (circa 1985) to offer voice, data, and video, all integrated into one service and generally distributed throughout the local area, as long as the on-premises equipment was not more than 3900 m (13,000 ft.) from the central office. The objective of ISDN was point-to-point digital connectivity and (in the United States and Canada) it was aimed at the small office, home office (SOHO) market. ISDN was conceived based on the sometimes-analog public telephone network. The twisted wire that now carries telephone service was a limiting factor. Later, as fiber optic loops fell in cost and became more readily available, so did ISDN. Note that (originally) ISDN presumed that you had an opposing endpoint to which you were to be connected. It was not a connection to the Internet, but to the phone system. Somewhere else on the phone system network the ISDN data traffic would be delivered, usually to a corporate mainframe or another SOHO termination.

Although the technology to implement ISDN has been around over 35 years, the implementation has lagged. Since many foreign governments have a monopoly over telecommunications and can dictate standards, many countries had previously established standards for ISDN. Some of these standards clash with the technology and rationale of the competitive systems found in the United States. One might assume (at one's own risk) that the political compromises necessary to realize an international standard could be achieved. However, they were not and other, newer systems have displaced ISDN as the system for connecting all the businesses and residences in the United States. ISDN did achieve (until DSL lines became widespread) a degree of popularity in the United States, not as an integrated service medium but for relatively fast Internet access from 1990 to 2000.

One of the key features of ISDN is out-of-band signaling. This means that the control signals are in a different medium than the messages. There were originally two ways to connect to an ISDN network: *basic* and *primary*. In the *basic* configuration, three channels are distributed on the wireline. These are two B (for bearer) channels, operating at 64 Kbps each, and one D (for data) channel, operating at 16 Kbps. This is called the *2B+D arrangement*. The original intention was that the B channels could carry either digitized voice or data on either channel and that the D channel would provide control signals and meet other low-speed signaling requirements. You can tell at what technical generation ISDN was conceived by the fact that it took 64 Kbps for subscriber-quality voice. In modern digital telephone systems, 32 Kbps provides toll-quality voice and 16 Kbps provides subscriber-quality voice. Toll quality is the highest quality voice signal (300 Hz to 3.3 KHz), subscriber quality, while acceptable, is not as high quality.

Message traffic on the D channel uses Link Access Protocol-D (LAP-D), which is essentially HDLC. The difference between LAP-D and HDLC lies in the address fields, where LAP-D uses a 2-octet address—one to identify its network and the other to indicate the end point. By using the D channel to control switching the B channels, a clear channel is established in which bit patterns on the data (B) channel do not affect transmission. At the time of ISDN's conception, 64 Kbps was chosen because that was the data rate needed to support voice digitization. That rate is high for voice given current technology and in many cases it is too low for some of the digital data services. It would be hard to use an ISDN line as a LAN port without having greatly reduced data rates (at least with the IEEE 802 types). In addition, some electronic switches will not support an ISDN channel at 64 Kbps. Instead, they require transmission at 56 Kbps because they extract the 16 Kbps signal by using 8 Kbps from each of the 64 Kbps channels. This results in two 56 Kbps channels and a 16 Kbps data channel that is in-band for signal control.

In *primary* access, many subscribers use common input trunks to a facility. These trunks are multiplexed together into 23 B channels and 1 D channel and are able to operate across one T1 carrier line. An example is the PBX facility that is fed by the T1 trunk and distributes data out to the subscribers.

Would ISDN be of interest to industrial users? It could provide a gateway between a LAN and the telephone WAN. However, on most LANs the data signaling rates might cause a severe bottleneck in data transfer. Some ISDN terminal adapter vendors make a modem that uses the entire bandwidth (128 Kbps) but the terminal unit will only have a throughput maximum of 112 Kbps, as 8Kbps are taken from each 64 Kbps channel to use as the D (control) channel.

Author Thompson's experience with ISDN was that it was not quite ready for prime time unless you lived in a major metropolitan area and all you wanted was fast access to the Internet. First, configuration is not simple. Second, many operating systems did not and do not have drivers for an ISDN terminal adapter through the serial port. Worse, if you wish to connect via ISDN and are a user in a less-traveled location, more than 4 miles from the central office, you may experience more than just a few problems attempting your original installation. Due to the lack of demand, equipment may not be in the central office for a while, or ever. Installation costs (depending on the state and the tariffs imposed) vary widely and ISDN lines are still subject to recurring line charges. Lastly, another ISDN facility must lie at the other endpoint in order to make good use of the ISDN features.

As with X.25 packet switching, ISDN (at least in the United States) is considered an obsolete technology, since DSL is readily available (as long as you don't mind that your traffic traverses the Internet) and offers higher data rates. In Europe it is still possible to obtain ISDN service.

In spite of its obsolescence, one of the good ideas that came out of the original ISDN, lives on: frame relay, which we discuss next.

Frame Relay

Frame relay is the connectionless version of X.25; instead of using LCN, it uses a data link connection identifier (DLCI). The carrier company (or companies) provides Layer 1 and 2 services for your data. In other words, you have a frame relay connection into which you send and receive data. Frame relay speeds typically start out at 64 Kbps and can go as high as you have the pocket money to afford. Although, in theory, frame relay offers switched circuits like X.25, currently most connections are not switched but are rather permanent virtual circuits (PVCs). These connections are point-to-point, or multi-point, as far as the customer is concerned but the actual routing and number of nodes vary and are unknown to the customer.

The customer is given a committed information rate (CIR) for which he or she pays a tariff. The user normally chooses the CIR at the average data rate of his or her traffic. As long as traffic is below the CIR, packets will not be discarded. As the data rate goes above the CIR, a best effort will be made to deliver the packets and the user's Layer 4 functions will discover the discarded packets. At some predetermined rate above the CIR, all the user's packets are discarded if there is congestion on the network. Discarded packets are the frame relay network's way of saying that it is becoming congested. Frame relay may be used to connect corporate LANs, and the present-day cost per single channel is low enough for industrial use, particularly if the frame relay Digital Data Service (DDS) is used. Frame relay DDS provides 56 Kbps with no CIR, using the DDS-type tariffs (actually fractional T1 now), and it provides this service typically at the monthly cost of a business telephone line. (The previous name for DDS, as offered by AT&T, was *Digital Dataphone Service*.)

Many industrial users of SCADA technology, particularly in urban areas, have paid for leased analog phone lines from the local telephone company in order to communicate with their RTUs. In the last decade, most of those telephone companies have either stopped supporting or stopped leasing additional analog lines, forcing SCADA system owners to migrate to alternate communication technologies. Various manufacturers produce frame-relay access devices (FRADs) that provide an analog phone circuit or RS-232 circuit but connect to a frame-relay network. Many SCADA systems have transitioned from leased analog phone lines to frame relay through the use of FRADs.

T1 Carrier

A T1 carrier (in North America) is a 24 digital-voice channel (DS0), time-division multiplexed signal. It multiplexes the outputs of 24 DS0 channels onto one 1.544 Mbps line, generally a coaxial cable but increasingly fiber optic cable. In the past, the user had to buy all 24 channels (or 1.544 Mbps) and this was expensive. Due to the falling price of multiplex equipment and increased competition, fractional data rates from 64 Kbps on up can now be bought. Each of the 24 channels that make up the T1 has a data rate of 64 Kbps; however, in most cases, the customer actually only gets 56 Kbps, since the provider uses bits (8 kb) from each DS0 for control.

A T1 channel is also known as a DS1 (digital signal 1) channel. In the extended super frame (ESF) method of signaling, there are 24 DS0 channels with 193 total bits: 192 bits for data and 1 bit for framing. Since the original voice-sampling rate was 8 KHz, a signal of 1.536 Mbps (24 × 64 Kbps) was produced, plus the one framing bit for an 8 KHz framing channel, which equals 1.544 Mbps. The framing channel is broken down as 12 m bits (4 Kbps for out-of-band channel management), 6 c bits (2 Kbps for CRCC), and 6 Fe bits (2 Kbps for framing). In South America, Mexico, and Europe the equivalent WAN channel is called *E1* and it has a data rate of 2.048 Mbps.

T3 Carrier

There are various hierarchies of T carrier, not all of which find current use in the United States. The T3, however, is used here. A T3, which is also an international standard known as DS3, consists of 28 T1 lines (according to the international hierarchy, it consists of seven T2 channels, each of which is composed of two T1C channels and each T1C is made up of two T1s). T3 is only an option for large corporations and government organizations since its cost per month is high.

Dataphone Digital Service

The two dataphone digital service terminations—a digital service unit (DSU) and its corresponding channel service unit (CSU)—offer digital transmission without the need for modems. These two units perform services analogous to a modem on an analog line. DDS was originally a leased-line (not switched) service that offered data rates from 300 to 1,200 bps, as AT&T Digital Data Service (same initials) it offers bit rates from 2.4 KBs and up (and is usually a fractional T1).

Fractional T1

Fractional T1 is a line that consists of one or more DS0 channels. Multiplexers are required to put the data on a T1 line. Various fractional T1 combinations provide different amounts of bandwidth:

DDS	1 DS0	64 Kbps
H0	6 DS0	384 Kbps
H11	24 DS0	1.536 Mbps (T1 without framing)
H12	30 DS0	1.920 Mbps

Most fractional T1 (FT1) lines run at 384 Kbps (6 DS0), 512 Kbps (8 DS0), or 768 Kbps (12 DS0).

Functions of Basic Telco Digital Services

Frame relay, fractional-T1, full T1, and T3 are all digital communication mechanisms for routing message traffic onto and back out of the telephone system. They do not lead to the Internet unless the user specifically asks for this when ordering the service. With most of these services, the assumption is that the user has multiple locations to network together and doesn't want to build or maintain the necessary infrastructure or have message traffic going over the Internet and opening corporate sites up for attacks from across the Internet. In some cases, a company may want to (or have to—e.g., along a remote pipeline or electric transmission line) build its own WAN and can do so using the same equipment and technology as the phone company. The following WAN technologies are, and have been, used by the various service providers and can also be used to build private WANs.

Fiber Distributed Data Interface

The original high-speed data bus, fiber distributed data interface (FDDI), is a fiber optic token-passing ring. It may use single- or multi-mode fiber. FDDI consists of a dual counter-rotating fiber ring (primary and secondary) where data may circulate in opposite directions. More than one frame will be on the line at a time, due to FDDI's high speed (this is the same method used with the 16 Mbps IBM token ring). This high speed is the result of a process known as early token release (ETR) in which the transceiver usually receives its own signal before it finishes transmission and marks the poll bit as a token on the fly. There is always data on the line. The dual ring design enables *self-healing* (it reconnects automatically after a transmission break) via loop-back as shown in figure 7-14.

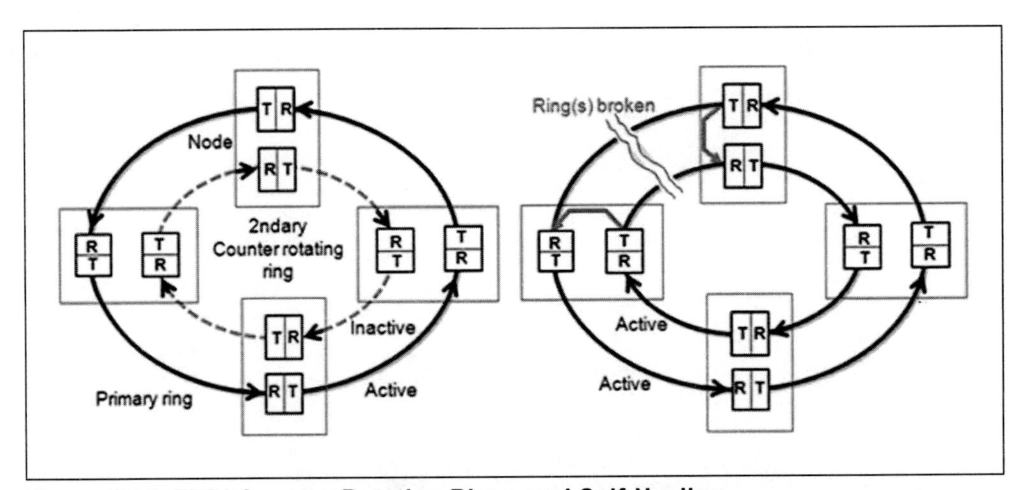

Figure 7-14. FDDI Counter-Rotating Rings and Self-Healing

FDDI allows up to 1,000 connections (not nodes). A node with both a primary and a secondary ring connection will count as two connections. FDDI does allow a 100-km (62.5 mile) network span (the length of the ring) with up to 2 km (1.2 miles) between any two nodes. (It can extend much farther—up to 20 times farther—if single-mode rather than multi-mode fiber optic cable is used.) Single and dual attachment stations (SAS/DAS) and concentrators are typical nodes on the ring(s) and those provide a bridge/gateway function to the local Ethernet LANs at each site. FDDI was much faster than the original 10 Mbps that Ethernet allowed when it was originally introduced, however

Fast Ethernet (100 Mbps,) which came about soon after, was as fast and subsequent iterations of Ethernet are much faster.

FDDI has been modified from its original packet-switching format to be able to handle circuit switching for voice and video transmission if necessary. This is FDDI-II. Both FDDI (the original) and FDDI-II require 125 megabaud (MBd) media for 100 Mbps transmission because they use the 4B/5B method of encoding, where 4 bits are transmitted as 5 bits. This represents an efficiency of 80 percent. (Conversely, 100 Mbps Ethernet requires a 200 MBd medium. Two cycles of clock per bit time [Manchester encoding] make for a 0 and 1 cycle of clock for a 1, which results in 50 percent efficiency of media capacity.) Encoding 4-bit (hex) groups using 5 bits is such that—for data—no consecutive combinations of 5-bit patterns will have more than three consecutive 0s.

When introduced in the 1980s, FDDI was intended to be a campus-wide (and metropolitan area) networking technology but it was made obsolete by subsequent technologies, such as ATM and SONET, that offer equivalent reliability and even greater speeds and distances.

Metropolitan Area Network

The metropolitan area network (MAN) discussed here is described in IEEE 802.6 (there are other MANs, but this is the only one we will describe). Installed in the United States (usually in government in centers Washington D.C.; Tallahassee, Florida; and Sacramento, California), MANs generally use ATM and the Distributed Queue Dual Bus (DQDB) protocol. MANs are currently operated at 155 Mbps, with accommodation for Switched Multi-megabit Data Service (SMDS) with speeds up to 600 Mbps when SMDS is standardized. The typical industrial user will probably not be concerned with a MAN, and if he or she were, it would be through a DSU-like device and at the data rate of the LAN (or near it).

Asynchronous Transfer Mode

Asynchronous transfer mode (ATM), also known as cell relay, is primarily a fiber optic transmission system and is highly suitable for that medium, as the small transmission packet, called a *cell*, is Layer 2 only (no routing, only switching), which requires it to use hardware switching and error detection. Data is transmitted in 53-octet packets (cells), with data taking up 48 octets while the remaining 5 octets are used for header data. ATM was originally scheduled to have a 155-Mbps data rate, although there are now different rates to suit differing applications, from 25 Mbps (ATM25) through 2.3 Gbps. When it was introduced, ATM quickly became the technology of choice for carriers because it handles voice, video, and data over WANs or LANs with a specified quality of service (QOS). ATM switches can be configured in a range of topologies, including trees, stars, meshes, and rings, and can support automatic re-routing around failed nodes and circuits.

As with frame relay, the service provider establishes permanent virtual circuits through their system to connect various locations into a point-to-point or multi-point configuration with a guaranteed continuous data rate and delivery latency. (Switched virtual circuits are also offered but are less common.) ATM technology is perfect for high-bandwidth applications, such as multi-site video conferencing. There are five basic service contracts offered on ATM networks:

- Peak Cell Rate (PCR)
- Sustainable Cell Rate (SCR)
- Maximum Burst Size (MBS)
- Cell Delay Variation Tolerance (CVDT)
- Minimum Cell Rate (MCR)

Because routes (virtual circuits and virtual paths) are pre-configured, routing cells (which are all fixed size and format) through the ATM network is very fast, with low latency.

For LAN users, ATM LAN emulation (ATM adaption Layer 5 [AAL5]) allows the ATM connection to appear to be a token ring or Ethernet connection. The token ring or Ethernet packet is then encapsulated into the cell structure of ATM. When it appears at the other end of the connection, it is appropriately reassembled into a token ring or Ethernet packet.

ATM is more of a technology for telecommunication carriers and service providers than a LAN technology; however, now that 40 GbE (Gigabit Ethernet) and 100 GbE are standardized (July 2010), carrier companies have displayed a great interest in moving services toward Ethernet. It was widely speculated (1994) that ATM would be the network of the future, particularly where deterministic services are required, yet this was when 10 Mbps Ethernet was standard. 100 Mbps Ethernet was two to four times as fast as Desktop ATM and was only a fraction of the cost. Even 1 GbE with duplex switches is deterministic and has a cost structure lower than ATM. While many network topologies and technologies were crowned successors to TCP/IP over Ethernet, 10 GbE is standard and here now. Moreover, 40 GbE and 100 GbE are standards (July 2010) and, along with IPv6, will provide a standard for QOS. Although ATM is the primary technology used in B-ISDN, SMDS, and MANs and is offered by other carriers providing WAN services, 10 GbE over fiber optic lines is becoming a serious contender in the long-haul (point-to-point) business. Of course, ATM supports topology and redundancy features that were not available with the original Ethernet, but many Ethernet vendors now offer redundant and constant availability schemes (particularly in industrial Ethernet applications). While Ethernet itself is not routable being a Layer 1 and 2 technology, using a routing protocol (Layer 3, such as IP of TCP/IP) greatly increases the ability to use mesh and other high availability networking schemes. Although ATM rapidly displaced FDDI as a MAN technology, the next advance (SONET) has largely displaced ATM as a base technology for the telephone system and Internet backbone. You may look for the higher speed Ethernet (10 GbE, 40 GbE, and 100 GbE) to do the same.

Synchronous Optical Network

A Synchronous Optical Network (SONET) network can be configured in differing topologies (such as a tree or a star), but it is most often used as a dual counter-rotating ring topology, similar to FDDI as shown in figure 7-15.

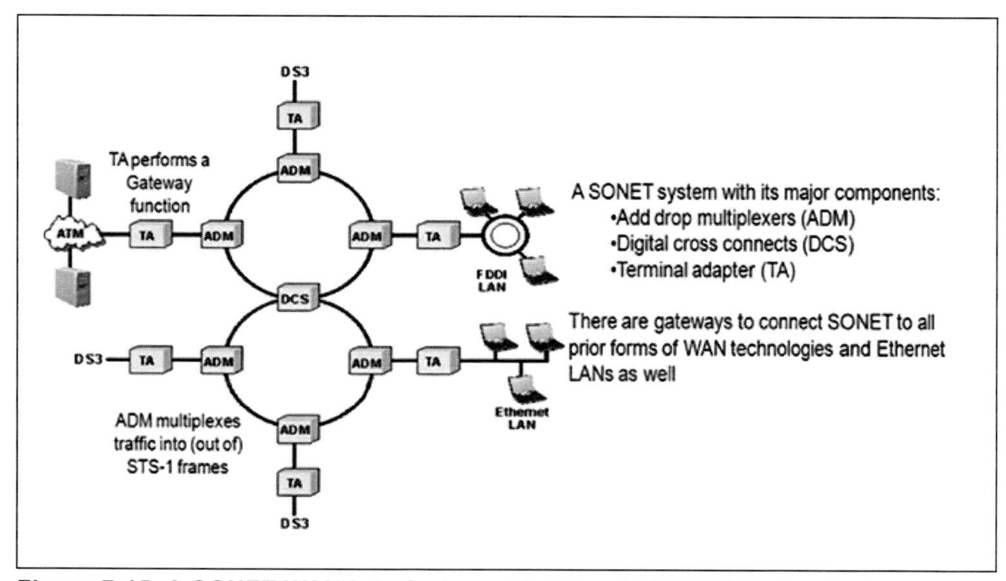

Figure 7-15. A SONET WAN Interfacing with Other LAN/WAN Technologies

SONET will encapsulate into its payload whatever data appears at the interface. SONET has four layers that are roughly analogous to the OSI model. Its minimum data rate is 51.84 Mbps and its maximum (at the moment) is 39.813 Gbps. Table 7-2 outlines the data rates available (all rounded off except for OC-1).

Table 7-2. Optical Data Rates

Optical Carrier (OC) Level	Transmission Rate Mbps
OC-1	51.84
OC-3	155
OC-9	467
OC-12	622
OC-18	933
OC-24	1,244
OC-36	1,866
OC-48	2,488
OC-192	10,000
OC-256	13,271
OC-768	39,813

SONET networks have been proposed as a *fiber-to-the-curb* infrastructure in which copper will carry the transmission into individual residences—something like having a central office in every neighborhood. Although it is nice to know that these technologies exist and are being pursued by carrier companies, for those in the industrial field, that information is all that is needed since almost any and all expansion beyond what a carrier company now uses is "proposed."

In situations where a WAN is needed and no infrastructure exists (typically SCADA applications—pipelines, transportation systems, electric transmission, etc.), it is not uncommon to see organizations installing fiber optic cables and using SONET technology to build a private LAN using WAN technology. In many such instances, the organization may have spare fibers and bandwidth and might sell or lease this to a telecommunication provider. Pipeline operators often bury fiber optic cables in the trench alongside the pipe. Electric utilities have strung fiber optic cables on their conductors. SONET turns that unused fiber (*dark fiber*) into a network trunk or network connection.

Digital Subscriber Line

Using the same copper pairs that bring the venerable voice-grade wireline to your residence or business, digital subscriber line technology, xDSL, has gained wide acceptance. It is considered a broadband technology because you can have concurrent telephone, video, and data services on the line. Unlike the digital connection services described previously (e.g., DDS, T1, frame relay), DSL does not specifically connect a user to the telephone system network, except for the existing analog phone service; it connects the user's home computer devices to the Internet and the phone company is the Internet service provider (ISP). *It should be noted that DSL, as offered, is not just a connection medium and it must be considered as a connection to the Internet.* ISP services are included in the price and connectivity is completed with one or more IP. When considering DSL, the user should, of course, ascertain if virtual private network (VPN) or other security techniques are used.

There are many varieties of DSL. Each requires that various modifications be made to the central office wiring and perhaps to the wiring at the user's premises, such as eliminating impedance-matching devices that were designed for voice frequencies, leaving just the bare copper.

Among the many kinds of DSL are these:

- ADSL – asymmetric DSL (what most of us have)
- HDSL – high-bit-rate DSL
- RADSL – rate-adaptive DSL
- SDSL – symmetric DSL
- VDSL – very-high-bit-rate DSL

Here, we will only discuss the most often encountered type: ADSL. In ADSL, the upload data rate is generally less than the download data rate. Depending on which type of ADSL is in use, the upload data rates may vary from 64 Kbps to 1.5 Mbps and the download data rates from 500 Kbps to 6 Mbps. These data rates are tariffed quite reasonably, with an ISP providing residential service with a cost of about $15—$24 a month (circa May 2014). This tariff will typically allow the user one IP address. Businesses, on the other hand, are generally given a more symmetrical rate (i.e., upload and download data rates are closer together) and several IP addresses for a cost of about $90 (circa May

2014) a month. Either service includes the capability of carrying on a telephone conversation at the same time that data is being transmitted or received on the DSL lines. This makes an attractive package: trading business telephone line costs for not only a business telephone line but also high-speed Internet and ISP access for the same rate.

Digital subscriber lines are not totally digital, in that they use modulation of sorts, either CAP (which uses encoded trellis modulation as previously discussed) or DMT (where 8 or more individual tones are phase modulated and—depending on the scheme and number of channels—may be encompassed in an overall modulation of a carrier). One DMT scheme has the data divided into 256 channels, each with 4 KHz capacity. Each data segment is assigned a unique ID and is spaced all over the allocated bandwidth. At the receive end, the segments are reassembled by ID and the packet is passed upward.

At some point in the near future, everyone could have a DSL to their business, to their residence, and so on. There will be challenges along the way, however, because DSL supplied by the telephone operating companies uses the telcos' installed copper voice pairs. The data rate varies inversely with the distance from the central office, and once the subscriber is past 18,000 ft. or so, DSL doesn't work well at all without the use of repeaters. If you aren't near a central office that supports DSL, you may obtain DSL-like speeds over radio links or even at a quite reasonable cost by up/down link satellite.

Cable Modems

Another Internet connectivity medium, although not deployed in a large number of industries, is the cable modem. The modem typically provides a 512 Kbps to 6 Mbps data rate to customers when the cable segment is not loaded; data rates drop as loading goes up because all customers are using the same shared media. As with the telcos, cable providers offer different rates and tariffs depending on the speed and service required.

Cable modems have brought one key benefit: they gave telcos the impetus to push for wider implementation of DSL. Unfortunately, neither cable modems nor DSL do rural and far-suburban customers much good. If cable TV is not available in a customer's area, then cable modems for Internet access are off the table as an option. These users are, at present, either stuck with 33.6 Kbps analog modems and dial-up telephone or they must use a satellite system with its somewhat increased costs (over a DSL or cable modem—although satellite link costs have been decreasing steadily). Some hope for the rural user or remote plant site is the adoption of WiMAX technology by cellular phone service providers for 4G support.

In the next section we introduce some solutions for these customers. Several wireless schemes are proposed for remote rural customers but the direct satellite link is already here (typically costing $49 monthly, circa 2014) and, with the exception of occasional weather-related outages, satellite can supply DSL data rates or better.

WANs for Mobile and the Hinterlands

Wireless WAN Technologies

Industrial users might wonder why they need to even consider a wireless WAN. (Not, of course, industrial users, whose processes require the application of SCADA technologies. They probably already know why wireless WAN technologies are useful, even if employed in a LAN, if not actually necessary.) The main reason industrial users with a single large facility would consider wireless is that nearly half the cost of any large networked installation, both initial and life-cycle costs, is in the wiring. The benefits of wireless can be great, although whether this is truly a wireless WAN or merely wireless LAN technology may be up for debate and depends on plant/facility size.

In the plant area, the technology developed for wireless may filter down to the LAN applications. Having unbound nodes, such as roving operator or maintenance workstations, allows a great deal of flexibility but also comes with a host of problems: walls and process units that are made of materials that are either highly absorbent or reflective of radio waves (especially at certain wavelengths or frequencies); large amounts of electrical noise; and many points for multi-path reception and the resulting interference problems. Yet spread-spectrum technology, smart antenna technology, and smart radio technology have overcome many of these problems. (See our discussion of basic wireless concepts in chapter 3.) Wireless WAN technologies can be useful for establishing communications throughout a geographically-large plant or industrial facility, for linking geographically separated facilities (possibly into a corporate/enterprise WAN), and for establishing communications where wires may be impractical (e.g., off-shore, or across a canyon or river).

Worldwide Interoperability for Microwave Access

Worldwide Interoperability for Microwave Access (WiMAX) is a wireless communications standard (IEEE 802.16e-2005 [with several subsequent revisions]) designed to provide 30 to 40 Mbps data rates to both fixed and mobile users with the 2011 update providing up to 1 Gbps for fixed clients or stations. The name *WiMAX* was created by the WiMAX Forum, which was formed in June 2001 to promote conformity and interoperability of the standard. The forum describes WiMAX as "a standards-based technology enabling the delivery of last mile wireless broadband access as an alternative to cable TV and DSL." WiMAX was originally seen as a means for providing wireless broadband Internet services in rural areas that did not have cable TV or DSL infrastructure due to distances and low housing density. It has been described as "Wi-Fi on steroids," due to its high speed (34 Mbps to 1 Gbps) and extended distance (1 to 50 km; 0.6 to 30 miles) specifications. Like Wi-Fi, the bandwidth will drop as the subscriber's distance increases and signal strength falls off.

WiMAX technology is also becoming the basis for 4G cellular telephone service in the United States. Because of its high bandwidth and low latency, WiMAX is being used to deploy so-called triple play services (phone, Internet, and streaming video) on 4G cellular phone networks. Figure 7-16 shows a simplified diagram of a subscriber location obtaining phone and broadband Internet services via WiMAX (4G) connectivity. WiMAX base stations can provide both service to local mobile and fixed subscribers, as well as make point-to-point inter-station connections in order to provide a high bandwidth trunk channel to connect to the Internet and wired telephone system. Because of their range and bandwidth, a large area can be covered by a few interlinked WiMAX base stations.

From an industrial point of view, WiMAX has been used to create a site-wide wireless cloud (umbrella?) so that suitably equipped mobile phones, laptops/tablets, and LAN gateways (*subscriber stations*) can all connect to a central WiMAX base station and be linked into a private, site-wide wireless WAN. Multiple base stations can be configured to provide both redundancy and alternate pathways around obstacles. WiMAX gateways are special subscriber stations that bridge WiMAX to conventional wired Ethernet and/or Wi-Fi; some may also offer analog telephone jacks and employ VoIP technologies. Some large offshore petroleum production and processing platforms have

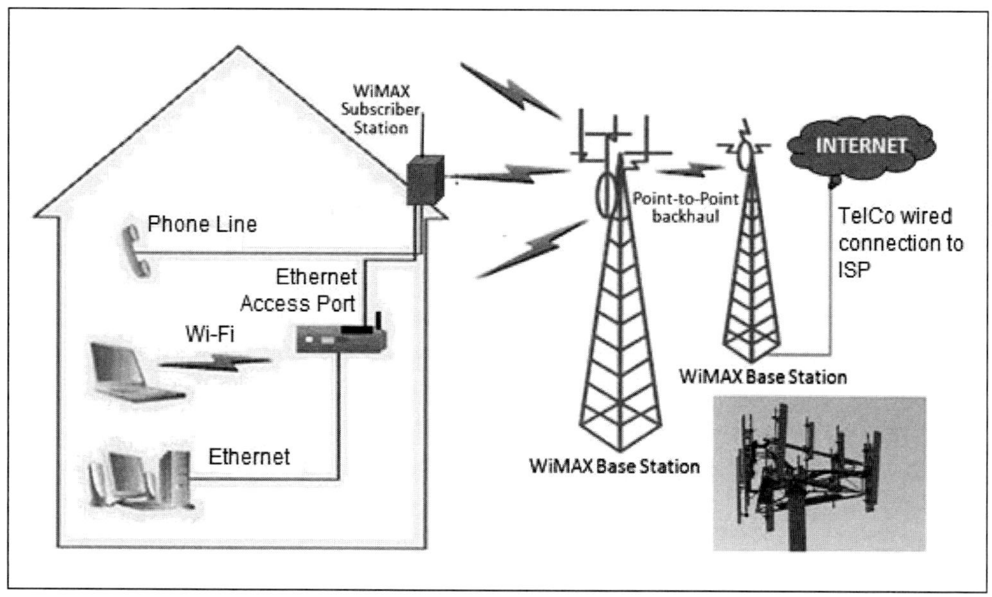

Figure 7-16. WiMAX-Based Internet/Phone Service

used WiMAX to create a wireless WAN for the main platform, as well as for nearby satellite platforms.

Wireless Mesh Networks

It is often handy to have wired Ethernet available around a large plant or industrial facility but it can be messy and even dangerous to run cable and install a wired Ethernet infrastructure. One alternative to wired Ethernet is to have Wi-Fi access points installed. WiMAX, as just discussed, is an option, however not all devices that can use wireless Ethernet also support WiMAX and adding that capability can be expensive or impossible. Another option is to create a wireless mesh network that allows Wi-Fi access points (APs) to be placed anywhere desired (as long as power is available) and to use a second radio in the access point to establish point-to-point links to other access points. This would enable message store and forwarding (routing) and eventual delivery to an AP connected to the wired plant LAN infrastructure. Companies, such as Aruba, Cisco, Tropos (ABB), and Sonos, manufacture wireless mesh routers (repeaters) and provide software to make them function in an optimal manner while providing overall network management; various levels of performance and latency guarantees; and cybersecurity functionality. Figure 7-17 shows a highly simplified conceptual diagram of a wireless Ethernet mesh network.

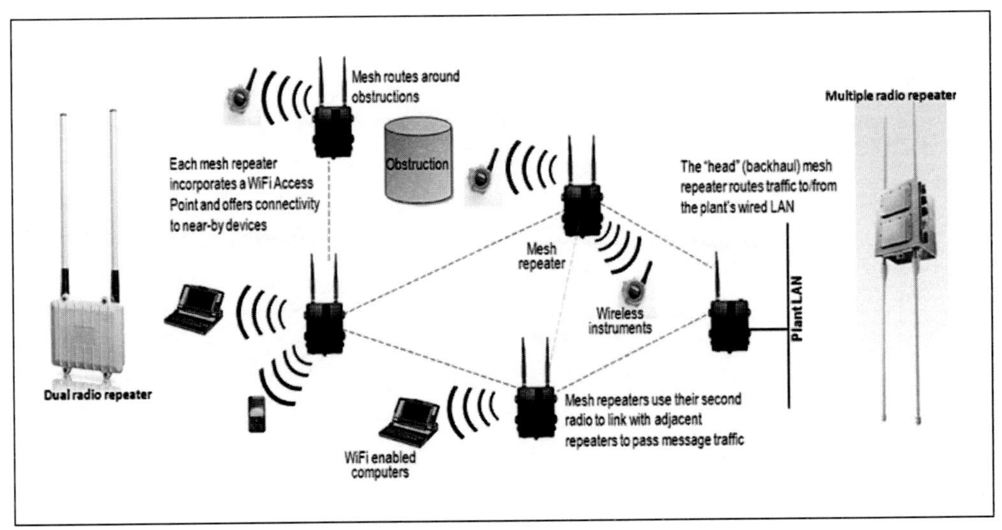

Figure 7-17. Wireless Ethernet Mesh Network

Great advancements have developed in wireless Ethernet networking over the last few years. Higher data rates have enabled wireless video monitoring and VoIP applications (which have been too costly in the past). Increased output power on the radio transceiver and outdoor mounting options have expanded the communication distances. Mesh technology provides wireless communication with multiple pathways that can address line-of-sight obstructions and single points of failure, while providing built-in redundancy and self-healing for network reliability. Wireless Ethernet mesh networks are built with wireless mesh routers that use multiple radios (one to provide AP services to local clients and one or more for inter-router message passing) and an integral computer that handles routing, message storage, and path failure recovery. Unlike using mesh networks to blanket a college campus or urban center, the typical industrial communications network is not as concerned with covering large areas with Internet access, as it is with the movement of critical data for control and system monitoring. The redundant/alternate mesh pathways can be automatically used as a backup if the primary communication path is lost due to router failure. The self-healing mesh network can be used to guarantee communication integrity between industrial controllers, smart devices, and systems; and the network setup can usually define primary (preferred) and secondary (alternate) paths between communicating nodes.

SCADA Applications

Wireless networks have historically been a perfect solution for supervisory control and data acquisition (SCADA) systems. One of the largest changes coming to the classic SCADA network is the requirement for higher security at remote locations. Remote video is an effective means of increasing security without wasting valuable manpower. Many of the security cameras are now Ethernet-based and most industrial controllers are now coming equipped with an Ethernet interface, therefore a wireless Ethernet solution that can support both the camera and controller on a single network is ideal. One requirement for remote video is increased bandwidth over the typically lower data rates required for the industrial controller. The distances from the master location to the remote sites are often significant when monitoring a large geographically dispersed area, so minimal latency (e.g., lowest number of hops) is necessary in order to make these long-range Ethernet SCADA networks function efficiently.

There are differences of both function and purpose between a business IT deployment and one used in industrial networks. This is true between IT mesh networking technology and industrial mesh networks in that most industrial communication networks support a much smaller overall number of concurrent client devices and a far greater percentage of fixed-location clients. Providing delivery reliability with minimal latency is the primary objective for the industrial mesh network. With the use of high-gain directional antennas, it is possible to make LOS links between routers that could be several miles apart, which means that a wireless mesh network could cover a large area or extend out to significant distances.

Network security in an industrial mesh network must be a priority today and the WPA2/IEEE 802.11i wireless security standards discussed in chapter 9 of this book are essential for such networks.

Using mesh routers/repeaters in an industrial environment requires that the hardware be packaged appropriately and be capable of dealing with the potential presence of corrosive gases and/or liquids, high levels of electromagnetic fields (EMFs)/radio frequency interference (RFI), and possibly even explosion hazards.

Digital Microwave

Microwave frequencies range from 30 GHz to 300 GHz, corresponding to wavelengths of 1.0 cm to 0.1 cm. These frequencies are useful for terrestrial and satellite communication systems, both fixed and mobile. In the case of point-to-point radio links, antennas are placed on a tower or another tall structure at sufficient height to provide a direct, unobstructed LOS path between the transmitter and receiver sites.

An LOS microwave is used for both short- and long-haul telecommunications to complement wired media, such as optical transmission systems. Applications include local loop, cellular backhaul, remote and rugged areas, utility companies, and private telco carriers. Early applications of an LOS microwave were based on analog modulation techniques but today's microwave systems use digital modulation schemes, where trellis modulation is used for increased capacity and performance.

LOS links are limited in distance by the curvature of the earth, obstacles along the path, and free-space loss. Average distances for conservatively designed LOS links are 42 to 50 km (25 to 30 miles), although distances of up to 167 km (100 miles) have been used. Microwave transceivers enable full-duplex, high-bandwidth digital communications from location to location, usually bridging a high-speed Ethernet LAN at each location as shown in figure 7-18.

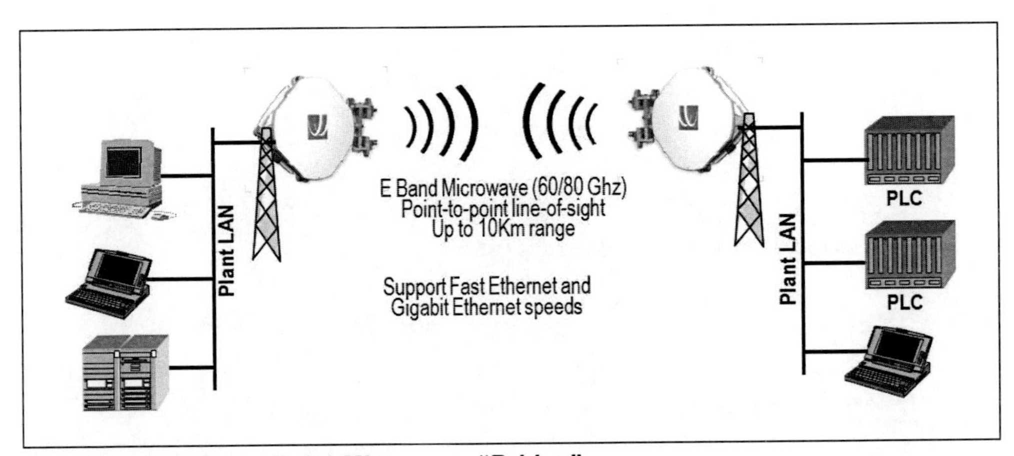

Figure 7-18. Point-to-Point Microwave "Bridge"

Digital microwave repeaters provide a means for creating a wireless LAN where no other infrastructure exists. Prior to the invention of fiber optics, it had been common to use a series of microwave repeaters to provide telecommunication services and SCADA polling connectivity in remote areas where no other telecommunication infrastructure existed (e.g., along a long-haul pipeline or electric power transmission line). Some large industrial facilities have used point-to-point microwave repeaters to link disparate DCS systems into a facility-wide system. Many offshore facilities use point-to-point microwave links to connect to both on-shore facilities and adjacent offshore facilities.

Satellite

While both 4G cellular and WiMAX continue to make inroads into far suburbia and rural areas, the primary option for rural areas that want high-speed communications access is still going to be satellite service. Corporations can lease transponder bandwidth on commercial satellites for point-to-point and point-to-multipoint networking. By placing private ground stations at each location, the satellite operator can provide differing levels of bandwidth guarantees and throughput. Such a scheme might be used to link multiple isolated locations, such as off-shore production platforms or pipeline pump stations, into a corporate network.

Residential and commercial users can contract with providers that offer low to moderate-speed, asymmetric uplink (transmitting) and downlink (receiving) satellite service, providing a bi-directional connection to the Internet. It is often the only high speed bi-directional option for many rural subscribers. (Earlier generation service providers used satellites for downlink communications, but used an analog phone line for uplink communications. This has mostly disappeared with the latest generation of transponders and ground station equipment.) Service providers include: Exede, Dish Network, DirecTV (HughesNet), and WildBlue—all of which are satellite-delivered Internet and TV service providers. Again, remember that this is an Internet connection, which may be acceptable as long as appropriate cybersecurity mechanisms (e.g., VPN technologies) are used.

As mentioned above, satellite service usually provides one data rate for uplink (transmission) and another for downlink (reception). Those rates range from 0.25 to 3 Mbps for uplink and 1.5 to 12 Mbps for downlink. There will usually be restrictive time-based bandwidth allowances (x GB per month), so that each user gets their fair share according to their plan. When a user exceeds their allowance, the provider may slow down their access, de-prioritize their traffic, or charge additional fees for the excess bandwidth used.

The majority of satellites used to provide communication services are in geosynchronous orbit about 37,000 km (22,236 miles) above the equator. Even at the speed of light, radio waves take about 1/4 second to make the trip up and down and double that time (1/2 second) for a response. This makes it almost impossible to use satellite links for interactive telephony or video.

Mobile Telephony

If there is to be a publicly available, high-speed data service (more than 56 Kbps) to outlying areas in the future, it will most likely be provided by satellite or in the form of cellular telephony services. Note that some of the end points when using mobile services are confined to point-to-point, others connect to the Internet. Routing would depend on the service or services desired.

One form of wireless is the Universal Mobile Telecommunications Service (UMTS), which is a mobile technology that has a data rate of up to 2 Mbps. Then, there are the other mobile technologies, such as the General Packet Radio System (GPRS), which offers a data rate of 56 to 114 Kbps, and the Enhanced Data GSM Evolution (EDGE), which provides a data rate of 384 Kbps. At one time, the most promising wireless technology for fixed (not mobile) customers was the Local Multipoint Distribution System (LMDS), which had a data rate of 155 Mbps. It used technology similar to cell phones, in that each transmitter covers a *cell*, typically 3 to 5 km (5 to 8.3 miles) in diameter, but the client is fixed and does not move. The technology is now used by IEEE 802.16 (fixed) and 16e (mobile) under the name *WiMAX*, which was discussed above. This is certainly a promising technology for rural and perhaps urban customers. Table 7-3 illustrates some of the typical data rates for mobile and wireless services currently available.

Table 7-3. Selected Wireless Data Rates by Service

Name	Medium	Speed
Plain old telephone service (POTS)	Unshielded Twisted Pair (UTP)	3 KHz; up to 56 Kbps
Mobile Telephone (GSM)	Radiated (RF)	9.6 to 14.4 Kbps
General Packet Switched Radio (GPRS)	Radiated (RF)	56 Kbps to 114 Kbps
Dish, DirecTV, Excede	Satellite RF	12 Mbps download (max), 3 Mbps upload (max)
4G	Radiated (RF)	100 Mbps (1 GBs proposed)
WiMAX	Radiated (RF)	128Mbps download, 56 Mbps upload

Summary

Our brief discussion of WANs has hopefully familiarized you with some of the services and terminologies. A variety of WANs are in service today—many of them packet-switched digital networks that use different protocols. Although some may have industrial applications, many are better suited to the data traffic required for database transfer and point-of-sale (POS) terminals. The data rates that are cost-effective today are relatively low compared to a LAN, although system evolution will change that. There are ways to make standard long-distance (more than 50 m; 165 ft.) media accept higher data rates and several standard networks do so. Packet switching is a way to take a large continuous stream of binary data (messages), break it up into smaller frames or packets, and transmit these packets through the public switched network at whatever speed the connection will support.

X.25 is the primary standard that covers packet switching for open systems, but it is slowly losing out to connectionless frame relay. Connectionless frame relay operates at much greater speeds and efficiency than X.25, provided that errors and congestion do not exceed the parameters for reception of reliable data.

ISDN is a concept that could have combined the digital phone along with digital data, however, its window of opportunity has passed. Now the competition is between ATM, SONET, DSL, and 1/10/40 GbE Ethernet.

Although little in this chapter may appear to be directly related to process control, it certainly is related to industrial data communications, in that all of the concepts and equipment now used in industry began use in general data communications. Many of the concepts presented here are already working themselves into the industrial area and, with the advent of corporate intranets spread over a large geographical area, adoption of WAN technologies in the process control and automation industries seems a foregone conclusion.

Bibliography

Note that Internet links were current at the time of writing but may change.

About.com. "Wireless/Networking. DSL vs. Cable Modem." http://www.compnetworking.about.com/od/dslvscablemodem/a/dslc

"A tutorial on Fourier Analysis – Fourier Series." http://www.gaussianwaves.com/2013/05/a-tutorial-on-fourier-analysis-fourier-series

"DSL." http://www.technical.philex.com/networks/sharing/dsl.htm (page no longer available)

Feibel, W. *Encyclopedia of Networking*. 2nd ed. San Francisco: The Network Press, 1996.

Fortune. "Broadband." http://www.fortune.com/fortune/sections/broadband (page no longer available)

Langley, G. *Telephony's Dictionary.* 2nd ed. Chicago: Telephony Publishing Co., 1986.

Shannon, Claude. "A Mathematical Theory of Communication." *Bell System Technical Journal* (1948).

Stallings, W. *ISDN: An Introduction.* New York: Macmillan Inc., 1989.

Whatis.com. "The speed of . . ." http://www.whatis.com/thespeed.htm

8 Internetworking

Internetworking—the seamless transfer of information throughout an enterprise, even if that enterprise has plants and offices throughout the world—is really the end goal of enterprise communications. There are a myriad ways of internetworking. Most internetworking deployments are generally based on established standards, or standards to be, and that is the rationale for this chapter. This chapter reviews some previously discussed information to ensure that key concepts are understood.

Layer 2: Internetworking Equipment

A Layer 2 (Data Link layer) device reads the data link information and uses it to perform some action. A bridge is a Layer 2 device and so is a switching hub. Both devices read the packet's Layer 2 destination address (media access control address [MAC], the physical address). A bridge or a switch, depending on its feature set, can identify the protocol by reading the type/length field and determine the 802.2 packet information or any other protocol information available for the purpose of filtering. Figure 8-1 illustrates a Layer 2 device.

Before continuing, the definition of some of the networking devices (or a review, depending on your point of view) is necessary.

Switch Definition

Recall that a switch reads the Layer 2 address (MAC address) and then uses an electronic switch to move the information to the addressed location. The

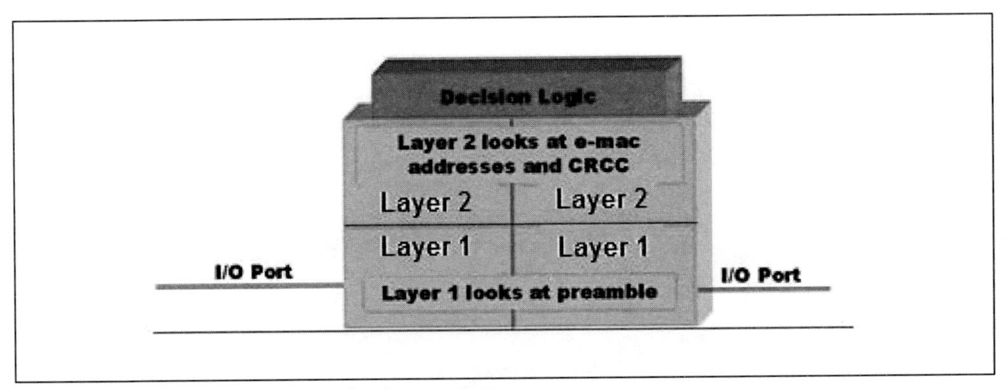

Figure 8-1. Block Diagram of a Layer 2 Device

switch consists of a processor, or processors, and an electronic crossbar that replaces the collapsed bus in a hub; this electronic crossbar is the switching fabric. The processor checks the address and uses the switching fabric (the electronic crossbar) to transfer the frame to the correct port. Modern Ethernet switches can auto-detect speeds and perform conversions for 10 Mbps, 100 Mbps, and 1000 Mbps Ethernet.

Bridge Definition

A bridge is a device that contains two sets of Layer 1 and 2 functions and connects network segments of the same type. It reads the Layer 2 addresses (source and destination MAC addresses) and transmits or receives the packet. If the bridge is transmitting, it develops a cyclic redundant check character (CRCC), also called a frame check sequence (FCS), and appends it to the transmitted packet payload. If the bridge is receiving, it computes a CRCC on the received packet and compares it to the transmitted CRCC. If there is no match, the next step depends on the protocol being used, the packet may be dropped or it may go through an automatic retransmission query (ARQ) procedure. The bridge uses an internal table, called a *bridge table*, to determine which side of the bridge the devices reside on. (Note that a switch may be considered a multi-port bridge that performs all of the same actions but typically for three or more ports.)

Regardless of the type, all bridges perform the following functions:

- Forwarding
- Blocking
- Filtering

Forwarding is when a packet is moved from the source port to the destination port(s). Blocking is when a packet from a particular source port is prevented from moving through a destination port or ports. If the destination device is on one network segment and the originating device is on another network segment, the packet is forwarded by the bridge to the segment with the receiving device. If the originating device and the destination are on the same network segment, the packet is blocked from traversing the bridge. Packets are filtered by the packet address using the source and destination IP to determine the path through the bridge connections.

Another function of a bridge is to limit collisions to specific domains. Two methods are used to reduce the number of collisions on a many-node shared network segment: (1) increasing the data speed (e.g., upgrade from 10 Mbps to 100 Mbps) or (2) using a bridge to break up the collision domain into smaller domains. Collisions are a Physical layer action; they do not have a MAC address (a Layer 2 function), hence they are not passed through the Data Link (Layer 2) and are not passed on by a bridge or a switch. The collision domain includes all the computers on the same physical network segment. By breaking up collisions into smaller domains, the collisions will be contained to only those devices in the smaller domains, therefore, limiting the effect a collision can have on performance. When using a half-duplex switch, the collision domain is limited to the switch port and the attached device; if the switch is duplex, there cannot be any collisions and there is no collision domain.

Types of Bridges

Several different devices are called *bridges*; some actually are bridges, and some are not. In general, bridges are classified as:

- Static
- Learning
- Transparent
- Translation
- Source route

Static Bridge

Static bridges, popular from the mid1970s to the mid1980s, are no longer in use, however it is helpful to understand how they worked. When using a static bridge (see figure 8-2), the system administrator had to physically enter the physical (MAC) addresses for each workstation into the bridge table. The table was a database containing each device's connectivity to each port (which may be affected by access control lists [ACLs]). The purpose of the table was to ensure that the bridge would *forward* packets across the bridge when necessary (e.g., in figure 8-2 workstation 06-fe-01-35-c2-ce sends to workstation 06-fe-01-c3-24-78) and *block packets* that did not need to be forwarded (e.g., in figure 8-2 workstation 06-fe-01-c3-24-78 sends to workstation 06-fe-01-10-00-f4). Physically entering the workstation MAC addresses into the bridge table reduced collisions across the whole network, leaving each network segment as a collision domain.

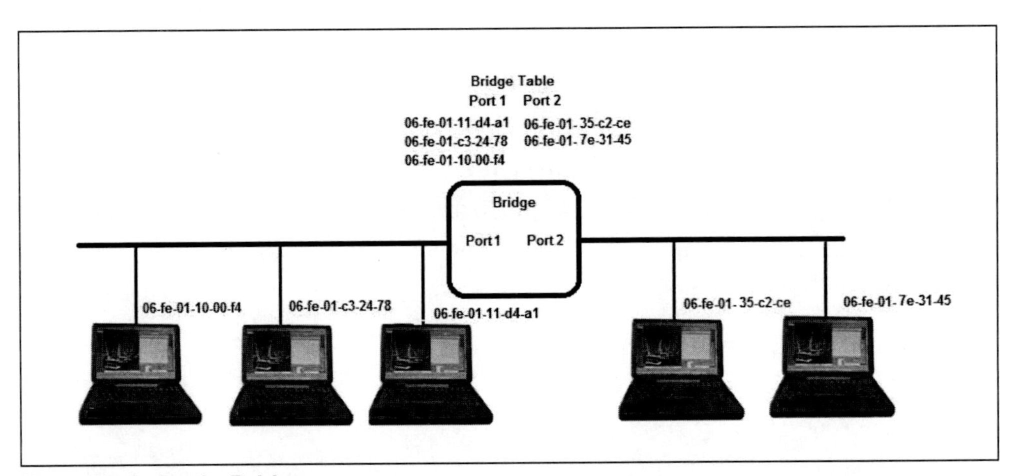

Figure 8-2. Static Bridge

To summarize, *a static bridge has several problems:* It is not dynamic in that changes are not reflected unless the administrator manually enters them. In a large network, manually computing a bridge table is time-consuming and always replete with errors.

Learning Bridge

The learning bridge is the contemporary model bridge. When first activated, the bridge listens to the network talking. Since the source address is part of the Layer 2 frame, the bridge quickly builds its own bridge table. If the administrator moves a workstation from one side of the bridge to the other, the bridge will soon determine which side the workstation is now on and forward or block accordingly. When it is first activated, or a change is made to the network, the bridge takes 30 to 60 seconds to determine the bridge table. During that time period the table might be incorrect: for example, a workstation that went offline may still be listed (not declared inactive) or a new workstation may not be included in the table.

Learning bridges are dynamic, in that changes (workstations coming on or off of the network segment) are identified and reflected within a short time period. They require no intervention on the administrator's part. All modern bridges are learning bridges. On systems running the Transmission Control Protocol/Internet Protocol (TCP/IP) suite of protocols, the Address Resolution Protocol (ARP) can be used to populate the bridge tables.

Transparent Bridge

Transparent bridging is the easiest from the user's point of view as there are no computations other than the spanning tree algorithm for the user to implement. In many cases, transparent bridging only applies to Ethernet and there are a host of other network protocols, particularly in industrial networking, that must be interconnected on a Layer 1/Layer 2 basis.

Translating Bridge

A translating bridge passes (or bridges) packets between two different networks, such as 802.3 (Ethernet) and 802.5 (token ring). It is actually a data link gateway and not a bridge per se. A translating bridge must change between big endian (802.5) and little endian (802.3) transmission, different packet sizes (17K octets versus 1,518 octets), and different bridging techniques (source routing versus spanning tree). Although this may be accomplished using Layer 2 information (that is why it is a bridge), a translating bridge functions much more as a protocol converter. The 802.5 network also uses source routing (Ethernet uses transparent), a different concept than transparent bridging, which requires translation between the two bridging techniques: transparent and source routing.

Source Route Bridge

Used in token ring networks, a source route bridge places the Layer 2 packet routing and loop detection on the client, rather than on the bridge. In source routing, the station transmits a discovery packet that goes to all routes. The first returning packet, which has all the hops and bridge numbers it used along the way, determines that packet's routing for the rest of the message. Source routing is not as efficient as the spanning tree algorithm because it depends on the client, rather than the bridge. Token ring networks can use the spanning tree algorithm instead of source routing, provided that all of the token ring network bridges/switches can also do so.

Remote Bridges

Many bridges, called *remote bridges*, have provision for wide area connectivity and enable bridging across the wide area network (WAN). One concern with remote bridging is its lack of bandwidth in most WAN applications. Even a T1 line may be a bottleneck on a 10 Mbps LAN, let alone a 28.8 Kbps connection. A bridge is not the answer to crossing a WAN; a Layer 3 device should be considered.

Switches as Bridges

A switch reads the data link addresses, just as a bridge does, and determines which port to connect. A switch should be considered to be a multi-port bridge. For both switches and bridges, there are some considerations that must be taken into account with complex connections. For example, a multiple path for packets at the Data Link layer is not allowed. Higher-end (higher-priced) switches have provisions for bridging, as well as switching, and contain the spanning tree algorithm. However, not all switches have equal features. In any case, it is essential that good wiring practices be followed when nesting (daisy chaining) switches, as opposed to stacking them. (When switches are stacked, external connections—not port-connections—are used to connect one switch to another, making the switch an expansion and not a hierarchy, as is the case when you connect the port of one switch to the port of another switch.) According to EIA-568, connecting two ports should require no more than three levels of switches.

Filtering

A bridge operates in "promiscuous" mode. In other words, it reads every packet on each network segment, not just the ones addressed to it. Filtering means the bridge matches a packet to its bridge table. This produces one of three results: the match is on the same port, the match is on a different port, or there is no match. Let us examine each of these results in turn.

Same Port Match

Referring to figure 8-2, note that when the bridge (any type) receives a packet from workstation 06-fe-01-c3-24-78 for delivery to workstation 06-fe-01-10-00-f4, it looks in the bridge table and it finds a match—on the same port. Then the bridge *blocks* the packet, preventing the packet from crossing the bridge. The bridge takes two actions on this packet: it prevents it from travelling across the bridge and it checks the CRCC.

Different Port Match

Referring to figure 8-2, note that if the bridge receives a packet from workstation 06-fe-01-35-c2-ce for delivery to workstation 06-fe-01-c3-24-78, it finds a match in the table—on a different port. The bridge therefore *forwards* the packet across the bridge and it checks the CRCC.

Flooding: No Match Found

Referring to figure 8-3, if the bridge receives a packet with a destination address that cannot be found in the bridge table (e.g., workstation 06-fe-01-01-00-f4 wants to send a packet to workstation 00-11-c1-33-2f-1a), the bridge *floods*—sends the packet to—all of the available ports (1 and 3) except the port where the packet originated (Port 2). A two-port bridge would merely forward to the other port; bridges with three or more ports would flood this packet to all ports in the hopes of obtaining an answer from one of them. This could conceivably cause a "duplicate packet" problem. Ethernet does not allow for duplicate packets, nor do most Layer 2 protocols, so in bridged (or switched) systems, some method must be employed to reduce the possibility of duplicate packets.

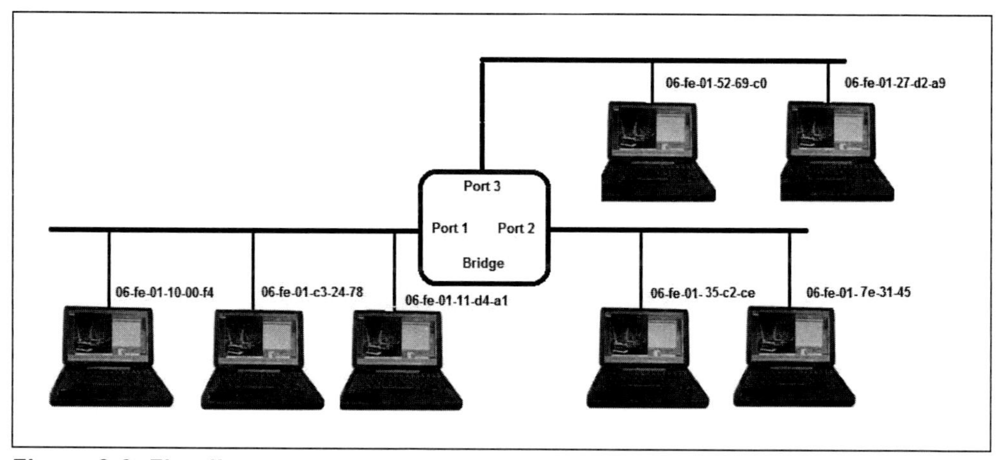

Figure 8-3. Flooding

The *duplicate packet problem* could arise if there are loops in the network. Referring to figure 8-4, packets sent between Workstation A and B can arrive via two different routes. This dual pathway is called a *loop*. Loops are not desirable because Layer 2 has no way to correct for duplicate packets. Particularly on an Ethernet network, *packets must have one, and only one path* between any two devices on the network.

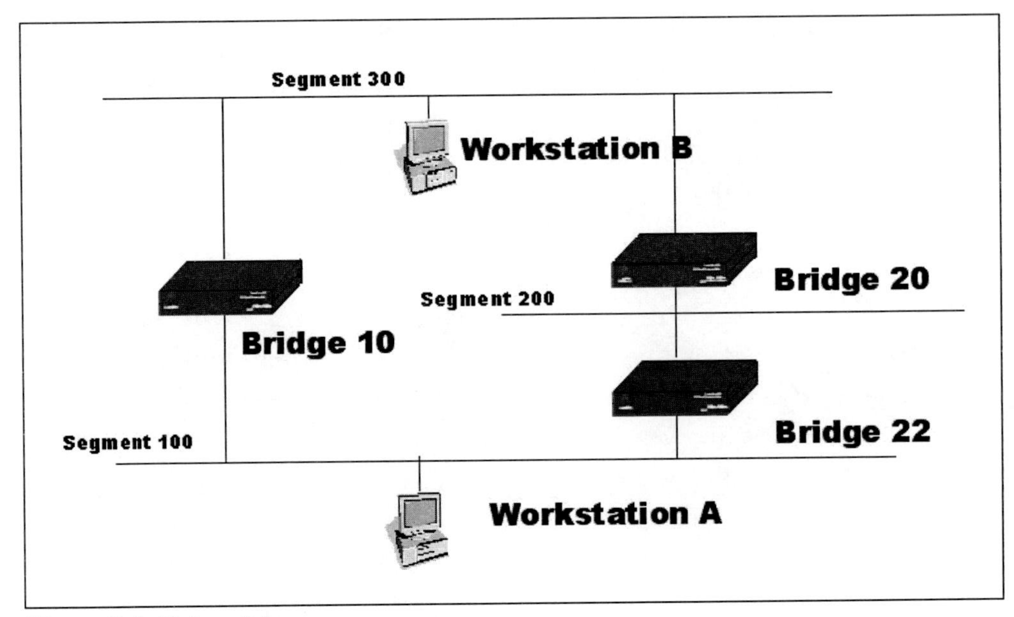

Figure 8-4. Network Loop

The duplicate package problem can be resolved by eliminating network loops. Ethernet in particular uses *transparent bridging*, where "transparent" means that the user or administrator takes no actions whatsoever and the bridge must ensure correct connectivity. It falls solely to the bridge to detect and eliminate network loops—actions taken to eliminate loops are transparent to the user. The method used to accomplish loop elimination in a transparent bridge is the spanning tree algorithm.

Spanning Tree Algorithm

The spanning tree algorithm (STA) is an IEEE 802.1d standardized method that enables bridges to detect and eliminate network loops. Although this adds more traffic to the network, it uses configuration messages between bridges to determine a single path between network segments. It does this by developing a tree structure—one root and many branches; the branches are paths to the network segments (bus systems) or ports (switched systems). The STA determines a root bridge and then opens the loops by blocking certain bridge I/O ports.

The logical tree design of the spanning tree algorithm is depicted in figure 8-5. There is only one path to every location. Bridges are assigned an ID. The algorithm used to create the tree and to locate devices within the tree structure uses this ID. For our present purposes, a configuration message should have three parts:

1. Root ID (the numerically lowest value ID)
2. Number of hops
3. Bridge ID

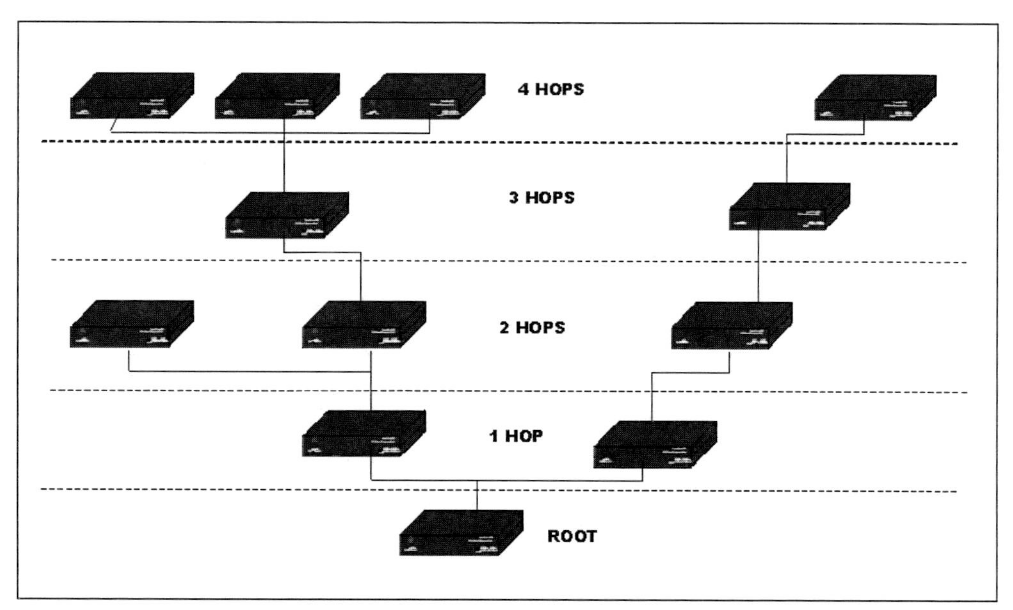

Figure 8-5. Spanning Tree Diagram

As mentioned above, the tree is set up using spanning tree configuration messages. When a bridge is initialized, it assumes that it is the root and broadcasts a configuration message. To express this, we will use a symbolized message (in other words, not the actual message but a representation) that has the format: root ID, number of hops, and the bridge ID. In figure 8-6, the root is bridge 2. The bridge's configuration message is "0002, 00, 0002."

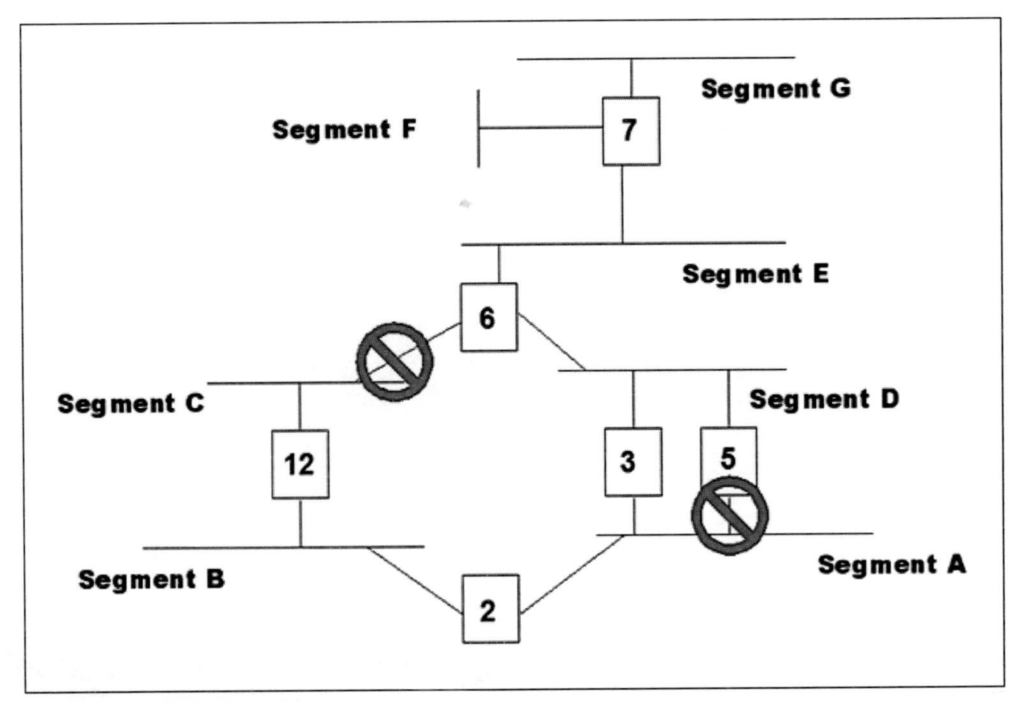

Figure 8-6. STA Example

When bridge 5 comes on line, it transmits the message "0005, 00, 0005," assuming it is the bridge. However, bridge 0002 has a lower ID, so when bridge 0005 hears bridge 0002, it drops its claim as the root and acknowledges 0002 as the root.

Once the root is established, loop detection starts. Both bridges 0003 and 0005 are one hop away from bridge 0002. Therefore, since one Ethernet hop is a cost of 10 (an arbitrary value for comparative costing) and both bridges have one hop from 0002, they each have a hop cost of 10 so their messages would be "0002, 10, 0003" and "0002, 10, 0005." Since both bridge 0003 and 0005 provide a path to segment D, a loop would be formed. In this case, the bridge with the highest ID (0005) would block its highest numbered port between the network segments. (Note that bridge 6 would also block the highest numbered connecting port.) In this way, the spanning tree algorithm automatically prevents loops.

Spanning Tree Summary

It is not necessary that you be able to compute a spanning tree algorithm. However, it is essential that you understand its basic function: to automatically prevent network loops. Crucial to your understanding of bridges, or any

Layer 2 device, are: the fact that the lowest ID is the root and the concept of configuration messages. All modern bridges (and most routers in bridge operation) support the spanning tree algorithm.

Most industrial bridges and switches use a slightly different algorithm called the *rapid spanning tree algorithm*. This algorithm does not need as much time to determine the root and tree upon startup or after loss of the current root.

Layer 3 Devices

Since we are dealing with industrial networks, redundancy is relevant. Network redundancy is common within industrial systems. Redundant industrial Ethernet networks, or any other redundant network dedicated to industrial requirements, are generally vendor-specific implementations. Even with redundancy, only one copy of a packet can arrive at the receiver since Layer 2 has no way of correcting for duplicate packets. For tying together multiple networks, or for organizing an industrial network into sub-networks, it is best to employ a Layer 3 device.

To address Layer 3 devices properly we will discuss the following topics:

- Layer 3 packet information
- IP router actions
- Router protocols
- Advertising
- Routing Information Protocol (RIP)
- Link State protocols
- Open Shortest Path First (OSPF)
- Bridges versus routers
- Multi-protocol routers
- Hierarchical routing
- Interdomain protocols
- VLANS

Now, is all of this essential to an industrial networking person? Absolutely! As we stated at the outset of this book, industrial systems are becoming a part of larger systems. This includes having their packets go other places—not simply remaining within the little islands of automation that they have been restricted to in the past. Layer 3 is, therefore, the heart of internetworking. Layer 3 is where you find the network you want to address and where you find out how to route to the location you want.

Layer 3 Packet Information

In this section we will deal primarily with TCP/IP over Ethernet: the LAN standard. You will recall that the type/length octet determines the protocols to follow. If the decimal value of the two Type/Length octets is fewer than 1500 decimal (the maximum number of information octets in a standard Ethernet frame), it is expected that the 802.2 control octets (2–3) follow. If the two octets are more than 1500 decimal, the frame is a non-802.2 frame and the Type/Length octets identify which protocol will be followed. If the two octets are 2048 decimal (0000 1000 0000 0000 = 0800 hex or 2048 decimal), then the next information will be an IPv4 header and, if TCP is the transport, TCP will be specified somewhere in the IP header; the TCP octets will follow the IP header. In a 20-octet IPv4 header, octets 12–20 will contain the source and destination address. Of course, all of this eats up some of the 1,500 octets reserved for data in this frame.

Layer 3 devices are protocol sensitive. A Layer 3 device must be designed to read (know what each bit represents) the protocol that it is handling. Since these protocols contain the addresses for different machines on different networks, they are essential to routing.

Figure 8-7 illustrates the general principle of routing.

Figure 8-8 illustrates an example of the industrial use of a router.

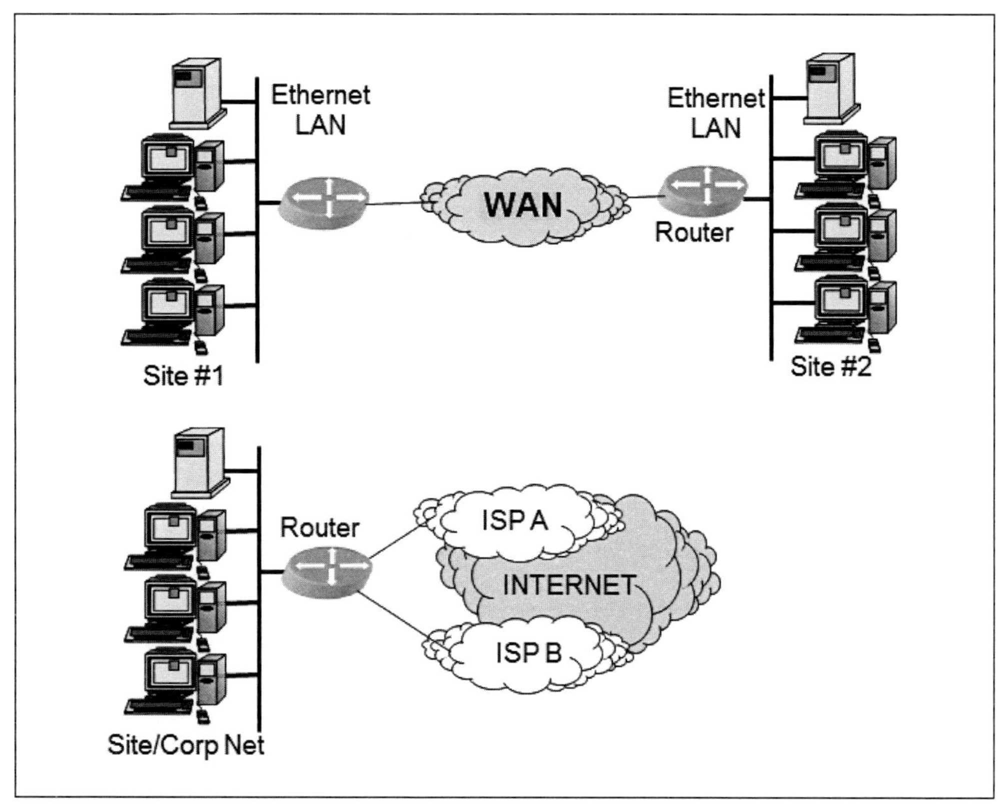

Figure 8-7. Routing in General

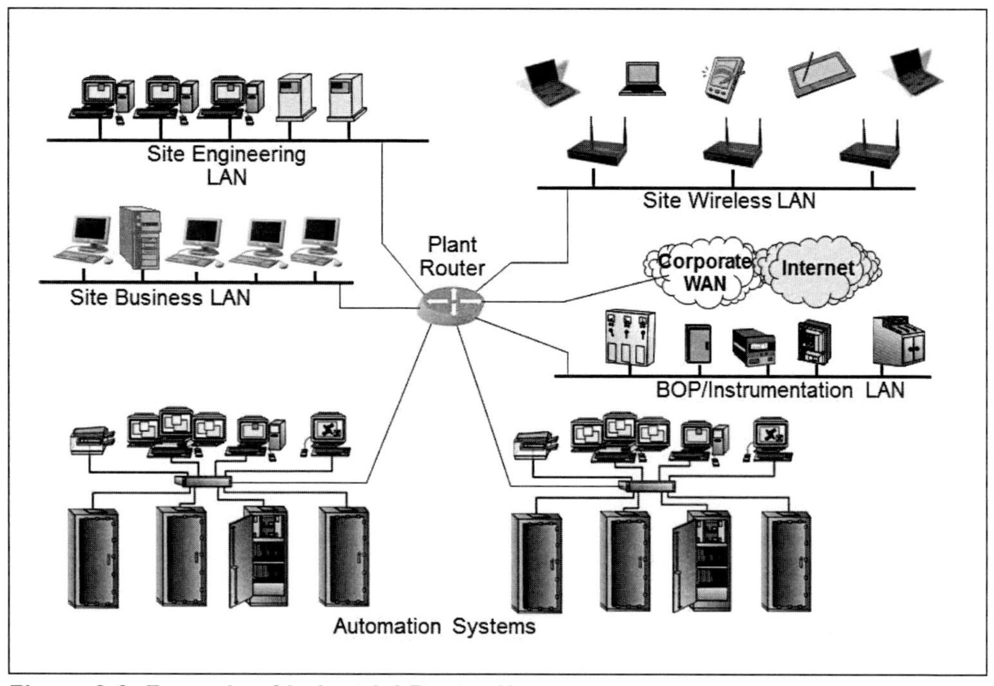

Figure 8-8. Example of Industrial Router Use

Router Actions

In order to simplify the explanation of routing, simple numbers, rather than complete addresses, are used in the following examples. Figure 8-9 uses a simplified notation of addressing. The decimal number by the workstation stands for the 6-octet MAC (Layer 2) address of the network interface card (NIC). The network address (Layer 3) is the decimal number with the "Network" prefix that identifies the network on which the workstation(s) resides.

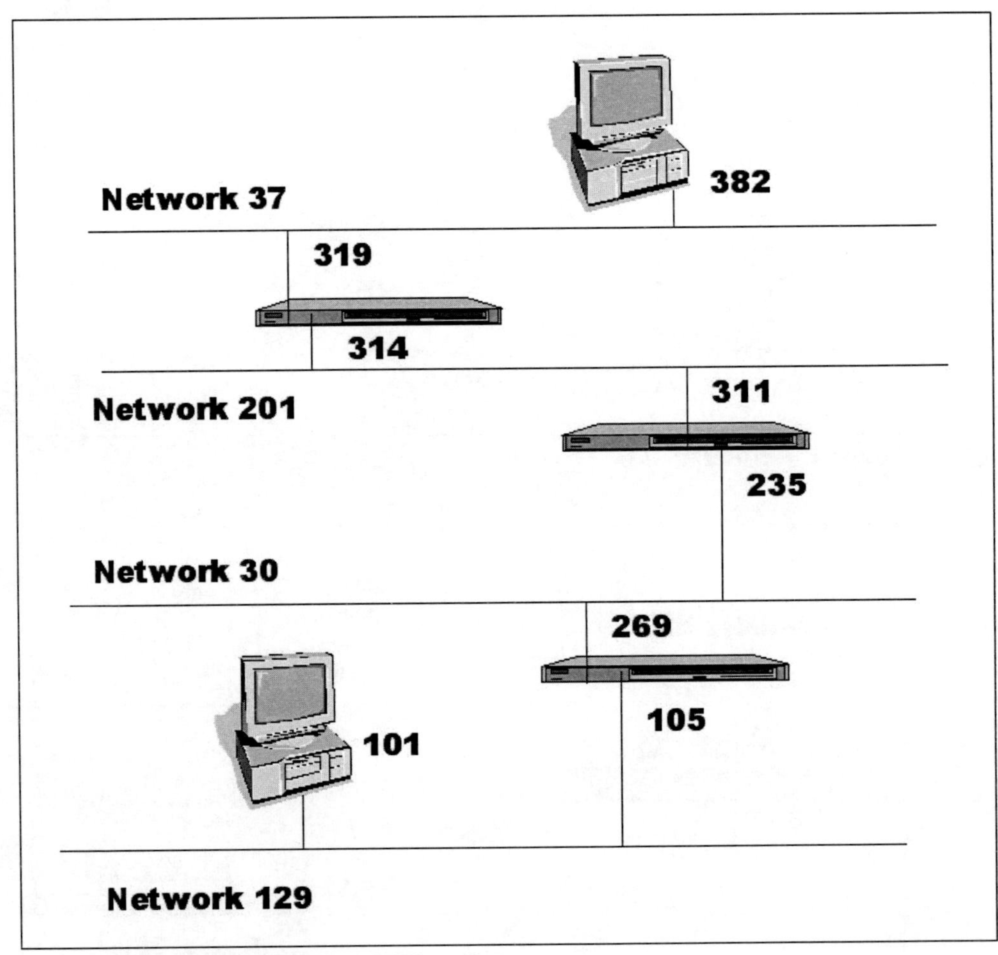

Figure 8-9. Simplified Router Addressing

When a router receives a packet that is destined for another network, it must create a route or path to the foreign network in the form of a list of router hops. Every time a router receives a packet and hands it off, it rewrites the data link address for the next router to which the packet will travel. Each router has to determine, on its own, the travel path for each packet.

Figure 8-9 is a simplified explanation of how a packet goes from a workstation with a Layer 2 address of 101 to a workstation with a Layer 2 address of 382. In figure 8-10, we have made the network address a combination of the network and the MAC address. It is not this simple in real life. Using the Internet Protocol (IP), version 4, involves using four decimal notations for addresses. Something (either a host file, a text file on the individual hosts, or the domain name system [DNS]) has to resolve the IP address as a decimal notation to the URL name (e.g., isa.org). Layer 2 uses ARP to resolve IP decimal notation to MAC addresses. For our purposes, let us assume that this is accomplished without further discovery on our part; that the workstation at node 101 knows the destination network address from previous communications.

DL Destination Address	DL Source Address	Type/Length	Network Source Address	Network Destination Address
105	101	2048	129-101	37-382

Figure 8-10. Simplified Frame Information

Note that the Data Link layer (DL in figure 8-10) is only aware of addresses on this network. If the MAC address it is looking for is not on this network segment, then it will plug in the MAC address of the default router, in our case 105. This frame (remember: this is an abridged and shortened frame model for purposes of discussion) goes to the router at 105. Router 105 would look in its routing table and see if a path exists to Network 129. Since we will discuss discovery later, let us assume that the routing table has that path. The router will move the packet across the router and change the Layer 2 addresses to reflect the next router in line, as shown in figure 8-11.

DL Destination Address	DL Source Address	Type/Length	Network Source Address	Network Destination Address
235	269	2048	129-101	37-382

Figure 8-11. Second Step in Network Addressing

You will note that the Layer 3 addresses remain unchanged but the Layer 2 addresses reflect moving the packet on the Network 30 segment. The router at 269 moves it across and transmits the packet on Network 201. See figure 8-12 for the Layer 2 addressing.

DL Destination Address	DL Source Address	Type/Length	Network Source Address	Network Destination Address
314	311	2048	129-101	37-382

Figure 8-12. Third Step in Network Addressing

The packet crosses this router, arrives at the destination network, and is transmitted from the router to node 382. Figure 8-13 shows the last Layer 2 addressing.

DL Destination Address	DL Source Address	Type/Length	Network Source Address	Network Destination Address
382	319	2048	129-101	37-382

Figure 8-13. Last Step in Network Addressing

Note that for Ethernet routing, it is Layer 2 that moves the packets on a network segment and it is the destination address that moves them to the final destination. If using TCP/IP, the addressing would be different. Figure 8-14 is an example of IP addressing.

In figure 8-14, all of the computers are on one /16 (old Class B) IPv4 network and the segments are separated by subnet masking. Subnet masking is a way to determine the network number (not the host address). All hosts (network devices, which may include workstations) must be on the same network address or there must be a router between different network addresses. A subnet mask is a contiguous set of ones that, when ANDED (a logic AND) with the IP address, determines the network number. The addresses looked at would be the Layer 3 addresses. But how does the packet move on its own network within the Layer 2 frame? There must be a mechanism for equating a four-dot notational IP address with the MAC address for that device. Enter the Address Resolution Protocol (ARP). The ARP compares the destination IP address with every outbound IP datagram to the ARP cache in the NIC card that will transmit the frame. If there is a matching entry, then the MAC address is obtained from cache and placed into the frame's destination address. If there is no match, then the ARP broadcasts an ARP Request packet onto the local subnet (in our case, 129, if 101 is the one wishing to transmit). This packet asks the owner of the IP to reply with its MAC address. If the IP address is not on this network segment, then the default router's MAC address (this is the default gateway) is used and the packet goes to the router.

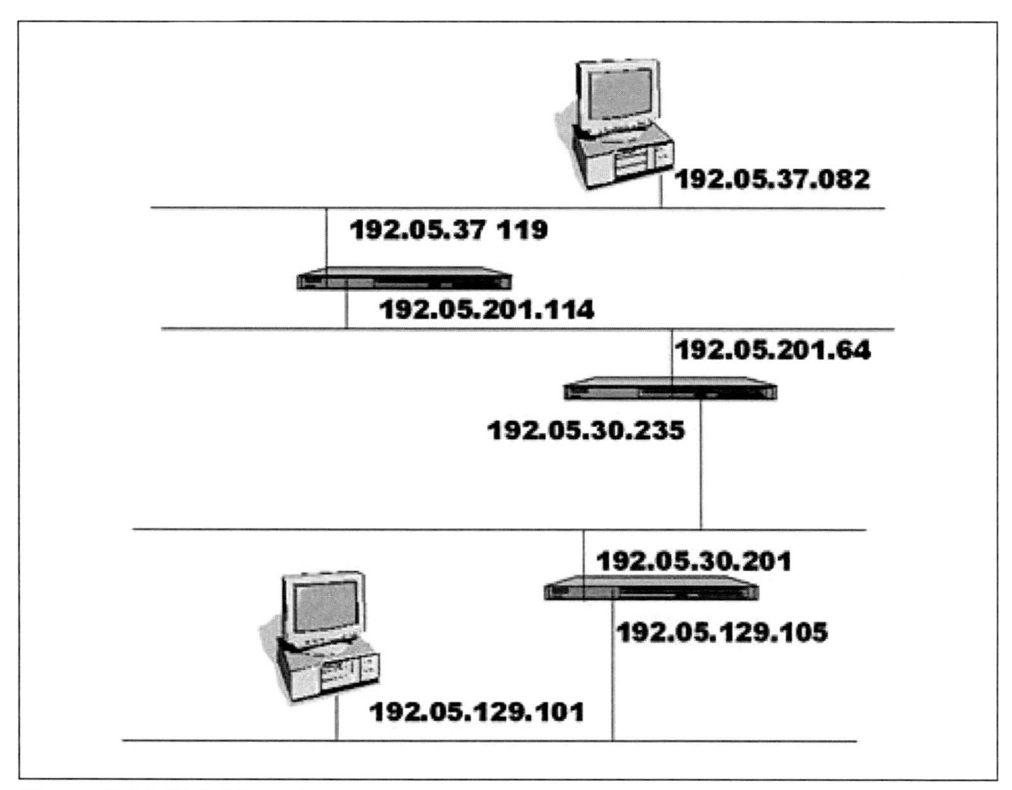

Figure 8-14. IP Addressing

This would happen for each segment the packet went through, until it arrived at its final destination. This is why the default gateway (which is actually a router) address is so important on a network.

Subnet masks are used to take an address space like the /16 (old Class B) and apportion it as separate segments, each with their own network. The IPv4 subnet mask itself is a 32-bit number that the receiving device uses to separate the network ID from the host ID. The easiest way to look at this is as follows: the network ID is assigned 1s and the host ID is given 0s. It should be noted that subnet masking does not generate more addresses; it generates fewer from a given space. Each device on a logical subnet must have the appropriate subnet mask.

If an end user wants to be able to browse (i.e., see the device in Microsoft Windows Network Neighborhood), they have to have either a DNS or a WINS server (Note: A WINS server is, as of Server 2008, no longer used in Microsoft Networking). If you lack either, you will need to install a "hosts" or "lmhosts" file (UNIX systems only require a "hosts" file). This file relates the four-dot notation IP address with the Windows computer name or host name (for UNIX), an essential item for Windows messaging.

Two assumptions were made to make this packet move from one device to another: (1) the originating node knew where the packet wanted to go, and (2) the router knew the path. These assumptions are not always true and that is the reason that routers (like bridges) must learn. One of the first things routers learn is which routers are their neighbors.

Routers use an algorithm to dynamically decide where to send the packet next, based on a cost calculation that can be determined by hop count or shortest path. Other router options, such as least congested or fastest link time are available, but for WANs they tend to cost more, both in terms of money and time because the least number of hops or fastest link may not be the highest performing route. For LANs, these options might cost more in response time and in number of hops to the destination.

Advertising

Computers on a network segment are aware of the destination MAC address of their router because it is the default gateway. The host acquires this address manually through static entry or dynamically through DHCP. Host addresses on multiple networks may be determined by the router by advertising, which means the router determines what addresses are available to it. The specific method of advertising varies with the protocol being used: some protocols have the hosts and the routers broadcast advertisements; some protocols have the router send out hello packets (advertisement) so it can obtain hosts' MAC addresses through the replies.

Router Protocols: Exterior Gateway

For a single connection to the Internet, the routing outside the Layer 2 network is simple: all messages not going to an internal host go to the default gateway that connects to the Internet.

Border Gateway Protocol (BGP) and other such "exterior gateway" protocols were defined to allow routers on the edge of adjacent networks to exchange information about the IP address ranges they know about (or can reach). To use BGP, the router must be on a /24 network (legacy Class C) and it must be assigned a 16-bit Autonomous System Number (ASN, an OSI terminology) by the American Registry for Internet Numbers (ARIN).

With BGP, the designated border router communicates with its peers, the border routers it directly connects with, and exchanges routing information, such as IP addresses in its network and routes it knows about from peer communications. Internal routers may use a protocol called *iBGP* to update the border router. This initially happens when the router powers up and then incremental updates (deltas) are exchanged when any parameter changes. In time, all routers know the architecture (if not the actual physical topology) of the interconnected networks and the routes to use to reach any given IP address. Due to the large number of addresses available as a result of communicating over the Internet, a border router needs a large amount of memory to maintain its routing tables.

With BGP, the router and its peers initially establish connections and authenticate; then they exchange positive and negative "reachability" information; finally they test connectivity and transmit updates as needed. Since route information can be extensive, and by necessity, error-free, TCP delivery is used between routers. BGP does NOT deliver or deal in metrics or performance, just *connectivity*. The ASN should only *advertise* the preferred routes using BGP.

Router Protocols: Interior Gateway

BGP is used to exchange routing information between autonomous systems and is the protocol used between Internet service providers (ISPs). Customer networks usually employ an Interior Gateway Protocol (IGP), such as Routing Information Protocol (RIP), Intermediate System to Intermediate System (ISIS), Enhanced Interior Gateway Routing Protocol (EIGRP), or Open Shortest Path First (OSPF) to exchange routing information within their networks. BGP can technically be used internally but it is more common to use the RIP or OSPF routing protocols.

Routing Information Protocol

RIP is a distance vector protocol. The protocol derives its name from the fact that it uses a single measurement of path length: the number of hops. It is a dynamic routing protocol that was developed for internal use on smaller IP networks. RIP uses User Datagram Protocol (UDP) port 520 for route updates and calculates the best route based on hop count. RIP requires less CPU power and RAM than other routing protocols but it does have limitations:

- **Metric:** Since RIP uses the hop count to determine the best route to a destination, based solely on how many hops it is to the destination network, RIP tends to be inefficient if there are multiple routes to the same location. The path with the lowest hop count might be over the slowest link in the network.

- **Hop Count Limit:** RIP cannot handle more than 15 hops. Anything more than 15 hops away is considered unreachable by RIP. This design is used to prevent routing loops, in which data could never get off of the media and would circulate forever, as though it were routed to nowhere.

- **Class Routing Only:** RIP v1 is a class routing protocol that supports IPv4 original classes: A, B, and C. RIP cannot handle classless inter-domain routing (CIDR), which supports the Variable Length Subnet Mask (VLSM), now used for IPv4 addressing. RIP v2 can handle VLSM but is not necessarily compatible with RIP v1. RIP v1 advertises all networks it knows as class type networks, therefore, it cannot subnet a network properly other than using /8, /16, or /24 subnet masks. There is a version of RIP for IPv6.

Routers running RIP broadcast the full list of all the routes they know every 60 seconds. There is a set of timers that RIP uses to manage and update route information. RIP presumes that constant updates are needed. When a router using RIP first comes on line, it sends out an update—who it is and who it is connected to. RIP updates are then sent every 60 seconds. The router listens to all the other routers that are sending updates and then builds a routing table from the updates. This table is continually being modified and is broadcast with each update.

RIP Routing Tables: A simulated RIP routing table is illustrated in figure 8-15.

60 Seconds			120 Seconds			180 Seconds		
Network	Router Address	Hop Count	Network	Router Address	Hop Count	Network	Router Address	Hop Count
12	244	2	12	244	2	12	244	2
15	315	1	12	312	1	15	315	1
62	420	1	15	315	1	62	420	1
			62	420	1	62	612	2
High hop count not used			62	612	2			

Figure 8-15. Simplified Routing Table

Note that the table in figure 8-15 has three entries at 60 seconds. At 120 seconds, the higher hop count entry would be erased from the table for both 12 and 62. This is because a RIP table will only indicate one route and this is the route with the lowest hop count (which, as mentioned, may not be the highest bandwidth route). If Network 12 address 312 becomes unavailable, due to link or router problems, it will be at least 60 seconds before Network 12 address 244 is reestablished.

RIP Problems: Below are two issues with using RIP.

- **Problem 1:** Routers using RIP transmit their tables periodically, whether any changes have occurred or not. This wastes bandwidth.
- **Problem 2:** Convergence (as used with routers, not voice-over-IP) is a measure of how long it takes all the routers to be notified of a change. A router does not know where it is in relation to other routers, only how many hops there are between routers based on RIP. This may cause the router to believe that routes are true when they are not and to believe an incorrect hop count as well; or as defined with technical elegance, it believes "gossip."

Gossip is illustrated in figure 8-16. If router 50 fails, router 10 will find out through the table broadcasts; however, this may take as long as 5 minutes. During that time, router 10 still believes that routers 20 through 45 can connect to Network 5 because each has broadcast that it can. As RIP only uses hop count, router 10 does not know the physical topology. The hop counter will count down to infinity for all packets routed to Network 5 (infinity is 15 hops in RIP) before router 10 realizes that whichever router it used did not connect to Network 5. If router 10 were the only router trying to send to Network 5, it would have to try each router. This will take some time. Because the RIP routers believe everything they hear, the result is called *gossip*.

Several methods have been worked out to compensate for this problem but they generally cause as many problems as they cure. For example, the *Hold Down method* keeps the last route update for a while before receiving any more route updates. Similarly, in the *Split Horizon* (poison path) method, the originating router cannot receive information it sent out. It is beyond the scope of this text to go into these methods in detail, but an Internet search will yield any number of explanations for these two corrections to RIP gossip.

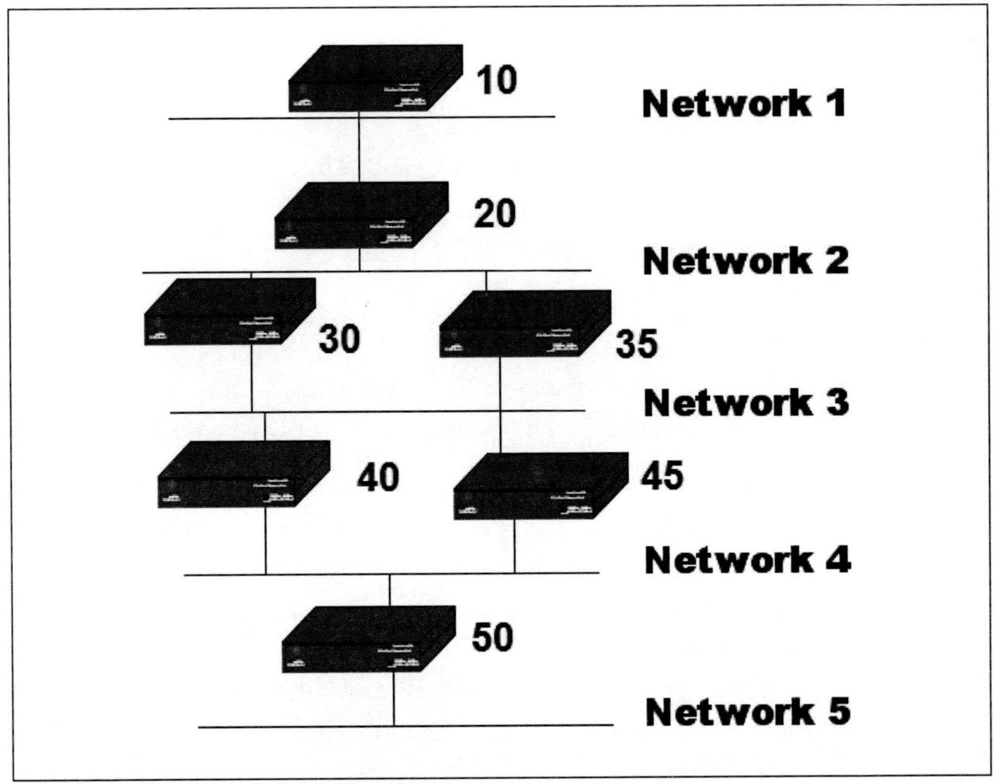

Figure 8-16. Illustration of Gossip

It takes a long time, relative to the speed-of-information rate, to propagate changes through the routers after a route change (convergence), so the information a router receives may or may not be accurate (gossip). Because of these faults, although some corrections are possible, inexpensive processing power has made available a different approach: link state protocols.

Link State Protocols

Link state protocols use more information than hop count, such as link speed, location, and the like. They develop a physical topology map in the router and, since they allow for multiple routing paths, they can provide load balancing. Link state protocols converge much faster than distance vector protocols. The protocol we will discuss is the Open Shortest Path First (OSPF).

OSPF: This routing protocol is used to determine the correct route for packets within IP networks. It was designed by the Internet Engineering Task Force to serve as an Interior Gateway Protocol, replacing RIP. It is becoming more widely supported and may eventually supplant RIP.

The advantages of OSPF are:

- Changes in an OSPF network are propagated quickly.
- OSPF is hierarchical.
- OSPF supports CIDR Variable Length Subnet Masks (VLSMs).
- OSPF uses multicasting within areas.
- After initialization, OSPF only sends exception updates, not the entire routing table.
- Using areas, OSPF networks can be logically segmented to decrease the size of routing tables.
- Table size can be further reduced by using route summarization.
- OSPF supports authenticating route updates (passwords and message digest).

OSPF uses *Link State Flooding* to tell other routers which LANs this router is directly connected to (accurate information). OSPF then performs the *SPF Calculation* to determine the shortest (fastest) path to a distant network. Lastly, OSPF performs *Neighbor Discovery*, which locates routers that are directly attached to the same networks as this router. Let us consider each of these parts in turn.

Link State Flooding: Using a process called *link state flooding*, routers using the OSPF protocol send out information concerning networks that they are directly connected to in the form of an advertisement. This is the link state (link status) which includes link speed, as well as where in the topology the routers are located. This information is distributed in a link state packet (LSP), which has a sequence number and a time stamp to prevent confusion. Routers collect this accurate information to build a topology map of the entire set of interconnected networks.

OSPF messages are in the form of IP datagrams packaged using protocol number 89 for the IP Protocol field. OSPF utilizes five message types for various types of communication:

- **Hello:** Routers use these messages as a form of greeting to discover adjacent routers on its local links and networks. Hello messages are used to establish relationships between neighboring devices (*adjacencies*) and to communicate key parameters that indicate how OSPF is to be used in the autonomous system or area.

- **Database Description:** These messages contain descriptions of the topology of the AS or area. Communicating a large link state database (LSDB) may require several messages to be sent: the sending device, designated as a *master* device, will send messages in sequence to the recipient (the slave) which will respond with acknowledgments.
- **Link State Request:** A router will send this type of message to another router to request updated information about a portion of the LSDB. The message specifically indicates which link(s) the requesting device wants updated information about.
- **Link State Update:** Update messages are sent in response to *Link State Request* messages. They contain updated information about the state of specific links on the LSDB and they are broadcast or multicast by routers on a regular basis.
- **Link State Acknowledgment:** This message acknowledges the receipt of a *Link State Update* message, which provides the element of reliability to the link state exchange process.

SPF Calculation: The shortest path first (SPF) calculation determines the shortest path (in terms of link speed and number of hops) from a router to any other router that it knows about. To ensure synchronization between the router databases, all routers broadcast at a set period, usually between 2.5 and 4 hours. This schedule resynchronizes and rebuilds the system topology map in each router.

Neighbor Discovery: Routers use these link state protocols to discover routers that are directly connected to the same networks to which they too are connected. OSPF brings in a whole new set of OSI system terminology regarding segmenting a network into a "hierarchy" with a "backbone" and then into lower layers and areas (all connected to the backbone), with terms like "stubby," "not-so-stubby," and "totally-stubby."

The backbone area is the core of an OSPF network. All other areas are connected to it. Inter-area routing is via routers connected to the backbone area and to their own associated areas. A stubby (stub) area is an area that does not receive route advertisements external to the autonomous system (AS). Routing from within the stubby area is based entirely on a default route (a default gateway); depending on the type of stubby, it may import information (routes) from other routing protocols. A not-so-stubby area (NSSA) is a type of stubby area that can import AS external routes (from other routing protocols) and send them to other areas but it cannot receive AS external routes from other areas. A totally stubby area is also similar to a stubby area; however, the totally stubby area does not allow summary routes or external routes. Inter-area (IA) routes are not summarized into totally stubby areas, the only way for traffic to get routed outside of the area is by a default route.

Routers are organized into logical spaces, called *areas*, *autonomous systems*, or *domains* (not to be confused with NT domains). With the newly defined router hierarchy comes different names for routers with different functions in the hierarchy and different descriptions based on where they sit in the hierarchy: designated router, backup designated router, internal router, backbone router, area border router, and autonomous system boundary router. Each router is assigned a 32-bit router number that is related to its IP address range and each area gets a 32-bit area identifier.

Multiple Protocols

When you are using a multi-protocol router (remember, routers are protocol dependent), you can keep protocol information in two ways:

1. Separate databases for each protocol ("ships passing in the night")
2. One database shared between protocols (integrated)

We will discuss these next.

Ships Passing in the Night: Keeping separate databases for each protocol ensures that problems that occur in one protocol do not affect the other protocol. The users of each protocol are effectively in two different networks and the synchronization and upgrading of link information occurs separately. In effect, you have two different routers.

Integrated: Using just one database for multiple protocols means that problems affect all protocols. Protocol updates are done simultaneously, which saves bandwidth. Using a shared database means you have only one set of users and one set of configuration settings—a fact not to be taken lightly if you have to administer many routers—which is usually the deciding factor in selecting between ships passing in the night or integrated routing.

How Routers Are Designated

Domains can contain multiple autonomous systems in an OSI view. One router is designated to be the reference router, to whom all other routers synchronize their topology maps. This reference router is called the *designated router*.

Most of the above router explanations were based on the assumption that it is an IPv4 world, which it will be for quite some time in the industrial areas. However, much inter-ISP messaging is done via IPv6. There is a RIP for IPv6; however, the OSPF routing protocol was designed to accommodate many protocols and IPv6 presents no problems for OSPF routers.

Bridges versus Routers

Should you use a bridge or a router? Figure 8-17 provides a quick comparison.

	Bridge	Router
Flat topology	●	
Segmentation		●
Multi-path		●
Reliability (data)		●
Inexpensive	●	●
Fast (data)	●	●
Protocol dependent		●

Figure 8-17. Bridge/Router Comparison. Note that a flat topology means a single Layer 2 network of multiple segments (no Layer 3).

The major and most important advantage of using routers over bridges is segmentation of the network: breaking it up into organizational units. With segmentation, troubles in one network organizational unit normally cannot affect other network organizational units in a routed system. Segmentation makes a robust (hard to break) network, and it allows a quarantine to be set up, if needed. Routers add reliability by allowing fast convergence after link problems and by providing alternative paths for data.

Routing Topologies

Most routers are found in a hierarchical topology. Hierarchies are established to help organize routers. As such, areas communicate with other areas using the *designated router* (Level 1). The routing rules are as follows:

- Level 2 routers talk to Level 2 routers at the backbone level.
- Level 2 routers talk to the designated Level 1 router at the area level.
- Level 1 routers only talk to Level 2 routers outside their area.
- Level 1 designated routers talk only to the routers in their area.

Figure 8-18 is an example of a hierarchical routing topology.

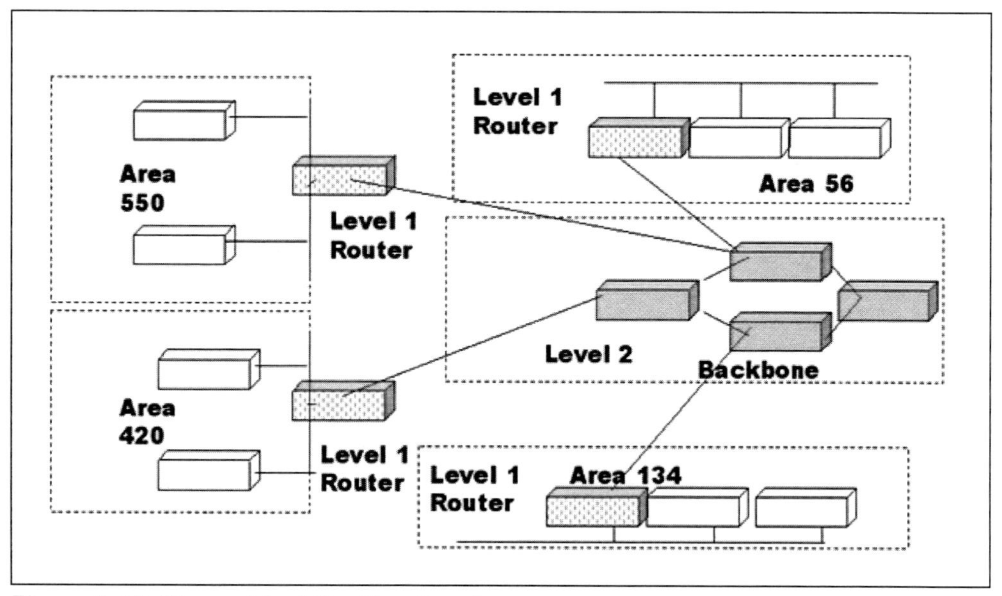

Figure 8-18. Hierarchical Routing Topology

Router Physical Connections

Routers may have one-to-one, one-to-many, or many-to-one I/Os. A departmental multi-protocol router typically has an Ethernet connection for the LAN side and a WAN connection (e.g., Frame Relay Access Device [FRAD], channel service unit [T1 CSU], data service unit [DSU], or fiber optic) on the routed side.

Sometimes hubs or switches are incorporated on the LAN side. When this is done, the routers are normally referred to as *brouters, departmental routers*, or (in the current lexicon) *Layer 3 switches*. These devices bridge on the LAN side and, when a packet is destined for another network, the router forwards the packet.

VLANs

If a traditional IP router is used to separate network areas, then the network is divided into subnetworks. If a Layer 3 switch (a brouter or divisional router) is used, the network can be divided into a number of virtual local area networks (VLANs). In either case, the router or Layer 3 switch is the main locus for all of the network traffic.

To create a VLAN, the routing switch is configured to define devices on different ports to act as if they are on the same LAN segment. VLANs can then be used to group arbitrary collections of end nodes on multiple LAN segments into separate domains. This is useful for zoning or dividing the plant floor organizationally.

Nodes may be accessed by multiple VLANs; however, the separate VLANs do not communicate with each other. This is identical to a firewall router DMZ (demilitarized zone). Packets sent between VLAN nodes are switched and packets sent between VLANs are routed. When two devices are defined as being on the same subnet or VLAN, the switch passes messages through without doing any filtering, just as if the devices were on the same physical segment. However, if two devices are not on the same VLAN, then the switch runs messages through its filtering software, passing or blocking each message as appropriate.

Remember that a router or switch can only filter traffic that passes through it. It cannot separate two devices that are physically wired to the same network segment. Thus, if it is important to filter traffic between two different groups of devices, it is necessary to ensure that they are attached to different switch ports. The best way to do this is to connect each device to its own switch port, creating a fully switched network. Figure 8-19 illustrates a typical fully switched, industrial VLAN.

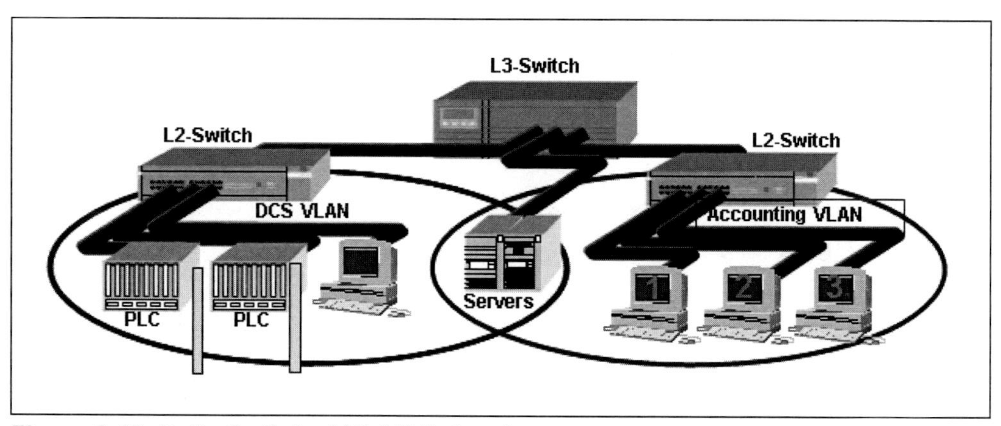

Figure 8-19. Fully Switched VLAN Network

Managed Switches

The features and configuration of managed switches vary by manufacturer, as well as by model (from the same manufacturer). The configuration interface used for managing a switch is generally some form of Web interface, although there will be limited hardware configuration elements (notably a reset switch).

Smart (intelligent) switches are managed switches that have a limited user-changeable set of features. The configurable features most likely found on a modern managed switch are:

- Port availability
- Port priority
- Spanning tree algorithm/fast spanning tree algorithm
- Simple Network Management Protocol (SNMP)
- Link mode port
- Link aggregation
- VLAN settings

At present, Layer 3 switches only examine packets and deliver them to the correct ports. Vendors are working to improve security and performance by adding functionality that would enable the switches to examine entire data streams and to take specific action upon them. Today, these different network tasks require several network devices to accomplish what the new Layer 3 switches will be able to do by themselves. This integration of function will reduce equipment count and complexity, as well as increase performance.

One of the problems with the present multiple-device approach is that infrastructure connectivity devices are often purchased by different groups within the organization. For example, IT purchases devices for load balancing; Security (usually, but not always, part of IT) purchases firewalls and the Intrusion Detection System (IDS); and Engineering obtains VPN, VLAN, and gateway equipment. If two or more devices end up conflicting with each other, a political battle that has nothing to do with technology may erupt and performance (and perhaps security) will likely suffer. Provided that it has the processing speed not to be a bottleneck itself, a full-feature, Layer 3 switch that is capable of all of these functions will make a network simpler and more efficient. However, the problem remains as to who will specify, acquire, and maintain this asset (particularly in industrial process control areas) as a full Layer 3 switch will fit the description of an inter-zone traffic flow control device.

In any event, managed Layer 3 switches have many more features and are capable of much more than just simple configuration and switching and routing packets. If a Layer 3 device is configured for advanced features (and this is no simple task), it can be considered as an Application Proxy Server (in addition to being an IEC 62443 Inter-Zone Traffic Flow Control) and—with packet examination (to Layer 7)—it may also be known as a deep inspection firewall.

Gateways

There are other devices that are used for internetworking, however, the only other one that we will discuss is gateways because of their interest to industrial users. The gateway types available include:

- Translating bridges/protocol converters
- Encapsulating bridges/tunneling gateways
- Network operating system (NOS) gateways
- WAN gateways
- Application layer gateways
- Site-to-site or computer-to-site VPN gateways

We discussed the translating bridge (protocol converter) gateway earlier in the chapter. Let us discuss the rest of the gateways in turn.

Encapsulating Bridges/Tunneling Gateways

To "encapsulate" or to "tunnel" means to surround the originating protocol with the transport protocol. As far as the tunneling (encapsulating) devices are concerned, everything is just a 1 or a 0. Physical devices put 1s and 0s on the media; they do not care what they represent. The Data Link layer is concerned with the frame organization yet, even in the Ethernet frame, the Data Link layer does not care how the 1s and 0s are arranged between the type/length and the CRCC. Layer 3 looks at its area, yet considers anything beyond its scope to be just 1s and 0s. Even TCP is only concerned with the packet length and sequence number; anything beyond that is just 1s and 0s. This means that when TCP/IP is used, it doesn't matter what combination of 1s and 0s is placed after Layer 4. It could be another protocol's entire Layer 2 frame. If the frame is longer than one packet, then it is broken up into fragments and delivered.

In the end, tunneling (encapsulating) is how Ethernet and TCP/IP will become the standard in industrial networking for Layer 1 through 4 protocols. By using a gateway, any protocol can be converted to Ethernet frames and delivered over the network. This will allow vendors to claim that they use "standard" protocols, even though the tunneled information (payload) may be usable only on that vendor's equipment. It is Layer 7 (Application) that determines what the payload 1s and 0s mean, and Layer 7 may be specific to a particular vendor, if it is not open source.

Network Operating System Gateways

Network operating system (NOS) gateways are also called *architectural gateways*. Most NOS gateways come with several protocols. Windows NT 4.0 had five protocols and could run all simultaneously. Windows Vista/7/8 has several protocols that must be separately loaded, with TCP/IP as the default and the only possible protocol if Active Directory is used. Novell (version 5.1 and up) was a gateway between native IP and IPX (Novell's Layer 4 network protocol). UNIX/Linux has TCP/IP and whatever the user loads as an additional protocol. All of the standard NOS gateways can run multiple communication protocols as communications stacks.

WAN Gateways

Given the widespread adoption of routing as the preferable method for networking multiple networks, WAN gateways have almost been superseded by routers. Most routers could be called *gateways*. They can mix and match to the WAN, for example:

- Ethernet to frame relay
- Ethernet to Switched Multi-megabit Data Service (SMDS)
- Ethernet to Asynchronous Transfer Mode (ATM)

Routers can connect through fiber-optic, wireless, or copper media, depending on their feature set and the Physical layer that they interface.

Application Layer Gateways

There are a multitude of application layer gateways, many more than can be discussed within the scope of this text. Some of the available application gateways are listed below, a few are discussed in this text in greater detail.

- Mail gateways
- File format gateways (NFS (Network File System)/FAT (File Allocation Table; FAT-16, FAT-32)/NTFS (New Technology File System))
- Multi-protocol networking
- Directory services
- VPN gateways

Virtual Private Network Gateways

There are a multitude of reasons why it is necessary to have communications traffic between and among computers that is confidential, source-authenticated, and content-verified, particularly in an industrial automation and control system (IACS). VPNs provide this across public or non-secure networks (such as the Internet) and restricted-access networks, such as a plant LAN. A VPN creates an environment among the connected devices that is virtually the same as being connected by a physically private, restricted-access network.

A VPN is a system of interacting computers that operates over a public infrastructure and achieves privacy by ensuring:

- ISO digital certificates (or local private key) are issued to each participant.
- Mutual authentication takes place prior to all other processes.
- Encryption is used for all communications.
- If there is no certificate (or private key), there will be no interactions.

VPNs may be described as providing encryption "tunnels" for data over the Internet. VPNs use the Internet for transport but the payload data is indecipherable to all but the intended group. The *gateway* term is applied to the VPN router that interfaces to a LAN (but not to the host); for example, a site-to-site VPN has two VPN routers that are called *gateways*.

There are three classifications of VPN topology (illustrated in figure 8-20).

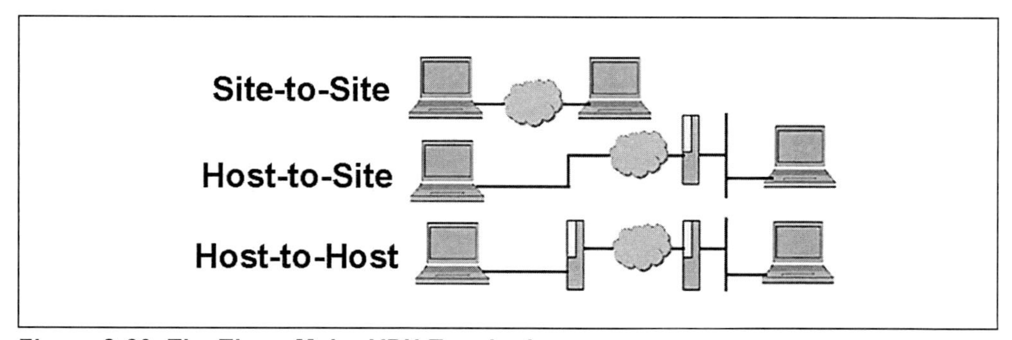

Figure 8-20. The Three Major VPN Topologies

One question for the potential VPN user to answer is whether to use hardware (router or security device) or software (a computer configured as a VPN host or site router) to create the VPN. Host software is more flexible but it uses host resources. If a software VPN is chosen, it will require CPU cycles from the host, although most laptops and desktops have plenty of spare cycles. In control systems, though, many prospective hosts have no spare CPU cycles or memory. Whether a software or hardware VPN is selected, a VPN is required at both ends of the network.

Host-to-Host VPNs

Figure 8-21 is one example of how a host-to-host VPN would work over a network.

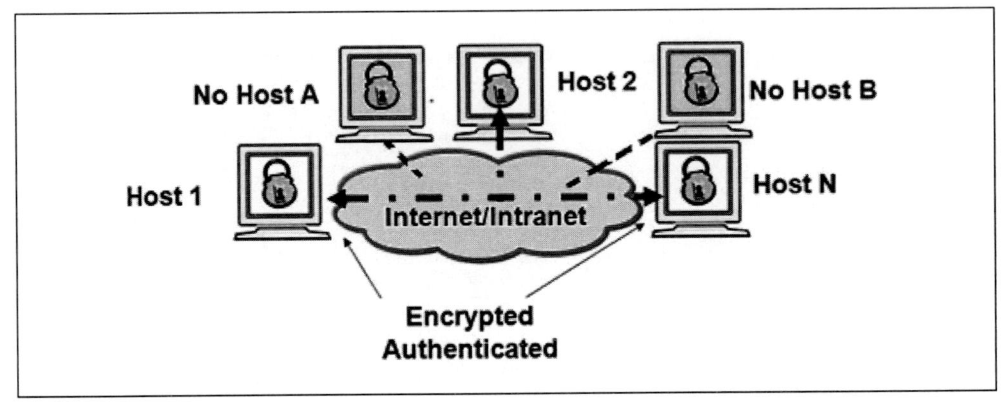

Figure 8-21. Host-to-Host VPN Example

In the above example, every host that participates in the VPN has VPN client software running. The host devices include software to do authentication and encryption. In the case diagrammed, No Host A and No Host B, while they are on the network, cannot communicate with those hosts running the VPN software or vice versa.

A problem with host-to-host VPN connections is that they may make it impossible for malware/attack detection technologies, such as a network intrusion detection system (NIDS), to do their job, unless they are also included in the VPN. A VPN does not prevent malware from being sent from one VPN member computer to another VPN member computer.

Host-to-Site VPNs

Figure 8-22 illustrates a typical host-to-site VPN.

Commonly called a *remote access (RAS) VPN*, a host-to-site VPN usually consists of adding VPN software to the remote client and using a router with VPN or a security appliance as the host gateway. This is a more secure way of connecting to a plant from a home or a remote office than a simple modem or DSL line as it provides encryption for authentication session and content.

Commonly used for remote or mobile workers to make a connection to the corporate systems from a remote location, the VPN connectivity may NOT

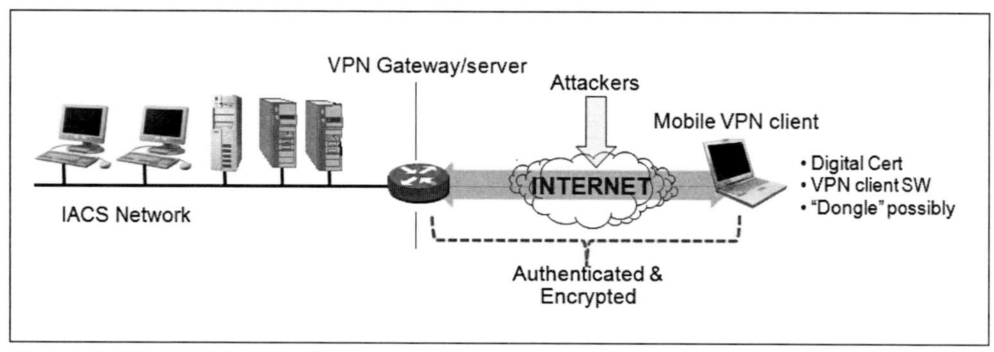

Figure 8-22. Host-to-Site VPN

eliminate the need for the remote user to go through a normal login once the VPN connection is established. For added security, some remote VPN client software requires a USB security module (USB dongle) to be plugged into the PC for additional user authentication.

Site-to-Site VPN

Figure 8-23 illustrates a typical site-to-site VPN, also commonly called a *LAN-to-LAN VPN.*

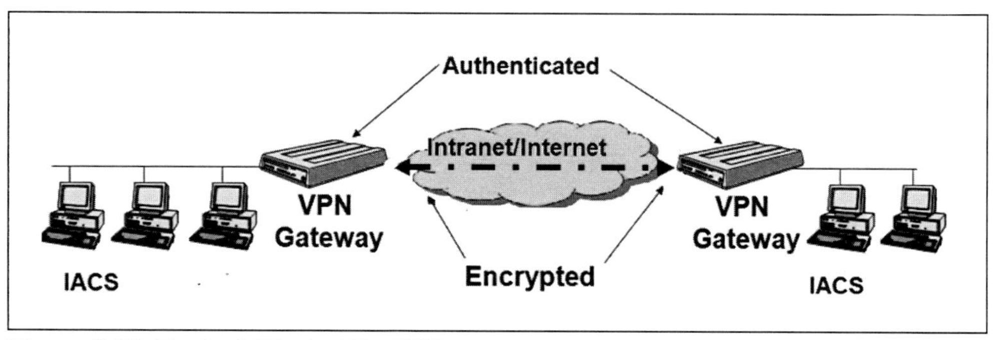

Figure 8-23. Typical Site-to-Site VPN

The gateways of the VPN are intermediary devices (routers with VPN software or security devices) that pass traffic from a trusted network via an un-trusted transport network to another trusted network, relying on VPN technology to secure the traffic on the un-trusted transport network. Authentication is a critical part, as communications are refused from any computer unable to authenticate. Keep in mind that site-to-site VPNs do nothing to add security to the local traffic on the LANs, only to traffic crossing the Internet/ intranets.

To ensure sufficient protection on a public network like the Internet, installing a VPN requires a detailed understanding of network security issues and careful installation and configuration. The reliability and performance of an Internet–based VPN relies on the ISP and its quality of service; these are not under an organization's direct control.

VPN products (hardware and software) and solutions (partial to turnkey) from different vendors are not always compatible, due to problems with VPN technology standards. Attempting to mix and match equipment may cause technical problems, although using equipment from only one provider may cost more.

SSL or IPsec

There are two contemporary technologies for building a VPN over the Internet or an intranet: Secure Sockets Layer (SSL) and Internet Protocol Security (IPsec). SSL is easier to deploy but is less standards-based; IPsec is harder to deploy but it is based on the Internet Engineering Task Force (IETF) standards. IPsec is included in many recent operating systems. The main difference between SSL and IPsec is that SSL makes each software application responsible for security by adding SSL logic, while IPsec builds it into the TCP/IP networking software (this is why it is called *IPsec*) so it is transparent to applications.

PPTP, L2TP, and IPsec

Many network protocols have become popular in VPN developments. We will discuss the following protocols in this section:

- **PPTP** – Point-to-Point Tunneling Protocol (favored by Microsoft)
- **L2TP** – Layer 2 Tunnel Protocol (favored by Cisco)
- **IPsec** – A collection of related protocols. It can be used as a complete VPN protocol solution or simply as the encryption scheme within L2TP or PPTP

A PPTP tunnel communicates to a peer on TCP port 1723. This TCP connection is then used to initiate and manage a second Generic Routing Encapsulation (GRE) tunnel to the same peer. The PPTP-GRE packet format is non-standard, including an additional acknowledgment field replacing the routing field in the normal GRE header. These modified GRE packets are directly encapsulated into IP packets (after the type/length octets) and seen as IP protocol number 47. GRE is a tunneling protocol (developed by Cisco Systems) in which the GRE tunnel is used to carry encapsulated PPP packets. GRE can be used to tunnel any protocols that can be carried within Point-to-Point Protocol (PPP), including IP. In Microsoft's implementation, PPP traffic can be authenticated with Challenge-Handshake Authentication Protocol (CHAP) or MS-CHAP v1/v2 (Microsoft's version of CHAP).

The Layer 2 Tunnel Protocol (L2TP) supports VPNs or is part of the delivery of services by ISPs. It does not provide any encryption or confidentiality by itself. Rather, it relies on an encryption protocol (typically IPsec) that it passes within the tunnel to provide privacy. The entire L2TP packet (including the header and payload) is sent within a UDP datagram. It is common to carry PPP sessions within an L2TP tunnel. L2TP does not provide confidentiality or strong authentication by itself. IPsec is often used to secure L2TP packets by providing confidentiality, authentication, and integrity. The combination of these two protocols is generally known as L2TP/IPsec.

IPsec is a protocol suite for securing Internet Protocol (IP) communications by authenticating and encrypting each IP packet of a communication session. IPsec includes protocols for establishing mutual authentication between agents at the beginning of the session and negotiating cryptographic keys to be used during the session. IPsec is an end-to-end security scheme operating in the Internet layer of the Internet Protocol Suite. Applications are automatically secured by IPsec at the IP layer.

So, which should you use: PPTP, L2TP, or IPsec protocols? It depends on what your equipment will support and upon the communications engineer or whomever is the engineering authority for your IACS. In industrial areas, with the increased emphasis on data transport security, the VPN is one of the easiest methods of providing that security. IPsec is mandatory in IPv6.

Summary

This chapter has been a brief tutorial on some of the facts you need to understand before starting out on the internetworking path. For an industrial user, this is a formidable path, particularly when you have a closed (proprietary) system that does not integrate well with the rest of the facility.

Although you can research particular aspects of internetworking on the Internet, most of the information available is either too general or is vendor-specific. Multiple reference texts are available but, because of the speed in which technology changes, some topics are not well documented. Continual study of the literature, upgrade training by vendors, and a good understanding of your requirements and how they are changing are the best ways for industrial users to understand internetworking.

Bibliography

Note that Internet links were correct at the time of writing but may change.

Cisco Systems Inc. *Cisco CCNP Preparation Library.* San Jose: Cisco Systems, 2000.

Doyle, J., and J. Carroll. *Routing TCP/IP, Vol. 2.* CCIE Professional Development series. Upper Saddle River: Prentice Hall, 2001.

Microsoft Corp. "Get Connected with Windows XP Networking." http://www.microsoft.com/windowsxp/using/networking/
(By July 2015 this should be updated to Windows 10 networking)

———. *Windows Internet Guide.* Redmond: Microsoft Press, 1996.

International Society of Automation. ISA Course: "TS13 Industrial Automation Cybersecurity: Principles & Application Research." Triangle Park: ISA.

Naugle, M. G. *Network Protocols.* Signature edition. New York: McGraw-Hill, 1998.

Wikipedia. Articles on PPTP, PPP, L2TP, CHAP, MS-CHAP, IPsec, and SSL. Accessed 2014. Bing.com.

9 Cybersecurity

Overview

Because bad things happen to good industrial automation control systems (IACS), we have included a chapter on the subject of cybersecurity and industrial networking. This chapter has, as its primary concern, industrial automation control systems. As the hardware and software used in the industrial arena have become far less proprietary and much more like commercial software, security problems have begun to multiply. It is not so much that commercial systems are more vulnerable but that they are much more widely known, which is the reason for their adoption in the first place. Due to the ever-present threat of viruses, hackers, backdoors, Trojans, and other malware (including spam and those wonderful phishing trips), as well as the need to test all software patches for operability among various applications, you might think that security was the full-time job of the industrial network technician—and you just might be right.

Types of Security

There are three major areas of security that persons working in and around control systems need to be aware of in order to participate in the maintenance of these areas. These are physical security, personnel security, and cybersecurity, as shown in figure 9-1.

Physical security consists of physically ensuring the security of information, personnel, and plant assets, including automation systems and networks,

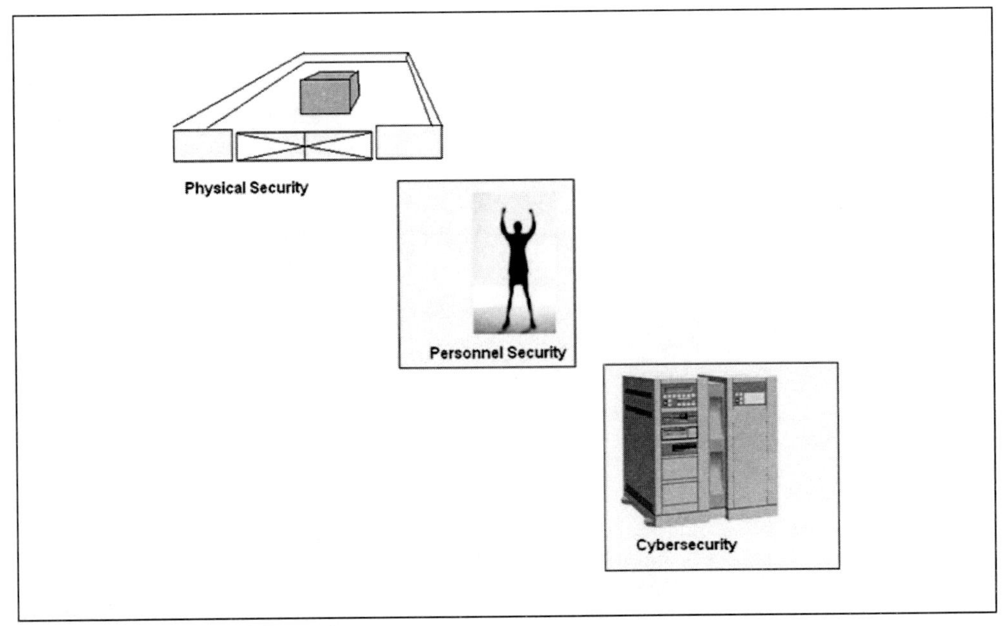

Figure 9-1. Types of Security

through the application of *physical controls*, such as guards, gates, locks, walls, or any other form of physical obstruction between the system and a potential intruder. If an intruder has physical access to your system, particularly to the servers, the intruder has your information.

Personnel security (also called *operational security*) is often achieved through the use of *administrative controls*, including policies, procedures, and training. It means ensuring that personnel are not security risks by conducting background checks on employees, having enforceable and enforced written policies for network use and staff conduct, and monitoring personnel for suspicious behavior, such as accessing the system when not on duty.

Cybersecurity has many aspects, including group policies (a term for policies on a domain controller, etc.), network use policies, firewalls, and password policies. Anything having to do with the computer system itself, such as access to files and applications, must be planned and must be consistent. Cybersecurity is often created and maintained through the use of *technical controls*, like firewalls, encryption, and passwords.

This division of security into three types originated within the security industry. The general public typically thinks of security as the first type, physical security, yet all three types of security work together—none is independent. If you only protect one type (cyber–physical–personnel), the potential invader may attack another type to determine if it is inadequately protected or unprotected. In addition, even though countermeasures (controls) are categorized as being physical, technical, or administrative, a security weakness (a vulnerability) may be best addressed by using any two (or all three) types of control. For example, to prevent unauthorized use of a system, you could use passwords (a technical control) and you could lock the system in a secure area and only give keys to authorized personnel (physical control.)

The model for modern information technology (IT) information security (for the business network, as opposed to the IACS network) is to preserve the following, illustrated in figure 9-2:

- **Confidentiality:** Ensure that information is accessible only to those who are authorized to have access and who have the need to know.
- **Integrity:** Safeguard the accuracy and completeness of information and processing methods. Integrity also includes non-corruption due to malware.
- **Availability**: Ensure that authorized users have access to information and associated assets as needed.

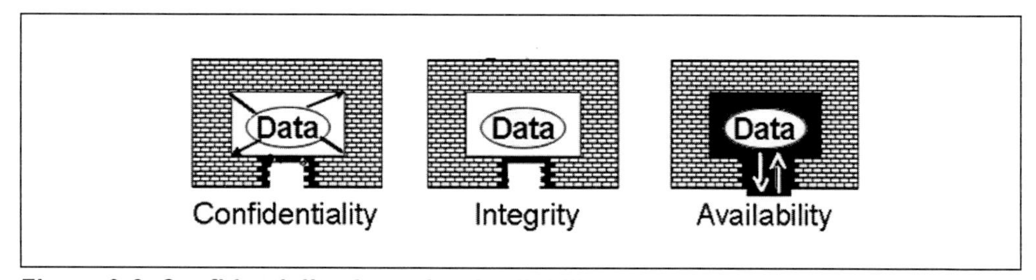

Figure 9-2. Confidentiality, Integrity, Availability

Although industrial and business users share these three concerns, physical security issues and the way that security is implemented and applied can be major differentiators between conventional security and industrial security. This is because industrial security must protect lives and facilities, as well as data. In the industrial environment, the priority is usually availability, integrity, and then confidentiality. Not that all three aspects are not important, it is just that availability tends to be the first priority, often because of safety concerns. Similarly, integrity is important because configuration settings, parameters, and information stored in various databases can impact the proper functioning of an automation system and, with it, the industrial plant. Note that safety and security overlap in an IACS, as shown in figure 9-3.

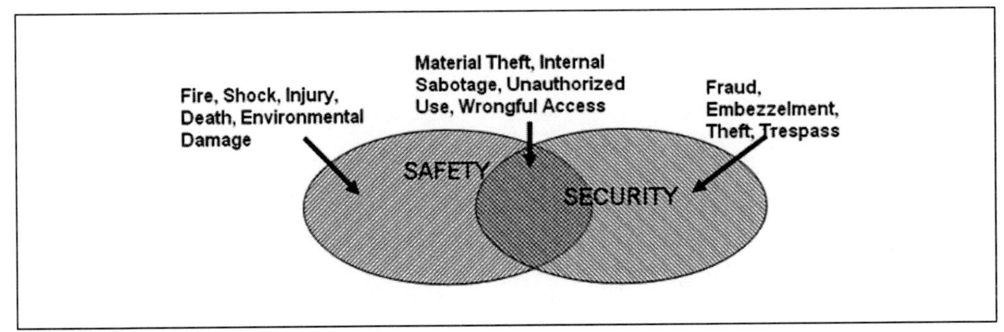

Figure 9-3. Industrial Safety and Security Overlap

All aspects of security are important, in fact, industrial cybersecurity is following the same implementation path as industrial safety (defense in depth, etc.), because cybersecurity vulnerabilities can be much more than just the theft of confidential data. Safety in industrial areas encompasses material for several volumes. Just the overlapping areas can lead to safety compromises, often in an urgent matter. The consequences of a safety breech (material theft, sabotage, and/or unauthorized or wrongful access) can most certainly result in dire circumstances.

Risk and Consequences

Any modern industrial control system may face some security risk and, although that risk may at times be difficult to estimate, it must still be accounted for. Risk cannot be ignored; however, there is no way to afford perfect security either. Good business practice dictates that a balance be struck between the cost of measures needed to mitigate a risk and the potential cost of that risk occurring. To correctly strike this balance, an industrial facility's risk managers must understand the factors for determining the security risk to the facility.

In a rapidly changing technological climate, a facility must have the capacity to continuously monitor the cybersecurity risk factors to determine if they are changing, and to determine the limits within which they may change before action is required. To be effective from both a technical and a cost perspective, the actions taken to minimize damage to a facility's assets must adapt to changes in threats, vulnerabilities, target attractiveness, and/or consequences. Let's define some of these terms:

- **Asset:** Anything—person, environment, facility, material, information, business reputation, or activity—that has unique value to a facility owner or an adversary.
- **Consequence:** The amount of loss or damage that can be expected from a successful attack against an asset. Loss or damage may be monetary, political, morale, operational effectiveness, or other business impacts.
- **Countermeasure (also called a *control*):** An action taken or a physical or technical capability provided whose main purpose is to reduce or eliminate one or more vulnerabilities or mitigate consequences.
- **Exploit:** A method, procedure, or mechanism used against a computer or system that takes advantage of (exploits) some known or hitherto unknown vulnerability.
- **Incident:** An occurrence resulting from an exploitation of a security vulnerability.
- **Local Exploit:** A type of exploit that is applied via having physical access to the target system. With physical access, many options are available that can't be pursued remotely, such as making use of CD/DVD drives or USB ports.
- **Remote Exploit:** A type of exploit that occurs via a communication pathway to the target system. Such exploits attack open ports and vulnerable services, usually by sending invalid messages that cause program faults and then injecting a payload.
- **Payload:** Malicious computer instructions carried by a worm, virus, or Trojan that are meant to be inserted into, or written over, valid program instructions.
- **Risk:** Potential for damage or loss of an asset. Risk is an expression of the likelihood that a defined threat will exploit a specific vulnerability of a particular attractive target or combination of targets to cause a given set of consequences.
- **Target Attractiveness:** An estimate of the value of a target to an adversary. Targets can be valuable to an adversary for a wide range of reasons, including terrorism or economic impact.

- **Threat:** Any indication, circumstance, or event that has the potential to cause the partial or total loss of, or damage to, an asset.
- **Vulnerability:** Any weakness that can be exploited to gain access to and compromise an asset.
- **Compromise:** To alter the functionality of an asset, including: causing total or partial loss of functionality, adding undesired or unauthorized functionality, and/or altering functionality.

Again, risk reduction must be balanced against the cost of the actions taken to reduce the risk. The goal of a *risk analysis* is to quantify risk. The reduction in risk for a given countermeasure should include a cost analysis of the countermeasure:

$$Risk = Threat \times Vulnerability \times Consequence$$

Risk and consequence, as used in this equation, are many times expressed in terms of dollars but could be expressed in product loss, personnel loss, or any detriment to a business. Threat and vulnerability are usually expressed as probabilities or percentages. As figure 9-4 illustrates, the optimal level of security is the point at which you get the most "bang for your buck." Here, the cost of security means what it will cost after the countermeasures (if any) have been employed. Note that there is a point of diminishing returns at which the cost of security rises exponentially, yet the level of security (risk reduction) does not. Unfortunately, the risk-consequence analysis cannot always be done strictly in terms of cost since the potential consequences can extend to loss of life, loss of limbs, or severe injury; damage to the environment; or loss of company reputation, to give a few examples. These factors are not easily (or even appropriately) assigned a dollar value.

The optimal level of security is the level you can afford balanced against the cost of the consequences at that level of security, in other words, the most security you can afford.

Another way of looking at threats and consequences of the threats coming to fruition is to use probabilistic risk assessment strategies in which consequences, especially those that are difficult to assign a monetary value, are given a general level of priority and threats are assigned a likelihood factor. In this type of scheme, the usual approach is to use three or four priority/likelihood levels. This approach is effective when precise numbers are not available

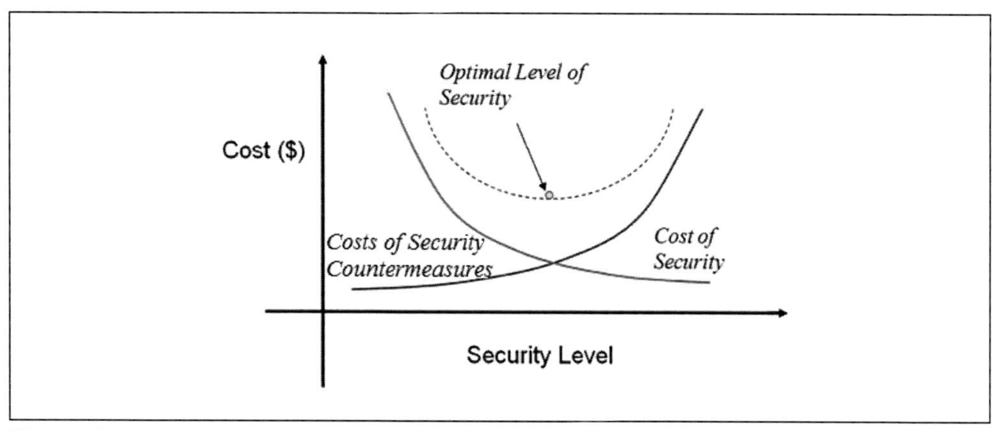

Figure 9-4. Security Costs vs. Level of Security

and when experience, common sense, and "gut feel" may have to be the basis for decision making. Table 9-1 shows an example of a consequence assessment that includes a range of potential consequences, each of which is assigned to a level of prioritization: high, medium, or low. High-ranked consequences are those that we want to avoid and we would be willing to invest in measures to ensure that they don't occur. Medium-ranked consequences are those that are worth making some additional effort and expenditure to avoid but only after implementing measures to address the high-ranked consequences. Low-ranked consequences are those that are tolerable to the organization and deemed not worthy of expending any significant additional effort to avoid. This example uses three priorities, however, four levels could be used instead. This approach gets too complicated if you try to devise too many levels.

The second consideration in a probabilistic risk assessment is the likelihood of an event occurring. Asteroids do strike the earth every so often and cause great damage, yet no one buys asteroid insurance for their home. This is because the likelihood of having one hit the earth is so low as to approach zero, although every few million years it does happen. As with all threats, cyber threats have to be given a likelihood ranking, in addition to a consequence priority. Table 9-2 shows various threats and assigns both a consequence and a likelihood ranking to each.

Table 9-1. Assigning Consequences a Priority/Category

Consequence								
	Risk Area							
	Business Continuity Planning		Information Security			Process Safety		Environmental Safety
Category	Mfg. Outage – One Site	Mfg. Outage – Multiple Sites	Cost	Legal	Public Confidence	People – On-site	People – Off-site	Environment
High	> 7 days	> 1 day	> $500 million	Criminal offense – felony	Loss of brand image	Fatality	Fatality or major community incident	Citation by regional/ national agency or long-term, significant damage over large area
Medium	> 2 days	> 1 hour	> $5 million	Criminal offense – misde-meanor	Loss of customer confidence	Lost workday or major injury	Complaints or local community impact	Citation by local agency
Low	< 1 day	< 1 hour	< $5 million	None	None	First aid or recordable injury	No complaints	Small, contained release below reportable limits

Table 9-2. Assigning Threats a Consequence and a Likelihood Ranking

Ranking based on a combination of likelihood and resulting consequences		Consequence		
		High	Medium	Low
Likelihood	High	Remote Operator Console support, using dial-in telephone access, is discovered by a war dialer program and hacker gets operator access to control system and trips a critical unit	Virus infects operator consoles and blinds operators so that process has to go to safe mode and production is halted for the entire day	Hacker gets into email server and causes email spam storm on plant business and engineering LAN
	Medium	Employee quits and dials into the Remote Operator Console modem from home and uses still-valid ID to operate valves and pumps so as to release toxic gas	Malware jumps from the business WAN onto the plant LAN and infects the DMZ-based historian causing loss of environmental reporting data	Malware jumps from the business WAN onto the plant LAN and attempts to locate and exfiltrate credit card information
	Low	Worm gets through all firewalls and into the primary and redundant process controllers and initiates an explosive process upset	Worm gets through all firewalls and into the operator console and modifies all operator inputs so that product is compromised and must be destroyed	Worm gets through all firewalls and into operator consoles and causes annoying messages to appear on the screen

If a threat is both highly likely and has a high consequence, this is clearly a threat to be taken seriously and one that requires the use of countermeasures to reduce its likelihood and its resultant consequence to acceptable levels.

Sources of Threats

What are some of the sources of security threats? They come in two varieties: external and internal.

External threats include:

- Script kiddies (those that modify or copy an original exploit by any of the below)
- Recreational hackers
- Virus writers and malware in general
- Activists/hacktivists
- Terrorists
- Agencies of foreign states
- Disgruntled former employees or contractors

Internal threats can be broken down into unintentional and intentional threats:

<u>Unintentional</u>

- Inappropriate access to systems or data
- Inappropriate manipulation of systems
- Incorrect configurations
- Conflicting software
- Infected software or hardware

<u>Intentional</u>

- Disgruntled employees (presently employed)
- Disgruntled contractor (presently contracted)

For external attacks, risk managers might well ask: "We have an industrial control system and it is not connected to the Internet, so how do any of those external threats affect me?" The system does not have to be connected to the Internet to be threatened. It can be attacked through any external connection: a support group or vendor modem connection or VPN access, an engineer using PCAnywhere, or even a remote monitoring workstation located within another corporate facility.

An internal survey of companies using industrial control systems would find that most managers believe their critical control systems are not connected to the business network, let alone to the Internet. However, an investigation would most likely show that every system is connected in some way to the enterprise (or management) network and that the business network is connected to the Internet using only the security needed to support the business processes. The security employed for the business process is usually not comprehensive enough to protect critical systems.

The ISA99 (International Society of Automation) committee has made numerous recommendations regarding the types of security strategies that ought to be deployed to secure industrial automation systems and facilities. In addition, the North American Electric Reliability Corporation (NERC) has issued a set of Critical Infrastructure Protection (CIP) guidelines for providing physical and cybersecurity for electric power grid facilities and assets. Organizations, such as the Interstate Natural Gas Association of America (INGAA) and American Gas Association (AGA), have published cybersecurity recommendations for protecting industrial automation systems. IEC/ANSI/ISA-62443 cybersecurity standards (formerly ISA99) is a series of published standards for implementing a cybersecurity program. We will look at the recommendations from the standards later in this chapter.

Security Vulnerabilities

The vulnerabilities that attackers can take advantage of reside in many systems and, hence, are well known. The reasons they are well known include:

- The use of common IT systems in supervisory control and data acquisition (SCADA) systems, such as Windows-based PCs for workstations and SQL Server for database and archival services

- The use of common networking technologies, such as Ethernet and TCP/IP, in distributed control systems (DCS), SCADA, and most industrial systems
- The use of embedded systems not originally designed for security
- The use of off-shore or outsourced development for critical software

Operating Systems

Security vulnerabilities in major operating systems (OSs) account for most instances in which a device is taken over. Most control systems rely heavily on Microsoft Windows or Linux and vulnerabilities (flaws) in these operating systems are well understood by hackers.

Security patches are a necessary evil. Given the complexity of modern operating systems and their interoperation with software applications, patches may fix a problem but they also may cause an application to fail, particularly if it is a little-known or custom application. Patch management is made much more complicated for the industrial user by control and safety issues. As an example of this complexity, author Thompson (author of this book) attempted installing Microsoft's Internet Explorer 7 (IE is now at version 11, so this was several years ago), a free download. It was found that some settings were irreversible (could not be uninstalled), except through registry editing, as they work well for most environments. Yet Thompson found several script errors, including one that wouldn't allow him to enter a zip code when ordering an item. Now this is not the end of the world, yet suppose the flaw involved a safety or control issue. It is certain that IE, in all of its revisions, was thoroughly vetted and went through numerous betas—a more rigorous process than most security patches go through—however, errors were still found. This is the reason that companies using standard OSs and/or applications have to ensure that, in every conceivable instance, the patch does not break their application. This is why you pay maintenance fees to DCS or SCADA vendors who do their own testing and issue their own patches. If you don't install the patches, viruses can allow Windows- or Linux-based HMI, programming stations, and the like to be taken over.

Unlike standard operating systems that are designed with security in mind (yet have security problems), SCADA and control systems were originally designed for performance, not cybersecurity. Likewise, the conversion of smart devices onto Ethernet-TCP/IP networking was generally done without any regard for cybersecurity. The IP stacks in such devices are usually minimal and lack the error checking and recovery functions found in a fully implemented stack. As a result, certain programmable logic controllers (PLCs), instruments, analyzers, and other devices fail, or at least cease communicating, while being vulnerability-scanned, which is a typical operation by a hacker or vulnerability checker. Such failure indicates a serious TCP/IP implementation problem. In addition, some PLCs have legacy commands still deployed on them that are dangerous from a cybersecurity standpoint and nearly all PLC/DCS/SCADA systems have authentication schemes that require only a password (possibly a single one for all users) for access.

Wireless Networks

As IEEE 802.11 (wireless Ethernet) has become more widely used in industrial settings, it appears that, all too often, these systems are deployed without any security features enabled (Wired Equivalency Protection [WEP], Wi-Fi Protected Access/Wi-Fi Protected Access II [WPA/WPA2], or 802.11i). Although this may reflect the "just get it to work, we will secure it later" attitude, somehow "later" never occurs.

IEEE 802.11 systems, which used WEP, had several well-known security flaws:

- The encryption scheme can be and has been cracked
- The authentication scheme does not protect the system as a whole
- The Media Access Control (MAC) layer is susceptible to simple (but effective) denial-of-service attacks

The interim solution was to use the WPA or WPA2 security schemes, both of which offer a much higher level of security then WEP (although WEP combined with features like MAC filtering is better than nothing). WPA was created by the industry trade group Wi-Fi Alliance. It was designed to be used with an 802.1X authentication server, which distributes different keys to each user (enterprise mode). WPA can, however, be used in the pre-shared key (PSK) or "personal" PSK mode, in which each user has the same pre-shared key.

WPA was introduced as a temporary fix to the security failings of WEP and was based on the preliminary IEEE 802.11i specifications, which are intended to be the currently-best solution to wireless Ethernet cybersecurity. With

WPA, data is encrypted using a 128-bit key with a 48-bit initialization stream. Another major security advantage of WPA/WPA2/802.11i over WEP is the Temporal Key Integrity Protocol (TKIP), which dynamically changes keys during system use (in WEP the same key is often used in perpetuity). However, when newer schemes utilized much larger key initialization stream, the inherent key weakness of WEP is removed.

WPA2 (with vendor certification by the Wi-Fi Alliance) fully implements the IEEE 802.11i standard. In particular, it introduces a new algorithm based on the Advanced Encryption Standard (AES) (in WPA2, AES is optional; in 802.11i, AES is the only choice). This algorithm is considered to be cryptographically *very strong* (it would be exceedingly difficult and time consuming to break). Official support for WPA2 began in Microsoft Windows XP (May 2005) and is supported in Windows Vista (Server 2008), Windows 7 (Server 2008 R2), Windows 8 (Server 2012), and Windows 8.1 (Server 2012 R2, July 2013).

Whether compromising an 802.11 network is considered an internal or an external threat is sometimes open to debate. We will treat it as an external threat in this book because the network's radio transmissions extend over the facility's boundaries, making them available to personnel outside the facility who may not have the company's best interests at heart. Although all the 802.11 variations have official distance limitations in their specifications, in the real world, hackers have demonstrated that they can contact access points (APs) from many miles away using amplifiers and special antenna designs. Hackers have a game called *war driving* where they try to locate and connect with wireless APs. The current distance record is over 20 miles.

Hardware

There are several other subtle ways in which outsiders can become "insiders." One is the use of laptops. Laptops often move between secure and insecure areas, and laptop software tends to be less controlled than software in desktop systems. A laptop can become infected while it is in someone's home and then can bring malware into the secure work environment. Once inside, the attacker/virus can take advantage of operating behind the company's firewall, spreading to other systems or transmitting information outside of the company using protocol tunneling (creating an outgoing web browsing session that hides the actual message content). Key logging programs will surrender user passwords and eventually, unless detected and removed, the malware infection will have compromised the entire system's safety, performance, confidentiality, integrity, and availability.

The *flash*, *pen*, or *thumb* USB drives, which are small in size and large in capacity (32 to128 GB), are as susceptible to infection by viruses and other types of malware as any floppy disk or hard drive. These drives are highly portable, which is the reason they are used, and a major system can be compromised by the use of these drives. When inserted into a PC or server, the file system is automatically mounted and any auto-execute file is automatically loaded into RAM and executed unless this function has specifically been disabled either by local security policy (on this PC) or a group policy on the domain server (in an Active Directory domain).

Ordinary CD and DVD media is also a commonly used means for planting malware into a computer. When people see an unlabeled DVD lying on the floor, they feel compelled to insert it into their PC to find out what it contains (the same is even truer for an unlabeled thumb drive). Leaving such infected materials in the public areas of a facility or an office building is a common social engineering ploy. Social engineering uses human interaction (social skills) to obtain or compromise information about an organization or its computer systems (according to the U.S. Cyber Emergency Readiness Team [US-CERT]).

Many portable electronic devices, particularly those with integral wireless and USB connectivity, can be accidentally or intentionally used to both deliver malware and to infiltrate files and data. There is malware that can leap from a smartphone to a laptop PC or to a network-connected printer via Bluetooth or Wi-Fi. An MP3 player, digital camera, or tablet can exchange files (including installing malware) via a USB connection to a PC. Organizations need to have comprehensive personnel security policies with regard to allowing such devices near critical automation systems.

Methods of Attack

What is the risk of an outside entity intercepting and acquiring process data? Doubtless, a reply could be: "What possible use could anyone make of some process data?" Unfortunately, hacking may take the form of industrial espionage. For example, given the fluctuation of gasoline prices from 2004 to 2014, it is not hard to imagine an operator of a fuel tank farm wanting to know what margin he can make when he has to provide a quote for an urgent large order request. Wouldn't it be useful to know the status of the other tank farm(s) in the region and what they are quoting? How useful would it be to get a daily printout of competitors' margins and wholesale prices? In the global economy, it is clearly as important to protect private process data as it is to safeguard commercial interests, sales, trade secrets, and anything that would aid a competitor.

Denial of Service

One of the most common exploits (because it is the easiest to do) is the denial-of-service (DoS) attack. This is an attack that is designed to render a computer or network incapable of providing normal services; it is deadly to a process control network. (The final results, due to process instabilities caused by unavailability of resources, are quite possibly deadly to all of those around said processes as well.) A DoS attack can be performed in several ways: *Bandwidth attacks* flood the network with such a high volume of network traffic that all available network resources are consumed and legitimate user requests cannot get through. *Connectivity attacks* flood a computer with such a high volume of connection requests that all available operating system resources are consumed and the computer can no longer process legitimate user requests.

The risk of being unable to view or control a process or system due to a DoS attack or any other intrusive, invasive, or performance-altering attack forces facilities to place great reliance on emergency and safety systems. These systems have traditionally been separated and made independent of the main control system and are generally considered to be "bullet proof." Regardless of the environment and situation, they will work as they are supposed to work. However, in following the current trend in designing the main control systems, emergency and safety systems are being based more on standard IT technologies, such as Ethernet and TCP/IP. The flexibility of software combined with faster data communications (not to mention the cost savings) is resulting in an increase in safety systems being connected to, or even combined with, the main control systems. This, of course, increases the potential risk of a common mode failure of both the main control system and the safety systems. For these reasons, the risks of cyberattack need to be considered not just when designing control systems, but when designing the safety systems as well.

Social Engineering in IACS

As previously mentioned, one form of external attack is *social engineering*, in which an attacker solicits information from the victim, who gives it to the attacker without suspecting its actual purpose. Social engineering methods may be used to gain entry into restricted areas, to gain access to information that can aid in staging a cyberattack, or even to trick personnel into inadvertently launching an attack. This can be significant to an IACS's personnel, as they are generally more isolated from the day-to-day travails of the office cyber environment. As an example of a social engineering attack: a firm e-mails you with an incentive (e.g., money, tickets, passes to a restaurant or play, a thumb drive) to answer an industry-specific questionnaire that asks you to divulge operating systems, types of equipment, firewalls deployed, and so on. Except for your facility, no one needs to know this information. Some legitimate vendor firms would like to know this information to tailor their bids but, by and large, most such requests should be ignored.

Another example of social engineering in an IACS: an attacker pretends to be an employee out in the field who has lost his contact information and calls another employee to get the remote access phone number. Does the employee receiving the call actually know who is calling him? What could the caller gain from using this remote access phone number? Why are employees giving out such private information without authentication?

Another distressful example (in effect everywhere, not just in IACSs) is *phishing*. Here, an e-mail arrives with the proper letterhead, web addresses, and so on, and says something to the effect that your account needs to be refreshed or substantiated and you need to respond with your user name and password. For phishing involving financial institutions, credit cards, or Internet banking, information such as user name, PIN, and credit card number are elicited. Respondents to these authentic-looking but entirely false e-mails will find themselves victims of identity theft or their accounts will be plundered. In an IACS, the e-mail would apparently come from corporate IT, asking for a user name and password. Many such emails provide a hyperlink that can be clicked to assist the victim with more information or offer some other plausible reason to click. Such hyperlinks invariably point your browser to a malicious web site that will then attempt to download malware into your PC.

May it be stated here (and repeated regularly) that no one needs to know your password. IT cannot see your password; they can only give you a new one. *No one EVER needs your password except to log on as you.*

Web-Based Research

One method attackers use to glean information from public sources is generally implemented before they decide what type of intrusion to employ. This method consists of using numerous sources of data to find out who owns your system and who has that IP. (Just type your name into an Internet search engine, tell it to GO, and marvel at the results.) Table 9-3 includes a list of common information resources that an attacker might use.

Table 9-3. Research Sources

Resource	Purpose
Ping (ICMP Echo Request) – built into most TCP/IP suites	Ping is used to determine if an IP is alive and what its response time is. Ping can use the DNS to return the IP address for the URL name.
Whois	Whois database lookups will provide information on the owner of an IP, when the IP was established, and a contact number.
Domain Name Service (DNS) – largest distributed database in the world	This database contains all the information about a given domain and can be accessed using nslookup and other hacker tools.
Traceroute (tracert) – usually provided with the TCP/IP suite	This resource lists router hops (with IP/DNS name) between the source and the target.
SMTP VRFY	This tool determines if an email address is valid. When used in a script with a file of common names, it can be used to gather valid user IDs (not passwords) from a system.
Research and Attack Portals	These are similar to Sam Spade software (which operates on an individual machine) but they enable users to access the portal via a browser. Traffic appears to come from the web server instead of the client machine.
Hacker Tools	Hacker tools are easily found by entering “hacker tools” in a search engine. There are many and diverse tools to assist in automating hacker attacks, as well as simple one-run exploits. NOTE: If you want to research these tools, it would be advisable to do that on a stand-alone machine that is not essential to the plant and has no useful information on it. Many times when you download from these sites, hacker insertions, viruses, Trojans, and malware galore come with your download.

Here are a few examples of ways that hackers might use the Whois Search command and Domain Name System (DNS) server sources to gather information:

Examples

Whois Search on the Internet Network Information Center (InterNIC)

Below is information that can be obtained by entering a domain name in the search box at www.internic.net/whois.html:

- Company name: Registrar used to register domain name
- Registrar website: (www.yournetwork.com)
- Company name:
- Telephone numbers for locating modems behind firewalls and those with unsecured remote access
- Contact and email for social engineering
- DNS names servers: DNS information

Whois Search on the American Registry for Internet Numbers (ARIN)

Below is information that can be obtained at www.whois.arin.net:

- Company name
- All IP addresses assigned to that organization

DNS Server Records

Entering an nslookup query in a command line utility will return:

- Domain name
- Specific IP address
- Host system type of domain name

Other Investigation Tools

There are many free tools available on the Internet that are intended for use in vulnerability testing, however, they can serve a darker purpose: most are also handy for hacking. The Sam Spade utility (figure 9-5) was designed as a useful tool for troubleshooting DNS and IP addressing problems, yet it can also be used to gather information that would be useful in staging a cyberattack (such as IP addresses) or a social engineering attack (such as names and addresses of personnel.)

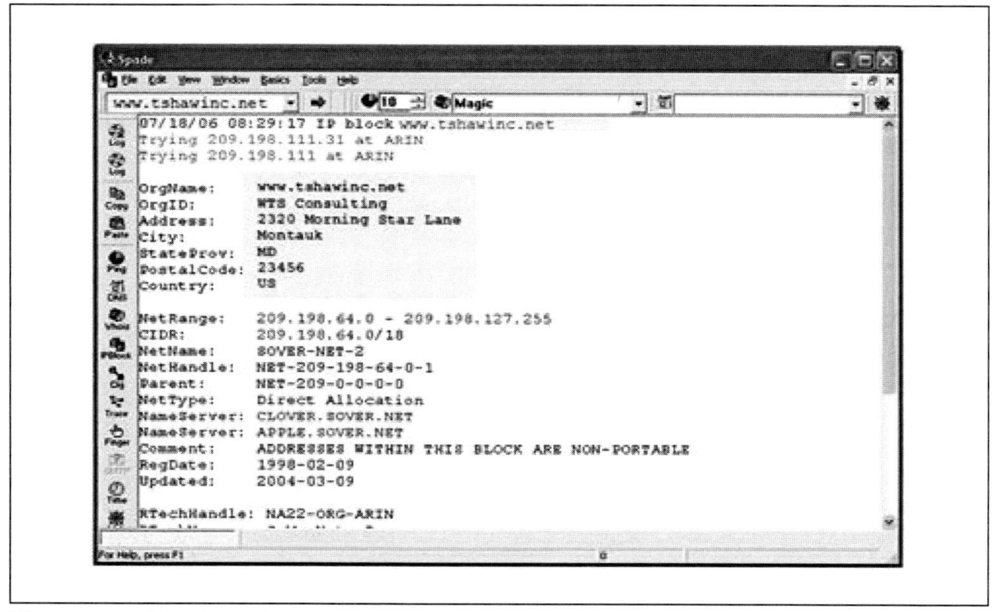

Figure 9-5. The Sam Spade Utility for Information Gathering

Password Cracking

A system's main protection from both external and internal attacks is the password. If you think passwords provide adequate security, you may also believe in the tooth fairy. The problem, simply stated, is this: people are human and attempting to have a person remember a strong password is a difficult task. A string of long passwords is next to impossible to recall. An effective password would be a string of 32 random upper and lower case alphanumeric characters with punctuation. But who can remember passwords of that length? If it is hard for the person who created the password to remember it, think how difficult it will be for an attacker to guess or even attempt to break. However, when a password cannot be remembered, it will be written down. And where would it be kept? On a sticky note on the monitor or, if the user is exceptionally security conscious, on the bottom of the keyboard?

The only way a secure password will be used is through personnel security dictates. If the user is allowed to generate his or her own password, it will be short and easy to remember. Password-cracking systems know this. Quite often passwords and user IDs are:

- Identical, usually the user's first or last name or the word "password"
- System defaults ("sa" for SQL Server, blank for XP Home users)
- Easy to guess, such as a company name, variations of user names, or TV characters (particularly Star Trek or other techie shows)
- Guest accounts or active terminated accounts
- Found on sticky notes attached to monitors

Automated hacker tools, such as Crack, L0phtCrack, and John the Ripper, tend to brute-force password cracking (the continued high speed attempt to go through tables of possible values of password characters—assuming there is no account lockout after x attempts fail) but can quickly determine short and common passwords. Another way to crack passwords is via *rainbow tables*, which are huge lists of pre-encrypted passwords that allow almost instantaneous cracking of even long, complex passwords. In all Microsoft Windows systems, up to and including XP, the user passwords were stored in a legacy LM (LAN Manager or LAN Man) format for backwards compatibility with old systems/workgroups, which turned out to be incredibly easy to crack. That legacy feature was removed as of Windows Vista.

Cain and Abel (often just abbreviated to Cain) started life as a password recovery tool for Microsoft Windows. It can recover many kinds of passwords using methods such as: network packet sniffing, cracking password hashes using various methods, and cryptanalysis attacks. Cryptanalysis attacks are done via rainbow tables, which can be generated by Cain and Abel see figure 9-6).

Vulnerability and Exploitation Tools

It used to be that to attack and compromise a computer, either across a network or via physical access to the computer, required extensive programming, networking, and operating system knowledge. Unfortunately, that is no longer the case. Incredibly powerful tools that require minimal computer proficiency are readily available. One of the most powerful is the Metasploit Framework, especially with the Armitage graphical user interface (refer to figure 9-7).

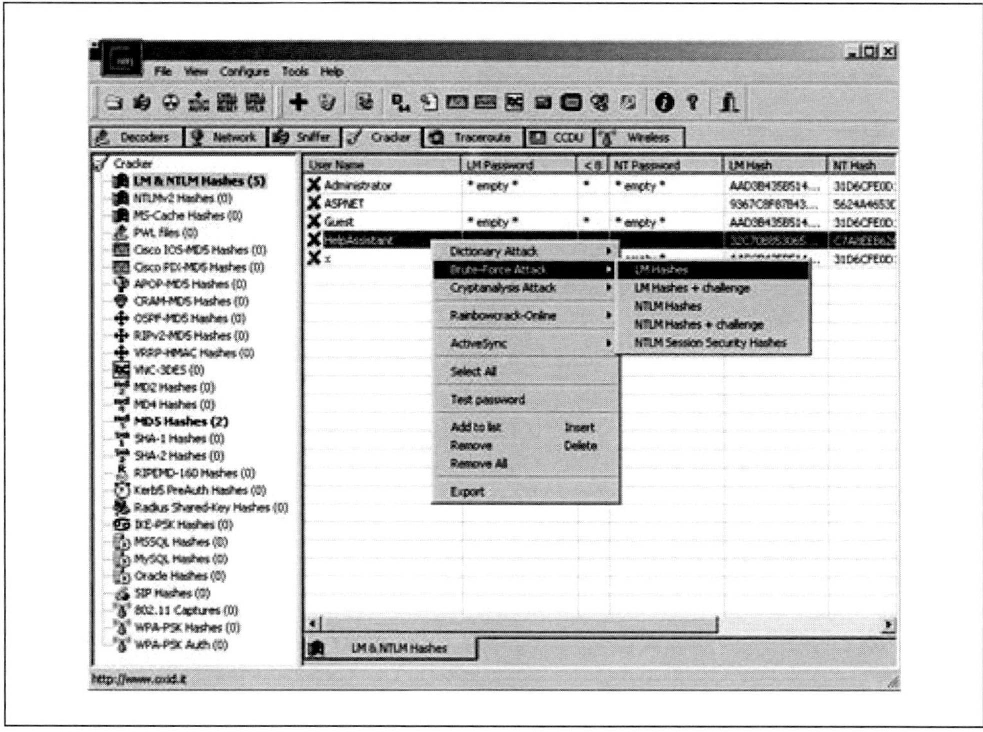

Figure 9-6. The Cain and Abel Password Recovery Utility

This tool can work with other tools, such as port scanners, and totally automate the process of locating, scanning, and selecting potential remote exploits to be used for prospective and likely cyber vulnerabilities. (Ostensibly, the Metasploit Framework is to be used by the good guys for penetration testing and not necessarily by hackers; however, this may not always be the case.) This is not a book on hacking (that would fill several more books), rather it is about communications awareness. Knowing that tools like Metasploit exist—and that they can be used enable attackers to locate and break into or otherwise compromise systems from across a corporate WAN and even the Internet—is critical to assessing cybersecurity risk.

Internal Threats

The greatest internal threat is a disgruntled employee. Most intentional attacks are primitive and include hacking behind the firewall, sabotaging files and applications, inserting time bombs (malware programs that initiate long after the employee has left), and password theft. Most of these attacks come from an employee who already has access to the system.

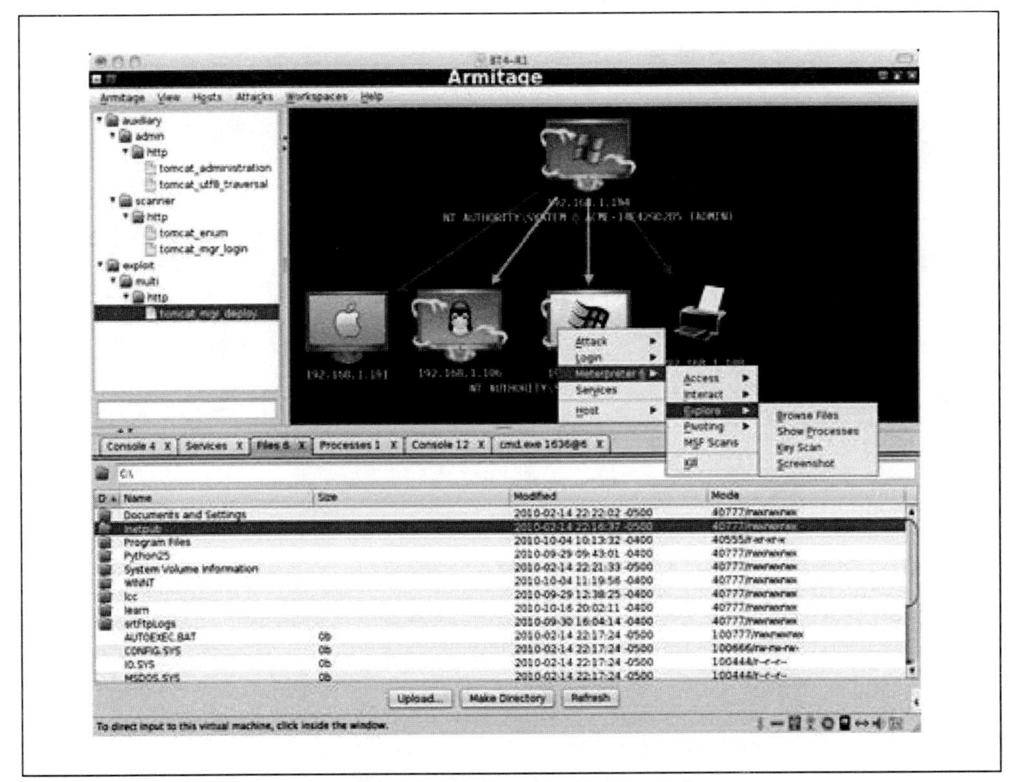

Figure 9-7. The Metasploit/Armitage Tool and User Interface

Other internal threats are not intentional and are generally caused by a faulty system design (i.e., not idiot proof), lack of communication, improper training, bad procedures, curiosity, and pure accident. These are minor concerns (unless an incident resulting from such a threat brings your plant down or has other serious consequences), and intentional threats are far more prevalent than unintentional threats—a point that should be kept in mind when you are performing risk analysis. Remember:

$$Risk = Threat \times Vulnerability \times Consequence$$

Risk Analysis

Assessing the value of an industrial cyberattack is not simply a matter of assigning a financial value to an incident. Although obvious direct financial consequences may be easily quantifiable (e.g., loss of production or damage to the plant), other consequences may be less obvious. For most companies, the consequences of an attack intended to damage the company's reputation may be far more significant than just the cost of a production outage. The impact of a cyberattack on health, safety, or the environment could be serious to a company's brand image. Even consequences as minor as a limited regulatory contravention may impact a company's reputation or, possibly, its operating license.

There are five steps to risk analysis:

1. List the assets to be protected
2. Identify threats
3. Determine vulnerabilities
4. Rate risks
5. Select countermeasures to mitigate risks

Step 1 is self-explanatory, and we have already discussed Step 2. We will now consider Step 3: Determine vulnerabilities. This step consists of determining what the IACS's vulnerabilities are and what they could be. The following are lists of typical system vulnerabilities—the lists are not inclusive. Most systems undoubtedly have other vulnerabilities.

1. Control system vulnerabilities:
    - Internet/intranet connections
    - Remote access
    - Laptops/tablets and USB drives
    - Modems (not used in a plant IACS, but may be found in some SCADA systems)
    - Wireless communications
    - Theft of handheld devices or laptops
    - OEM, vendor, and third-party access
2. System environment vulnerabilities:
    - Poor password management
    - Gullibility of people (social engineering)
    - System default configurations

- Lack of computer security policies
- Lack of communications with Human Resources, Contract Management, and other affected departments
- Lack of computer security training for personnel
- Inappropriate trust relationships between domains

In order to rate the risk of these vulnerabilities, you must determine their possible consequences. Below is a list of typical consequences, it is not inclusive. The consequences for the system being evaluated will depend on the system, processes, process materials, and environment, among other factors.

The possible consequences of a compromised system may include:

- Endangerment of public or employee health and safety
- Customer safety risks
- Interruption of production or other economic loss
- Violation of regulatory requirements
- Damage to company reputation
- Lowered corporate stock price

The lists of vulnerabilities and consequences provided here are based on ISA course IC32E, "Cyber Security for Automation, Control and SCADA Systems." Anyone involved in security for automation systems should attend that, or an equivalent, course. Table 9-4 is a typical small segment of a complete risk analysis table developed from examples in the ISA course. In this table, only one threat is listed, threat number 1.001, out of a possible many. There are two vulnerabilities listed that could bring about this threat. There are four consequences listed, two for each vulnerability, and there are four listed safeguards. Also shown are recommendations to enhance the safeguards and reduce the vulnerabilities, and the departmental responsibility for taking action on those recommendations.

Table 9-4. Example of Risk Analysis

Threats	Vulnerabilities	Consequences	Safeguards	Recommendations	Action
1.001 Manipulate PV/ SP by remote access, causing process upset	1. VPN from vendor	1.1 Possible on-site fatalities	1.1 SIS independent of main system	1.1 Restrict physical access to VPN equipment	HR
		1.2 Possible off-site fatalities (if any)	1.2 Process upset does not have potential energy or toxicity to cause off-site damage, directly or indirectly		IT
	2. Remote access by modem	2.1 Possible on-site fatalities	2.1 same as 1.1	2.1 Eliminate modem access	
		2.2 Possible off-site fatalities (if any)	2.2 same as 1.2		

IACS Countermeasures

Some of the actions that may be taken to help protect an automation system from an internal or external attack include the following:

- Train users (employees, contractors) how to identify/recognize, how to react to, and how to report social engineering.
- Locate modems, wireless access points, and wireless-enabled devices before attackers do. Ensure that security is enabled and that modems use callback; or better, remove all modems behind the firewall (best: use VPN technology for all remote access).
- Define and implement a password policy, enforce it, and train users in it.
- Develop Help Desk procedures to authenticate users who are requesting password resets.
- Use router-based firewalls to deny most incoming and some outgoing connections through a defined policy.
- Use a port-scanning tool to scan systems to identify open ports. Close all unnecessary ports (TCP/UDP) and uninstall unused software.
- Configure router services to allow Internet Control Message Protocol (ICMP) Echo Requests (pings) only from the ISP's management systems.
- Use an Intrusion Detection System (IDS).

- Regularly run a vulnerability scan against your own network.
- Do a scheduled update of user lists. Delete user information for personnel who have been terminated (at the time of notification if possible), who are retiring (effective date), and who are undergoing changes in their organizational responsibilities that give them more or fewer permissions.
- Restrict Domain Name System (DNS) information leakage:
  - Make sure there is no DNS HINFO (system information record) and no text files concerning Internet-accessible machines.
  - Restrict zone transfers (forward or reverse—a mechanism available to replicate DNS databases across DNS servers) to appropriate servers.
  - Employ a split DNS to exclude internal-only systems information.

Firewalls

Of the technologies available to secure and protect critical communications, one of the most basic is the firewall.

A firewall is somewhat like a filter placed in a liquid stream to block undesirable chemicals, particulates, or compounds by capturing them. A firewall is placed in a communication path; it inspects each message passing through—in either direction—and it blocks the message or allows it to pass. How that decision is made and what information is used in the decision making has to do with the differing types and capabilities of firewalls. A firewall is a program. It may run as a service on the computer or device it is protecting, such as the Windows 8 firewall or a basic firewall in a router or L3 switch, or it may run on a separate computer platform as a security appliance.

All firewalls are not equal. Firewalls may perform various functions including standard packet inspection, stateful packet inspection (SPI), application proxy, application layer filtering (ALF), and even deep packet inspection. If a firewall provides the interface between a corporate WAN and the Internet, it is an enterprise firewall and it may support many additional functions and features (e.g., NAT, VPN gateway, NAC). More sophisticated firewalls (especially enterprise firewalls) may run on a commercial operating system, such as Windows, UNIX, or Linux, and, thus, may themselves need to be hardened (hardening will be discussed later in detail) and patched. It should be noted that there are exploits that have been and can be used against many firewalls that run on commercial operating systems, so patches and security fixes must be current on these devices.

Standard packet inspection firewalls do most of their work at the Network layer (Layer 3) and they are usually called *packet filter firewalls*. They:

- Examine only the headers of each packet of information
- Accept or reject each message based on the packet's sender address, receiver address, or TCP port
- Perform firewall checks on each incoming or outgoing packet for its source address, destination address (to match against source/destination rules), and the desired function/protocol/program (based on the IP port assignment)
- Accept or reject packets based on a comparison of the above packet Layer 3 information to several predefined rules called *access control lists* (ACLs)

Standard packet inspection firewalls are typically Layer 3 only and are often router-based.

Pros: Very fast

Cons: Examine packet headers only and not the overall session or packet content (the message data)

Stateful packet inspection (SPI) firewalls surpass the effectiveness of the standard packet inspection (packet filter) firewall by:

- Tracking the relationships between packets in a session
- Tracking (and possibly timing) the state of the TCP protocol
- Tracking expected responses to UDP messages

They can be either router-based (e.g., Cisco ASA 5505) or server-based (e.g., Microsoft Internet Security and Acceleration Server).

Pros: Relatively fast
Flexible
Improved security over previous firewall types

Cons: Do not check message content, still check only packet headers

Application proxy firewalls do most of their work at the Application layer (Layer 7) but they also do all the previous firewall functions, from Layer 3 on up. They perform SPI (Layers 3 and 4) and:

- Handle packets for each Internet service by interpreting the command at the top layer and issuing a command for the function to be performed (if it meets pre-defined conditions)
- Act as an intermediary that accepts connections and requests from a client and then responds
- Interpret every incoming packet up to the Application layer, check it, and then reissue it to the target device

A more powerful version of this is **application layer filtering (ALF)**, which tracks the state and content validity of Application layer message traffic. This enables detecting tunneled protocols (e.g., using HTTP to send messages that are not HTML).

Application proxy firewalls are usually server-based:

Pros: Very strong security model

Cons: Slow and requires many CPU cycles
Only handle well-known services/protocols

Today it is also possible to have **deep packet inspection** firewalls that generate a copy of the entire TCP/IP message stream and perform malware scanning prior to re-transmitting the message to the destination. These firewalls are usually server-based:

Pros: Very strong security model

Cons: Slow and uses many CPU cycles; adding significant delays in message traffic
Must keep malware signatures up to date

By introducing a firewall between the enterprise and process control networks, the user can achieve a significant security improvement. This is illustrated in figure 9-8. Most firewalls on the market today offer stateful inspection for all TCP packets and application proxy services for common Internet Application layer protocols, such as FTP, HTTP, and SMTP. Correctly configured, firewalls significantly reduce the chances of a successful external attack on the process control network.

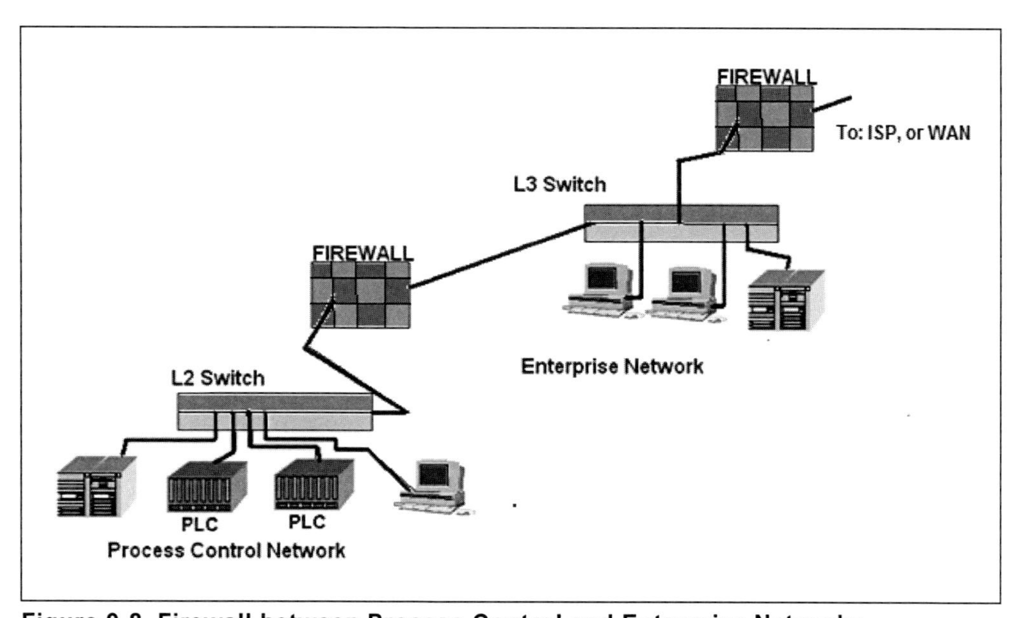

Figure 9-8. Firewall between Process Control and Enterprise Networks

Devices such as a data historian and mail server are considered shared between the Process Control Network (PCN) and the enterprise network and, therefore, can be placed in a **demilitarized zone** (DMZ) firewall between the enterprise and process control networks. Each DMZ holds a shared asset, such as the data historian, the wireless access point, or remote and third-party access systems. Incorporating a DMZ-capable firewall allows the user to create an intermediate network, often referred to as a Process Information Network (PIN).

In order to create a DMZ, the firewall must offer three or more interfaces, rather than the typical public (external) and private (internal) interfaces. One of the interfaces is connected to the enterprise, the second is connected to the process network, and the remaining interfaces are connected to the shared common asset. Using two separate, back-to-back firewalls to create a DMZ is considered to be more secure than using a single three-port firewall.

Many organizations think that a DMZ is somehow intrinsically secure and that they can put any number of systems in the DMZ and still remain secure. **This is not true**. Systems in the DMZ are visible to the unsafe side of the firewall—the outside—and can, therefore, be attacked. Once compromised, the DMZ systems can be used as a platform from which to attack the systems on the safe side of the firewall—the inside. Since the assets in the DMZ may be equally accessed by either side of the firewall (inside network and outside network) and they can communicate out to systems on both the inside and outside, these assets need to be protected and hardened. Therefore, any system in the DMZ must be stripped to minimal functionality (severely hardened), kept patched, and preferably provided with some form of host intrusion prevention (HIPS) technology such as "whitelisting." This is definitely the case for a DMZ facing the Internet, yet it is still a good idea for an enterprise-to-process control DMZ. The term "bastion host" is often used to describe such a hardened, DMZ-based system. Figure 9-9 illustrates a DMZ Firewall.

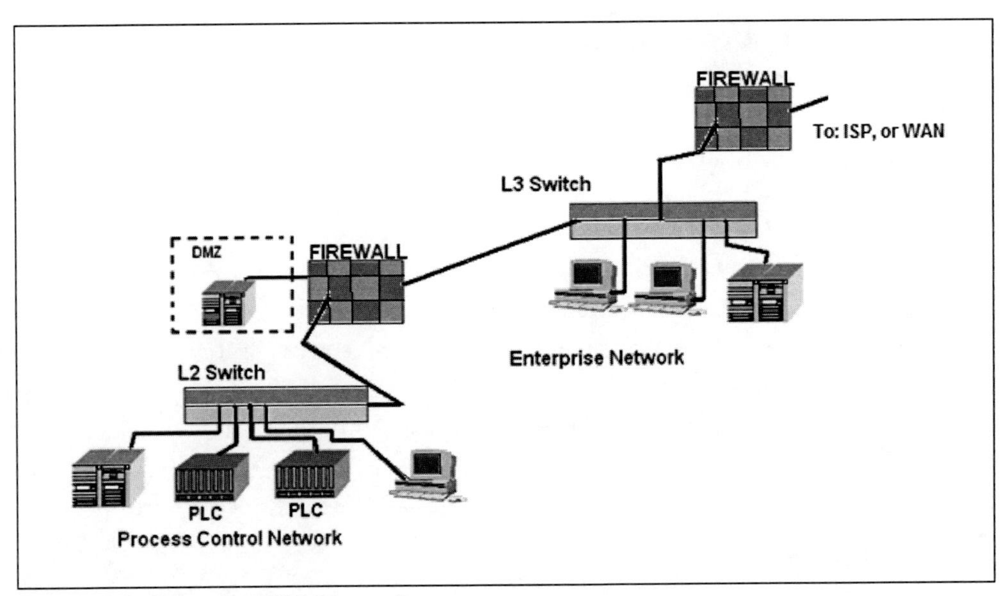

Figure 9-9. Use of a DMZ Firewall

Firewall Rules

Firewalls, regardless of type, are usually only as effective as the rules or ACL with which they have been configured. Firewall rules are an ordered list of conditions that must be checked to decide whether a message is to be passed through or discarded. The level and complexity of these conditions, and the data they utilize, has a lot to do with the type of firewall, but poor rules make for a poor firewall.

Rules consist of an "if-then" statement where the "if" is looking for specific data (like an IP address or port number) and the "then" is usually either to permit or deny the passing of the message. A major problem with firewalls is that people add rules out of order or append new rules without eliminating prior rules that conflict with the new rules. When assessing a message, the firewall starts at the top of the rules list and stops at the first rule that applies to the message; it then permits or denies the message based on the rule type. The problem occurs when there are multiple rules that may apply to the message. If the first rule the firewall finds in the list allows the message to pass, there will be no further checking of subsequent rules, even if a (misplaced) later rule would have denied passing the message.

Good firewall configuration uses specific "permit" rules for the handful of message traffic the user wants to allow, and then applies "deny" for everything else. Most firewalls have a factory default ruleset that "permits all" in either direction, essentially making the firewall a wire. Users must eliminate that default ruleset and configure a set that matches their needs.

Network Address Translation

One technique supported by most firewalls and routers when they are acting as the default gateway for message traffic leaving a local LAN is Network Address Translation (NAT). Originally designed as a way to conserve IPv4 addresses, NAT is now used to hide networks from prying eyes so that private networks can continue to use IPv4 even though the Internet itself is now running IPv6. There are several non-routable IP address ranges that can only be used on private networks; that is, if any of these IP addresses is used as the source or destination address in an IP datagram, the first router receiving the datagram will discard the packet. IPv4 addresses 10.xxx.xxx.xxx/8, 172.16.xxx.xxx/12, and 192.168.xxx.xxx/16 are all non-routable addresses. The NAT device merely maps the source computer's private address and source port number to a unassigned port number at the upper end of the range and transmits packets using the single valid IP address owned by the facility (usually its own IP address). In its transmission, the NAT device notifies the receiving station of the reply TCP or UDP port used.

Figure 9-10 illustrates one method of performing NAT. The device performing the NAT function maintains a lookup table that allows it to take returning messages and reconstruct the proper IP and TCP/UDP headers with the local IP address and source port, so it can deliver the message to the originator via the local network.

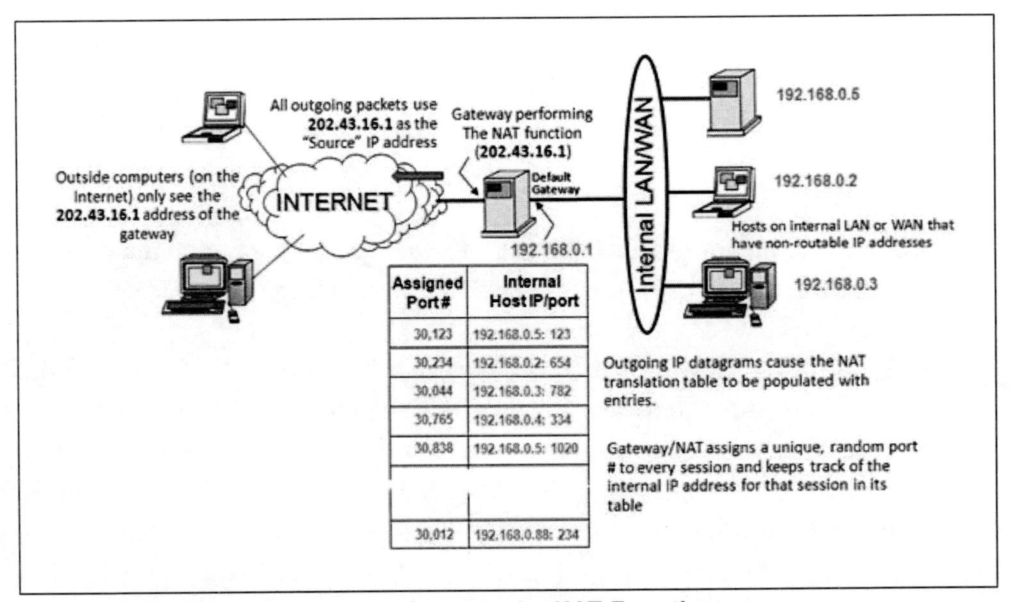

Figure 9-10. Default Gateway Performing the NAT Function

Although the NAT process "hides" the internal computers by not allowing an external computer to initiate communications directly to them, one form of attack is to compromise the NAT device itself and use it as a platform for attacking the internal network. NAT should not be thought of as a security mechanism and the device performing the NAT function (as with **any** Internet-facing device) should be hardened and kept patched.

Monitoring Network Traffic

We mentioned that a firewall acts like a filter to remove undesirable messages from a communication stream, yet sometimes things still get through. A plant may still need something to sample the stream, looking for malicious, anomalous, or otherwise undesirable content, and provide a warning. An in-situ analyzer would serve that role for a real process stream but a network intrusion detection system (NIDS) provides the equivalent functionality on a network.

A NIDS is a separate application or set of applications that monitors communications on all major entry and exit points of an internal network and processes that traffic to identify malicious, anomalous, or otherwise undesirable traffic. A NIDS can use three detection methods: signatures, statistical analysis, and anomalous behavior detection. It is important to use all three, since new malware (so-called Zero-day or 0day malware) won't have an identifiable signature until some organization spots the malware and develops one. A NIDS can monitor traffic originating both on the outside and on the inside and, if properly designed, can spot malware propagating between systems and LAN segments.

There are commercial NIDS products available from a number of vendors and there is also a shareware NIDS called *Snort* that can be implemented on an existing server or run on separate hardware. The Snort community has developed a huge library of signatures and anomaly detection libraries, including some for industrial protocols, such as Modbus/TCP and the IP version of DNP3. Figure 9-11 shows a conceptual model of a NIDS together with the various components used to build one.

Depending on how a NIDS is configured, it can even act like a firewall and block message traffic that it deems as malicious, dangerous, or anomalous. In that mode it is called a *network intrusion detection and prevention system* (NIDPS). The National Institute for Science and Technology (NIST) has an excellent free publication (SP 800-94) that provides an overview of NIDS technologies.

Using Data Diodes to Protect Systems

The government and military have computer systems that contain top-secret and other classified information and, yet, these systems need to be connected to a network to allow selected data to be exported. In those cases, they often place a *data diode* between the high-security system and the rest of the network. A data diode is a pair of computers with a simplex channel between them that permits data to flow out of the sensitive system, but *no message traf-*

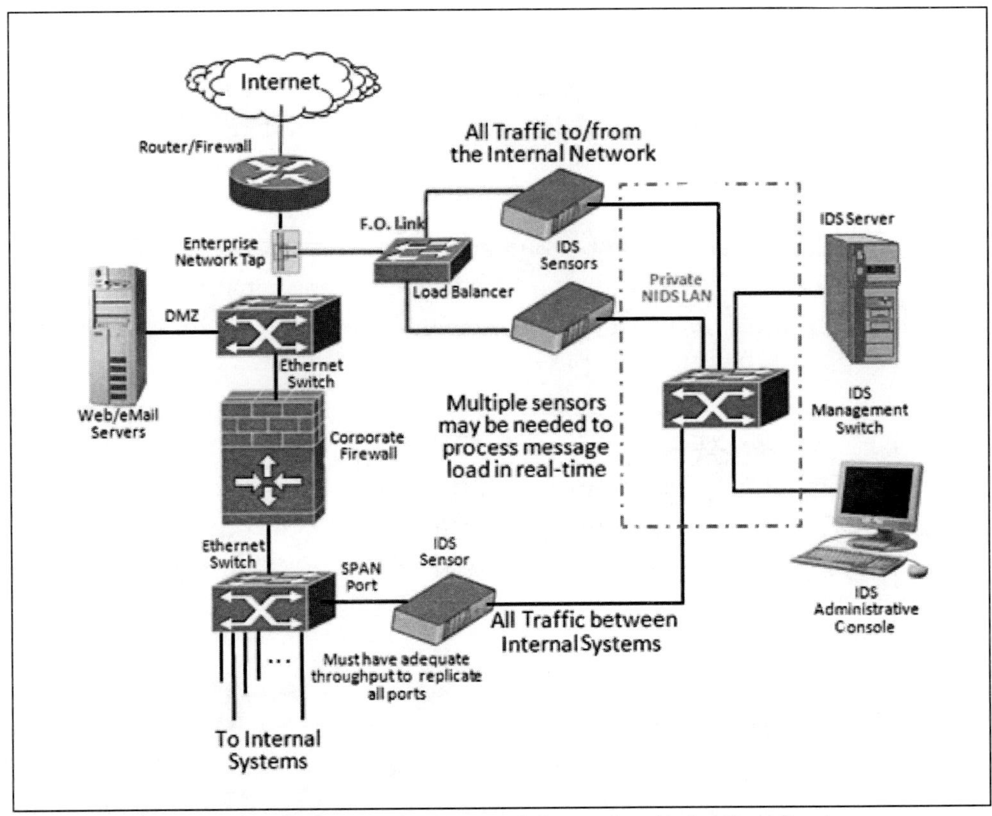

Figure 9-11. Conceptual Model of a Network Intrusion Detection System

fic of any type (even the handshaking and message acknowledgments used by many protocols, including TCP) can flow back. The data diode completely eliminates the possibility of an external attack, but it also makes some services impossible to support and it eliminates the possibility of remotely administering and supporting the system(s) behind the diode.

Data diodes are not plug-and-play devices; they have to be configured for the specific protocols they will pass and the systems to which message traffic will flow. However, data diodes do make external attacks impossible as long as there are no other communication pathways that bypass the diode. Figure 9-12 illustrates the use of a data diode. In the figure, SBC means single board computer. Of course SneakerNet, the manual movement of files or data between systems using portable media and devices, is still a threat.

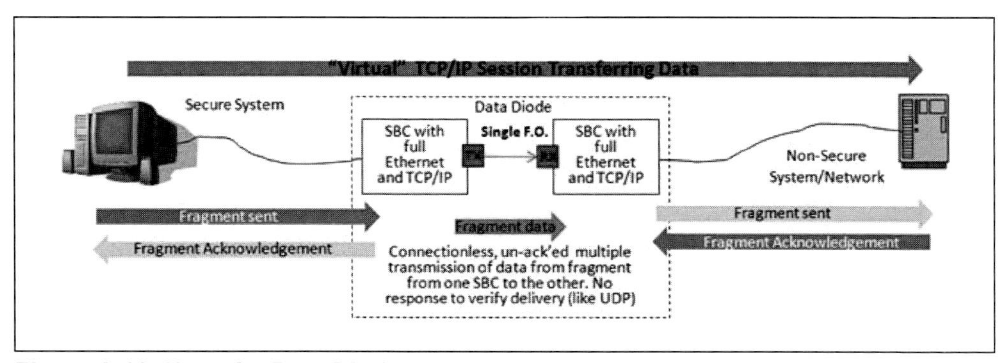

Figure 9-12. Use of a Data Diode

Hardening

Another strategy that can be used to greatly reduce the likelihood of being the victim of a cyberattack is hardening. Hardening a network asset entails a great deal of planning. The services, ports, applications, and privileges that are necessary for proper operation and the security policies that are or will be in use must be identified; then the implementation actions must be determined. It is not a simple process, but it is one that is employed to reduce the number of vulnerabilities presented by a network asset to an attacker. This is something that can be applied to individual assets, as well as to the plant networks.

Operating System Hardening

The standard "out of the box" distribution of Windows, from 3.11 through Vista, came with lots of features, functions, and capabilities included and enabled, just in case someone might need them. (The same can be said about OS-X and many Linux distributions. Fortunately Windows now comes relatively tightened down and the server must have ports and features installed first to work). The reason is simple: It was done so that people did not have to become computer-savvy in order to use PCs. Unfortunately, that means that a LOT of stuff the user will never need or even want—such as games, fax software, and drivers for printers that aren't even manufactured anymore—comes installed and enabled. In addition, all the TCP and UDP ports and all possible protocols are allowed. By default, the security settings are mainly disabled or are set to the least secure level possible and there are default administrative ("admin") accounts installed.

This provides attackers with an astounding number of vulnerabilities that can be exploited, both locally and remotely. Of course, as an automation system administrator, you can't just blindly go about deleting software and blocking ports. Your IT personnel may have experience in hardening Windows- and Linux-based systems. There is lots of advice on hardening out on the Internet but it is usually aimed at personal or home PCs or business and IT systems. Some things that might work for those cases might break an automation system.

Your automation system vendor should be aware of the need to harden their systems and ought to be able to recommend hardening practices, as well as things you should not do. A clear and obvious step in the hardening strategy is to apply vendor-supplied/recommended security patches and updates. NIST has free publications that offer a lot of suggestions, however, your vendor must confirm that any planned hardening won't harm the system.

Network Hardening: Components

LANs, which are mainly what is found in a plant or industrial facility, are composed of both media (e.g., CAT-5/6 cable or fiber optic cable) and active components (e.g., Ethernet switches, gateways, routers, and possibly even internal firewalls). There may be repeaters, patch panels, and junction boxes as well. A basic means of hardening a LAN is to use physical means to prevent access: locking cabinets for all components, conduit for all media, locked enclosures for patch panels, etc.

Network Hardening: Port Use

The easiest means of getting onto a LAN is to plug into a port on an Ethernet switch, but there are many ways to prevent access. Modern switches can be configured to reject a connection by an unauthorized device by checking its MAC address. Unused ports can be turned off via configuration settings. VLANs can be created to isolate groups of switches or ports from others. None of these actions guarantee that a LAN can't be successfully attacked, still, they can make it much harder for the attacker to do so.

Network Hardening: Remote Administration

Of course, with intelligent network components you must use strong passwords and avoid enabling remote administration (administering a device other than the local device over the network) unless it can be done in a secure (encrypted) manner. Many switches, firewalls, and routers support access, configuration, and administration via telnet, SSH (secure shell), HTTP (web browser), HTTPS (secure web browser), and possibly SNMP v3 (Simple Network Management Protocol). Most such network component devices also support local connection via a "console" port. If you can telnet into a network component device, *then an attacker can possibly do it as well;* the same concept applies when a browser can be used to access a device.

It is important to configure the devices to prevent outside access. Figure 9-13 shows an example of possible configuration setting options for such devices.

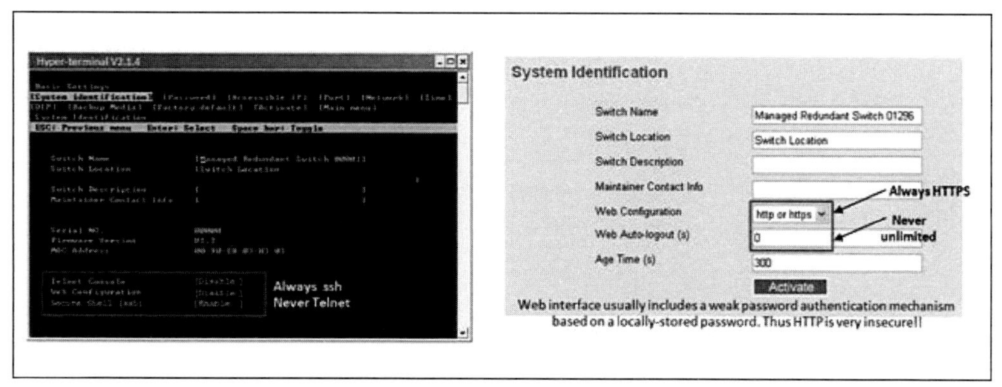

Figure 9-13. Selecting Secure Remote Administrative Options

Evolution of Threats

As technology changes and as countermeasures become effective, the threats **will** (not may) evolve into a different form. It is much like the influenza virus. Vaccines are prepared for the most likely strain and people are vaccinated, yet the flu virus seems to be able to morph into different forms against which the vaccine is not as effective as it could be or is not effective at all. Cyber malware is the same. It is the price we pay for connectivity to the larger world.

The Internet and VPN Countermeasures

We have included the Internet in a book about industrial communications because it has become an indispensable part of business and industry. The Internet is a world-spanning wide area network that is ubiquitous and is extremely cost effective. The use of higher bandwidth access devices, such as DSL or cable modems, allows a corporation to effectively, yet inexpensively, create a corporate WAN over the Internet that links multiple sites, each with their own LAN.

In the past, a corporation would have paid significant amounts of money to lease private point-to-point connections (e.g., T1 or frame-relay) for this purpose. Nonetheless, using the Internet creates problems, the most paramount being security because the Internet is shared by literally billions of others, not all of whom have your best interests at heart. No one wants to think that a 13-year-old could manipulate a set point in your plant; but, theoretically, that could happen if you connect your plant LAN to the Internet without adequate protection. However, with the advent of effective and secure encryption, Virtual Private Network (VPN) technologies allow corporations still using leased, private, point-to-point connections to greatly reduce their recurrent communications cost, while also reducing the threats described in this chapter.

Encryption

In this section, the discussion will center on encryption (and decryption) and how it is utilized to secure network communications.

Codes and ciphers are two ways of making information secret. *Codes* are just substitute words or phrases that are generally understood by both the sender and the receiver. *Ciphers*, on the other hand, are generally based on using the smallest unique part of the information. For a written letter, a cipher would be based on the individual characters; for an electronic transmission, the cipher would be based on each bit.

A cipher may be of the transposition or the substitution type. In the *transposition* type of cipher, the individual bits are retained and are only moved around according to a key held by both parties. The *substitution* type of cipher substitutes a cipher bit for the data bit and then replaces the cipher bit with the original bit upon deciphering. First developed by Vernam in 1917, the basic method of enciphering and deciphering (a polyalphabetic substation cipher) is illustrated in figure 9-14.

A *block* cipher encrypts a block of data (i.e., it encrypts multiple information units—in this case bits—at one time). Once a block is encrypted, it will then be stored until the transmit path is available (a process called *store and forward*). Block ciphers are fine for bulk transmissions, such as email or file transfers, where a bit of time delay is not a problem. *Stream* ciphers, on the other hand, encrypt the next bit to be transmitted one bit at a time, according to the key.

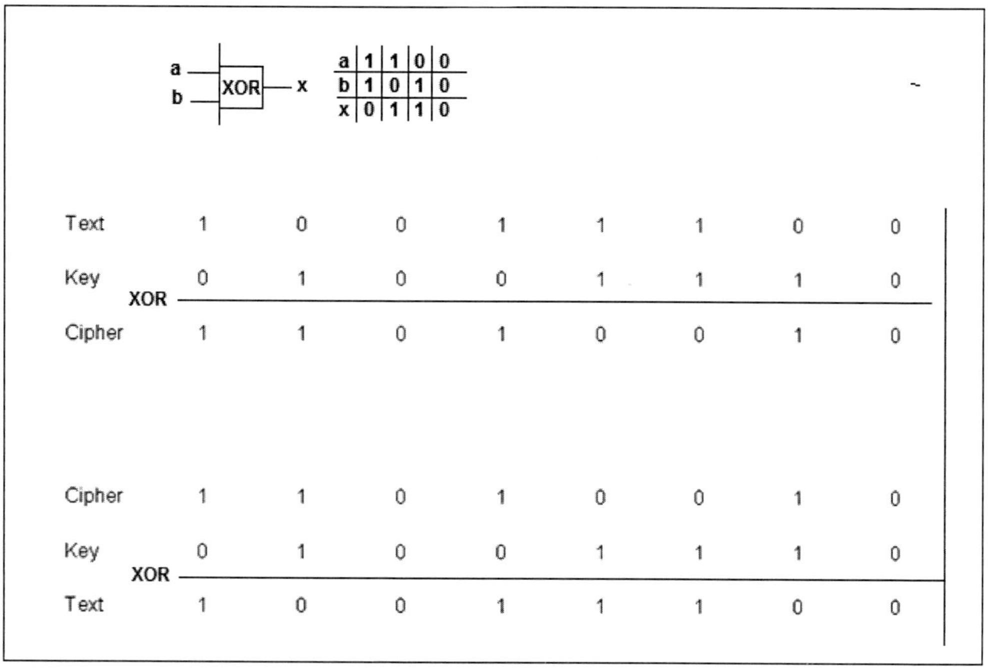

a, b → XOR → x

a	1	1	0	0
b	1	0	1	0
x	0	1	1	0

Text	1	0	0	1	1	1	0	0
Key	0	1	0	0	1	1	1	0
XOR								
Cipher	1	1	0	1	0	0	1	0

Cipher	1	1	0	1	0	0	1	0
Key	0	1	0	0	1	1	1	0
XOR								
Text	1	0	0	1	1	1	0	0

Figure 9-14. Encryption/Decryption Example

XOR (eXclusiveOR) is a logic gate, also known as a *binary half adder* or a *Modulo 2 circuit*. The rules are simple. For a two-input gate (which will have one output), if the two inputs are the same (either 1 or 0), the output will be a 0. If one input is a 1 and the other a 0 (the order is not important as long as the values are opposite), the output will be a 1. *Encryption* is performed by having each individual data bit as one input to an XOR logic and the key bit as the other input to the XOR. The output (of the XOR logic) is a cipher bit. A stream of cipher bits will not look like either the stream of key bits or the stream of data bits that produced the cipher stream. *Decryption* is the inverse of the encryption process. That is, each bit of the cipher stream is XORed with a key bit (the same value as the one that encrypted it) and the output will be the data bit (see figure 9-14). This requires more capable circuitry, in terms of speed and processing, than when using a block cipher. Stream ciphers use pseudo-random number generators (PRNG) to create a key-stream of bits to be used for the XOR process. Both the sender and receiver must use the same PRNG algorithm and starting (*seed*) value so the order and value of the bits in the key stream are identical at the sender and receiver (which is why the word pseudo is used—the bits appear to be random but are duplicated at the sender and receiver).

What is not simple is generating the key. To be a perfect key, it must be non-repeating or take a long time to repeat. If you (as the encryptor) lack both the time and the equipment to prove that a number is non-repeating, the next best alternative is to choose a large prime number (a prime number can only be factored by itself and 1). If the prime number is large enough, the results of its repeated division will not repeat themselves.

This non-repeating number is duplicated electronically as a pseudo-random number string. It is "pseudo" random because it has to be reproduced in the receiver as it is produced in the transmitter and the chance of two long, non-repeating numbers being the same by chance is slim (much like winning the lottery), yet it is not zero.

Stream ciphers are used to perform real-time encryption on stream data such as audio (e.g., VoIP telephony) and video, where delays would cause noticeable jitter (shaky voice or video) and viewing or listening problems. Wireless Ethernet (IEEE 802.11) uses stream ciphers to encrypt Ethernet frames as they are transmitted. The cellular telephone system uses stream ciphers to encrypt your conversation, but just between your phone and the cell tower.

In a simple point-to-point circuit or link, the key is limited to both ends; no one else should have it. It will probably be distributed by physical means. Keys must be changed after a specified period of time or they are vulnerable to being broken (compromised). This is because the key really isn't an endless non-repeating number and because the patterns for most primes are known. Irrespective of the complexity with which the key was modified, if the key is used for a long enough period of time, it is vulnerable to attack, particularly by computer. This is especially true if a weak encryption system was used to start with, as with WEP in the original IEEE 802.11a wireless Ethernet scheme.

Breaking a cryptographic cipher has many facets and the use of ever-more powerful computers and networks of computers has greatly aided the task. Letter frequencies, business environment, network functions, traffic flow patterns, and cultural uniqueness all contribute to the success of cryptanalysis of an encrypted message. However, the single most effective item in compromising a cipher is plain-text (message to be enciphered) source data whose time and location of transmission can be identified. With this information and the data of only a few packets, all messages sent or received during that key period (or in the case of a public key, many periods) can be disclosed.

Although most business data requires time stamps, if your network data has been encrypted (or will be) and the data is to be stored before transmission, it should be encrypted to deny attackers the opportunity to compare the plain-text stored data with the encrypted transmitted data.

Encryption today is performed using one of a few well-known algorithms, such as AES, 3DES, and Blowfish. Such algorithms needed to be published and peer-reviewed to ensure that they actually work, so they are not secret. The secret is the key used with the algorithm. In digital cryptography, a key is just a big binary number (e.g., from 128 bits to over 64 kb [kilobits] in length). Breaking encryption is usually a matter of guessing the key. Using many computers in collaboration, it is possible to make a lot of guesses, breaking the Digital Encryption Standard (DES) only took a month, and today it is possible to get a massive number of computers across the Internet working together on the problem. Given enough time and enough computing power, any key can be guessed. By making keys big (cryptographically strong) and changing them on a periodically-scheduled basis (e.g., once every 48 hours), it is possible to greatly lessen the likelihood of a key being guessed and the encryption being broken.

Cryptography is primarily designed to provide message confidentiality by making it (almost) impossible for someone without the key to decode the message. Properly used, cryptography can provided additional benefits. It can be used to verify message integrity; that is, determine if a message has been tampered with, modified, or forged. It can be used to confirm the alleged identity of the sender of a message (authenticate the sender), and it can be used to prove that a message originated from a particular sender (this is "non-repudiation"). In industrial control applications, the confidentiality of a message may or may not matter but, if the message is changing a critical parameter or issuing a critical command, it is essential to know that the message is authentic and was sent from an authorized operator workstation. Unfortunately, most current SCADA, PLC, and DCS products are somewhat lacking in their support for such capabilities.

In the encryption and decryption processes, the key is centrally important. The question arises: how can the keys on a LAN be distributed securely? As on a SCADA network, what if there are a lot of remote sites? The question becomes even more complex if each site has multiple levels of access.

Physical distribution of a key is costly and inefficient because there is no electronic way to transmit it on the same network as the previous key, for there is a chance the previous key may be compromised. So, typically, a courier is used to deliver the keys physically in some format. Keys that must be distributed in this manner are known as *private* ("symmetric") or *secret keys*.

Keys distributed over a network are known as *public keys*. A combination of private and public keys is required for electronic distribution of a public key. In a private key system, the same key is used to encrypt and decrypt the data. In a public key system, pairs of different but mathematically-related keys ("asymmetric") are used. These keys, a public and a private key at each node (typically generated at the node location), enable a useful encryption scheme. One key of the pair (i.e., the public key) encrypts (transmission end), while the other of the pair (i.e., the private key) decrypts (reception end). This is the basis for digital certificates, Secure Sockets Layer (SSL), and other Internet security methods. A message may be encrypted using *either* of the keys of the pair but, once encrypted, it can only be decrypted using the *other* key of the pair. Windows, Linux, and OS-X all have software to create a public-private key pair. Public keys can be transmitted over an insecure network to anyone since we don't care who has a copy. When two computers want to communicate securely, they each generate a key pair, keep the private key, and send the other computer their public key. Figures 9-15A, B, and C show various ways the sets of keys can be used to provide communication security.

Figure 9-15A. Sending an Encrypted Message

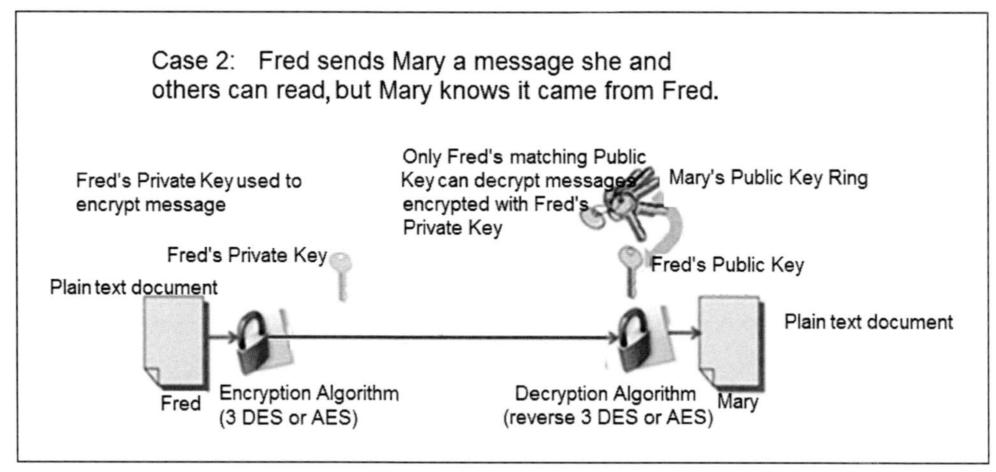

Figure 9-15B. Using Fred's Private Key to Prove Sender's Identity

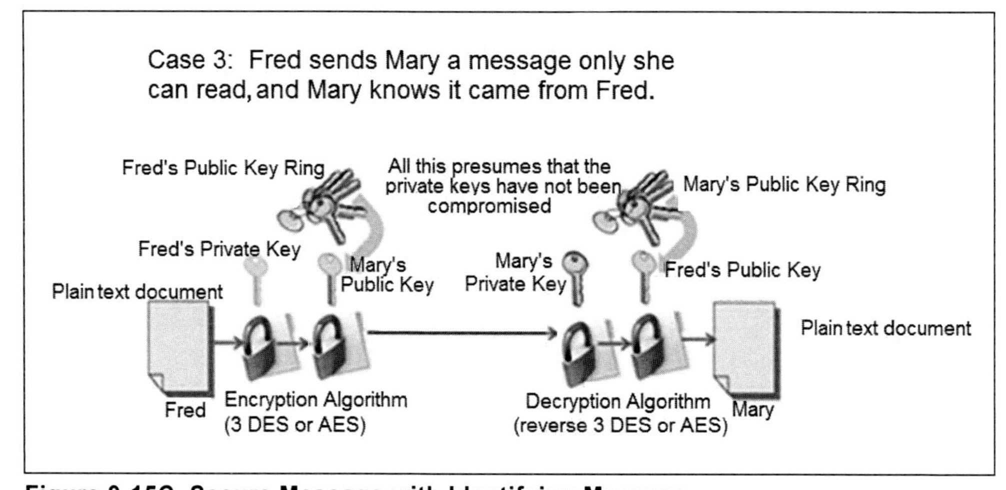

Figure 9-15C. Secure Message with Identifying Message

Double encryption, using both public and private keys (the last example in figure 9-15C and a message digest (*hash code*), provides the highest level of security and is how technologies such as SSL and VPNs function. A hash code is a special calculation that can be performed on a block of data (e.g., a file) to generate a unique numeric value that will be different if even a single bit in the data is altered. Well-known algorithms, such as MD5 and SHA-1, are used to create hash codes for messages (hash codes are also called *message authentication codes* or *MAC*—not to be confused with the Ethernet term MAC). Adding a hash code to the end of a message, prior to encryption, makes it possible at the receiving end to verify that the message has not been altered. It serves the same sort of purpose as a CRC calculation appended to a message, such as on an Ethernet frame.

There are numerous hash code algorithms but the sender and receiver must agree on and use the same one. The size (number of bits) of the hash code value is fixed for each of the algorithms but some produce larger hash values (more bits) than others. The larger the file or data block to hash, the larger the size of the hash code needed to ensure that the message has not been altered.

Some vendors who post software on their websites for download will also post a hash code value for the files. This is so that you can verify a valid download by re-computing the hash code once you receive the file(s). Figure 9-16 shows the hash code value outputs generated by hashing a small text file using several popular hash code algorithms.

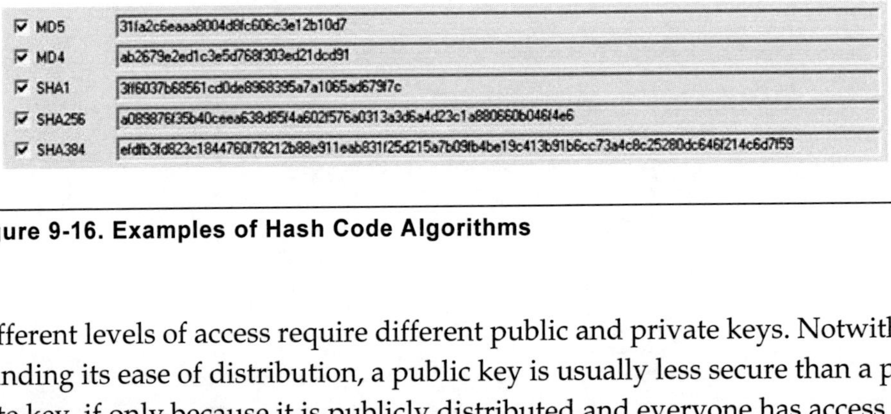

MD5 31fa2c6eaaa8004d8fc606c3e12b10d7
MD4 ab2679e2ed1c3e5d768f303ed21dcd91
SHA1 3ff6037b68561cd0de8968395a7a1065ad679f7c
SHA256 a089876f35b40ceea638d85f4a602f576a0313a3d6a4d23c1a880660b046f4e6
SHA384 efdfb3fd823c1844760f78212b88e911eab831f25d215a7b09fb4be19c413b91b6cc73a4c8c25280dc646f214c6d7f59

Figure 9-16. Examples of Hash Code Algorithms

Different levels of access require different public and private keys. Notwithstanding its ease of distribution, a public key is usually less secure than a private key, if only because it is publicly distributed and everyone has access to that distribution. That being said, one has to wonder what information most companies that are not involved in the defense sector would have on a network that would prompt such an expensive attack, since the cost of breaking the public and private keys would be high. Paranoia aside, a public key system, properly administered, will efficiently and securely protect most industrial and business internal LANs from external threats.

Many software companies offer encryption protection for files and transmitted data. This protection usually takes the form of hardware adapters, which are quicker than software, although effective software does exist and is increasingly being built into operating systems' environments. To make them hard to break, as was mentioned above, public keys are usually huge (several thousands of bits long; see figure 9-17 for an example). This makes them computationally excessive for continued use. Public/private keys are used to establish a secure connection over which a more conventionally-sized *session* key can be negotiated and exchanged (128 + bits in size); the session key is then used for all further encryption. Session keys are symmetric (both parties use the same key) and temporary (you only use them until the end of the session or for 24 hours, whichever comes first.) When a site-to-site VPN is created (discussed at length in chapter 8), the two gateway devices use public/private keys and key exchange to negotiate a session key. They will typically repeat this process periodically (e.g. every 24 hours) to make sure that a given session key is not used for an extended timeframe, thereby lessening the likelihood of the key being broken.

Figure 9-17. Example of a Public Key

The DES was a national encryption standard that has now been replaced, due to its lack of cryptographic strength, by algorithms such as 3DES (triple DES) and AES. DES was a block cipher; that is, an output block that was enciphered as a function of the input block and the complete key. In the DES, a 64-bit block was encrypted by a 56-bit key. An adapter with a very large scale integration (VLSI) chip usually performed this enciphering because software (until the advent of faster microprocessors) was just too slow. Most financial transactions by financial institutions and Federal Reserve banks were conducted using the DES.

One problem for those considering using DES was that it required a synchronous data stream. The PC is asynchronous. Some method of conversion was therefore required, usually an adapter card that had a number of hardware registers and a synchronous modem for communicating over dial-up lines.

The level of security offered by DES was always held in suspicion by some people because of its origin—IBM and the U.S. government. This could normally be dismissed as the usual outcry of anti-government people, except that the U.S. government does continually press for the integration of backdoors (a method of defeating the encryption process) into any encryption process. This requirement was ostensibly for maintaining law and order; using the backdoor could only occur by court order. Indeed, at one time the U.S. government was considering requiring the installation of an encryption chip with a specified backdoor in every computer used by government or private business facilities.

DES was quickly broken by a collaboration of computers on the Internet, so it gave way to a new standard encryption method: 3DES (1,024 key bits). The reason for the large number of bits is to make the key so difficult to break that an attack will be time consuming and cost more than the information is worth. The U.S. government actually sponsored a challenge for developing the standard algorithm for an AES that drew several notable contestants; 3DES was one of the proposed algorithms but it did not win. The government stated that any of the competing algorithms were sufficiently safe for modern data transmission.

Message encryption for one or two parties is quite simple; on networks it becomes complex. If you would like to encrypt a single file, do the following: use one of the many file encryption utilities, use the maximum length key, and ensure that the key is a random assortment of alphanumeric characters. After encrypting your file, encrypt it again with a different maximum length, random, alphanumeric key. No one can recover your file through cryptographic analysis if the file isn't too long in relation to your key size, so in this example, a file of about 4 kb and a key length of 256 octets, would be secure against all but a supercomputer attack. The problem is this: did you write the keys down? If not, you won't be able to recover the file either.

Now consider a LAN with a large number of nodes. How are all of these keys to be distributed, changed, and so on? This is the problem with ensuring that your data will withstand all attack. Less secure systems cost less, are easier to implement, and will probably meet the realistic threat to your data. You can spend a lot for security that is not needed, as long as your network is not connected to the Internet. However, when your data is placed on the Internet, then it is fair game to all the denizens of that world. Strong encryption is a necessity. SSL, PGP (Pretty Good Privacy), and digital certificates all have one thing in common: long (cryptographically strong) keys.

Internet technologies (n-tier) are now being employed on intranets in industry. One candidate for a fieldbus gives each device (sensor, actuator, etc.) its own web server and web pages. The information available to the operator is real-time and all the operating specifications are there to see. The need for protection, whether encryption or other means, is vital. Encryption offers the best and most secure option (provided it is selected for the case in hand and implemented correctly).

Internet Engineering Task Force Security Solutions

The challenges posed by insecurity and cyber threats were not lost on the Internet Engineering Task Force (IETF)—the people who control the standards and technologies used on the Internet. As part of proposing an IP revision (IPv6) that would solve the problem of address depletion, they also designed better security features and added enhancements to support streaming.

IETF also recognized that IPv4 would be around for quite a while, at least on private networks, and so someone needed to come up with a temporary solution for IPv4 that allowed authentication and confidentiality until conversion to IPV6 was complete. The result was two new IPv4 configurations (borrowed in part from the IPsec concept and also referred to as IPsec) that allowed communications to be authenticated and/or confidential.

One option in IPsec is to insert a new authentication header (AH) between the IPv4 header and the TCP or UDP header in an IP datagram (refer to figure 9-

18). This new header carries some security parameters and a hash code (message digest) for the entire IP datagram, except for the few fields in the IP header that are dynamic. The message receiver can re-compute the message hash and verify that it has not been tampered with, modified, or forged. The hash can also confirm the alleged identity of the sender of a message (authenticate the sender) and that the message is not modified in transmission, but this scheme does not provide message confidentiality.

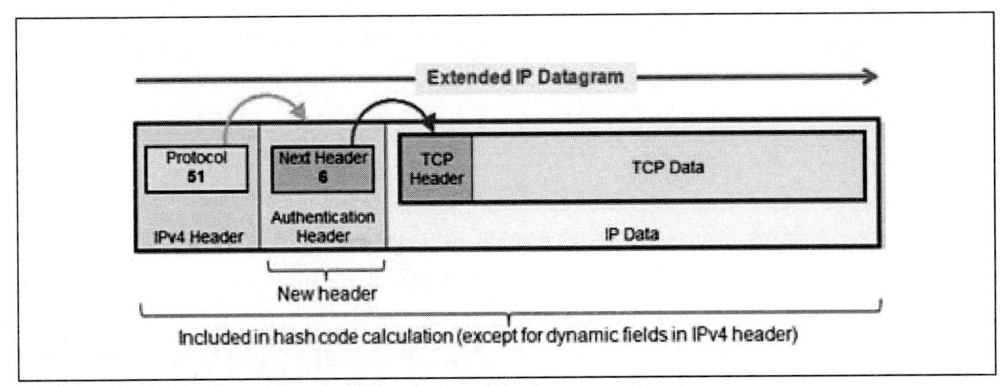

Figure 9-18. IPsec (IPv4) Authentication Header

Encapsulated Security Payload (ESP), a second option in IPsec, is provided to enable encrypting messages to provide confidentiality. As an option, this can also include authentication information to verify message integrity. Figure 9-19 shows an IPv4 datagram as it looks with the authentication header added.

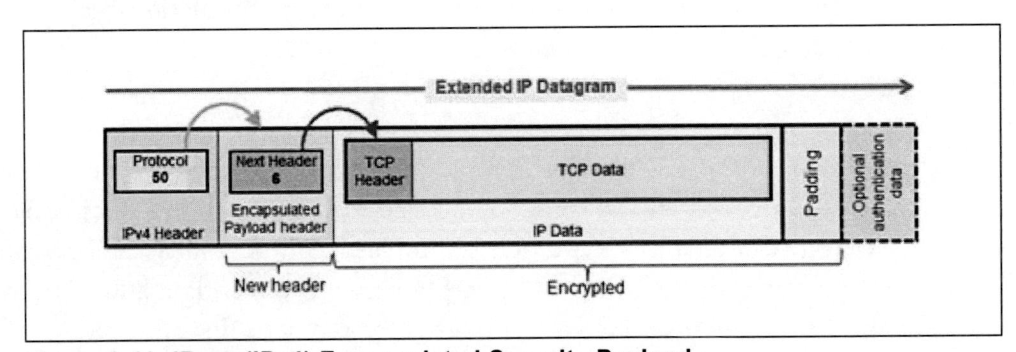

Figure 9-19. IPsec (IPv4) Encapsulated Security Payload

For both the AH and ESP versions of IPsec, the two communicating parties have to go through an initial handshaking sequence to agree on the encryption key, the encryption algorithm, and the hash code algorithm. This is done using the exchange of public keys and then using the encryption of messages, based on those keys, to perform the negotiations. The IPsec standard requires all computers to support a basic minimum set of encryption (3DES) and hashing (MD-5) but additional types can be added as long as all participants support them. IPsec is considered to be a network layer function that is transparent to communicating applications, as well as to the TCP and UDP transport layers. This means that enabling IPsec does not require any changes to existing application software.

Network Management and Security

Good network operation depends on good network management. Network management could be the subject of its own book and many automated management programs are available. To be accurately and completely performed, network management requires software assistance. In the following pages we'll look at these aspects of network management:

- Configuration
- Security
- Error/fault handling
- Accounting
- Performance

WAN/LAN management is a must for managing network performance. Configuration, security, performance, and available resources (error/fault handling, redundancy) are all listed as aspects of network management.

All network management programs will add overhead traffic to a network. Although many management programs are in general use, most provide only partial implementations of most companies' total network management goals, which would include all six topics listed above.

Integration of IT Practices with Network Management

DCS, SCADA, and PLC-based automation systems used to be isolated and were based on proprietary networking technologies. That is no longer the case. We have discussed the migration of such systems and devices onto conventional Ethernet-TCP/IP LANs and WANs, as well as the conversion of industrial protocols to Application layer (OSI Layer 7) protocols transported by TCP/IP. In addition, the engineering, operator, and programming workstations and servers for these systems are now running either a Windows or Linux operating system. This means that these systems can be monitored and administered using the same methods and technologies as are used by IT for the corporate business systems. In fact, a growing trend is to have corporate IT provide administrative support for such systems. Regardless of whether you think this is a good or bad idea, it is happening as corporations look for cost savings. Some of the IT practices that are making inroads into plant automation systems and networks include:

- Centralized monitoring and configuration updating using SNMP
- Centralized user authentication and policy enforcement using Microsoft Active Directory
- Centralized administration of patches, updates, firewall rules, AV (antivirus) signatures, and NIDS/HIDS signatures/libraries

A general problem with these practices is that they often either preclude the placement of a DMZ between the corporate WAN and the plant networks or they require any intervening firewalls or some routing host in the DMZ to have "holes" (permitted ports and protocols/services) through which such functions can be performed; this is the classic trade-off between security and convenience.

Network Management: Security and Configuration

Configuration is essential to implement network security and connectivity; to change, add, and delete users; and to change the logical order of station addresses and routing assignments. In addition, configuration includes adding or deleting access and/or functions to a station, bridge, router, or gateway, as well as managing equipment and node features, including software provisioning.

Security also involves user authorization, which is used in multilevel security and is generally accomplished using passwords. Different users in an organization have differing needs. Not everyone should be an administrator (although that does simplify access, it is hardly a suggested procedure). Users should have only the level of access they require, no more. Most major network operating systems provide differing levels of access, as well as designated times that a user can log on and frequency limits to control how many log-ins a user may have at one time. The administrator should be cautioned that most network operating systems or directory services use the lowest security assigned to the user, regardless of which groups the user is placed in, because a whole group may be denied if a member who is disallowed is assigned to it.

Key management is how a company stores and distributes encryption keys and digital certificates. Key management, which is essential to the encryption process, is a well-thought-out set of processes that has just now become adequately cost effective and non-limiting to system performance to be acceptable to use in industrial facilities. A company's liability would be high if its network security (which includes key management) were to be compromised through its data transmissions; the resulting damage could cause great monetary loss, or injury, or loss of life. If a company does not effectively manage keys and their distribution, then their system may well be compromised.

Access management is generally the task of the system administrator. There are and will be both technical and managerial problems as industrial networks become integrated into commercial and office networks. For a start, who will have access control over the industrial network? Wrong settings or poor response time to a request for access changes could be extremely detrimental to system operations.

Additionally, as a matter of interest, who will determine the network's technical requirements? An industrial measurement and control system has a host of requirements and commitments that are different from any office network. Who will allocate bandwidth and connectivity, and who will dictate platform?

Network Error/Fault Handling

A *network error* occurs when the "ought to be" condition becomes an "is not" condition, while a *fault* is the detection of that error and the manner in which the error, or set of errors, is handled by the network software. Since errors and the resulting faults will occur, it is necessary to have the following information to ensure that errors are handled correctly:

- Node status
- Backup and restore procedures
- Methodology, reporting of station alarms, server fault handlers, and system error handling

- Procedures and the location for logging and auditing errors
- System error tolerances

In general, some form of software program will provide this information. In most cases, industrial systems will normally have, at a minimum, these five types of information available as a safety and performance requirement.

Network Accounting

Network accounting is used to:

- Apportion traffic cost
- Determine license use
- Determine software use
- Determine user costs

The system administrator performs most of these tasks. It should be understood that the system administrator must have sufficient experience, training, and background to understand the industrial system requirements before setting policy.

Network Performance

Network performance is monitored to:

- Determine instantaneous network status
- Document historical records
- Analyze traffic
- Correct bottlenecks
- Track users' apparent speed (the response speed to a user's keystroke is important for eliminating multiple key strokes, user interventions, user panic, or other debilitating user actions taken when he or she thinks the system is inoperable when it is just slow)

IEC/ANSI/ISA-62443 Cybersecurity Standards

Due to the limited understanding of the different aspects of cybersecurity by most industrial instrumentation and control personnel and the profound extent of threat to a nation's infrastructure due to a cyberattack, the ISA99 committee designed a cybersecurity management standard that considers the industrial side of cybersecurity. Other agencies and countries have the same cyber threats to their critical infrastructures and international standards were in development. ISA99 is a set of standards, technical reports, and other related information defining procedures for implementing and deploying secure electronic IACSs. ISA99 standards apply to end-users (i.e. asset owner), security practitioners, system integrators, and control systems manufacturers responsible for designing, manufacturing, implementing, deploying, or managing IACSs. In 2010, the ANSI/ISA99 standard series adopted the IEC numbering, becoming the ISA-62443 standards. The IEC 62443 and ANSI/ISA-62443 standards are identical.

ISA addresses industrial automation and control systems whose compromise could result in any or all of the following situations:

- Endangerment of public or employee safety
- Loss of public confidence
- Violation of regulatory requirements
- Loss of proprietary or confidential information
- Economic loss
- Impact on national security

The concept of electronic security in industrial automation and control systems is applied in the broadest possible sense, encompassing all types of plants, facilities, and systems in all industries. An IACS includes, but is not limited to:

- Hardware and software systems such as DCS, PLC, SCADA, networked electronic sensing, and monitoring and diagnostic systems
- Associated internal, human, network, or machine interfaces used to provide control, safety, and manufacturing operations functionality to continuous, batch, discrete, and other processes

Physical security is an important component in the overall integrity of any control system environment but it is not specifically addressed in this series of documents. The ISA-62443 standards, as designed, have four working parts, which are illustrated in figure 9-20:

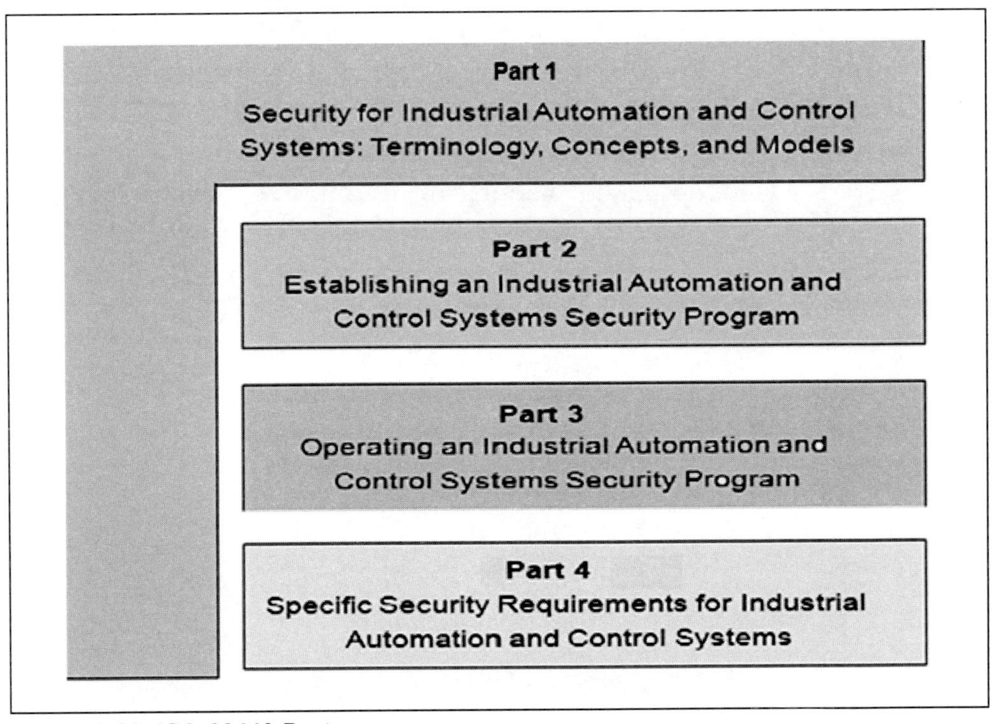

Figure 9-20. ISA-62443 Parts

Part 1 establishes the context for all the remaining standards in the series by defining the concepts, terminology, and models needed to understand electronic security for the industrial automation and control systems environment. It also establishes certain security metrics and security life cycles for an IACS.

Part 2 focuses on how to create a security program for control systems. The program is integrated into an overall Cyber Security Management System (CSMS). Details of the program are described by elements and requirements for each element. The elements and requirements are organized into three main categories:

1. Risk analysis
2. Addressing the risk with CSMS
3. Monitoring and improving the CSMS

Part 3 addresses how to operate a security program after it is designed and implemented. This includes the definition and application of metrics to measure program effectiveness.

Part 4 defines the characteristics of IACSs that differentiate them from other information technology systems from a security point of view. Based on these characteristics, the standard establishes the security requirements that are unique to this class of systems.

Cyber Security Management System

ISA-62443 standards define the elements necessary to establish a CSMS for an IACS and provide guidance on developing those elements. This document uses the broad definition and scope of what constitutes an IACS, as described in ANSI/ISA-62443-2-1 (99.02.01)-2009, *Security for Industrial Automation and Control Systems Part 2-1: Establishing an Industrial Automation and Control Systems Security Program*.

The elements of a CSMS described in this standard are mostly policy, procedure, and practice related, describing what shall or should be included in the final CSMS for the organization. Other documents in the ISA-62443 series discuss specific technologies and/or solutions for cybersecurity in more detail. For additional information, consult the references listed in the bibliography at the end of this chapter.

The ISA-62443 standards use specific examples to provide guidance on developing a CSMS. The examples represent the ISA99 committee's opinion on how an organization could develop the elements of a CSMS; because they are examples, the elements may not work in all situations. The user of this standard must read the requirements carefully and apply the guidance appropriately in order to develop a fully functioning CSMS for their organization. There may be cases where a pre-existing CSMS may be in place and the IACS portion is being added, or there may be some organizations that have never formally created a CSMS at all; this standard does not attempt to create a solution for all cases.

Figure 9-21 illustrates the elements of a CSMS.

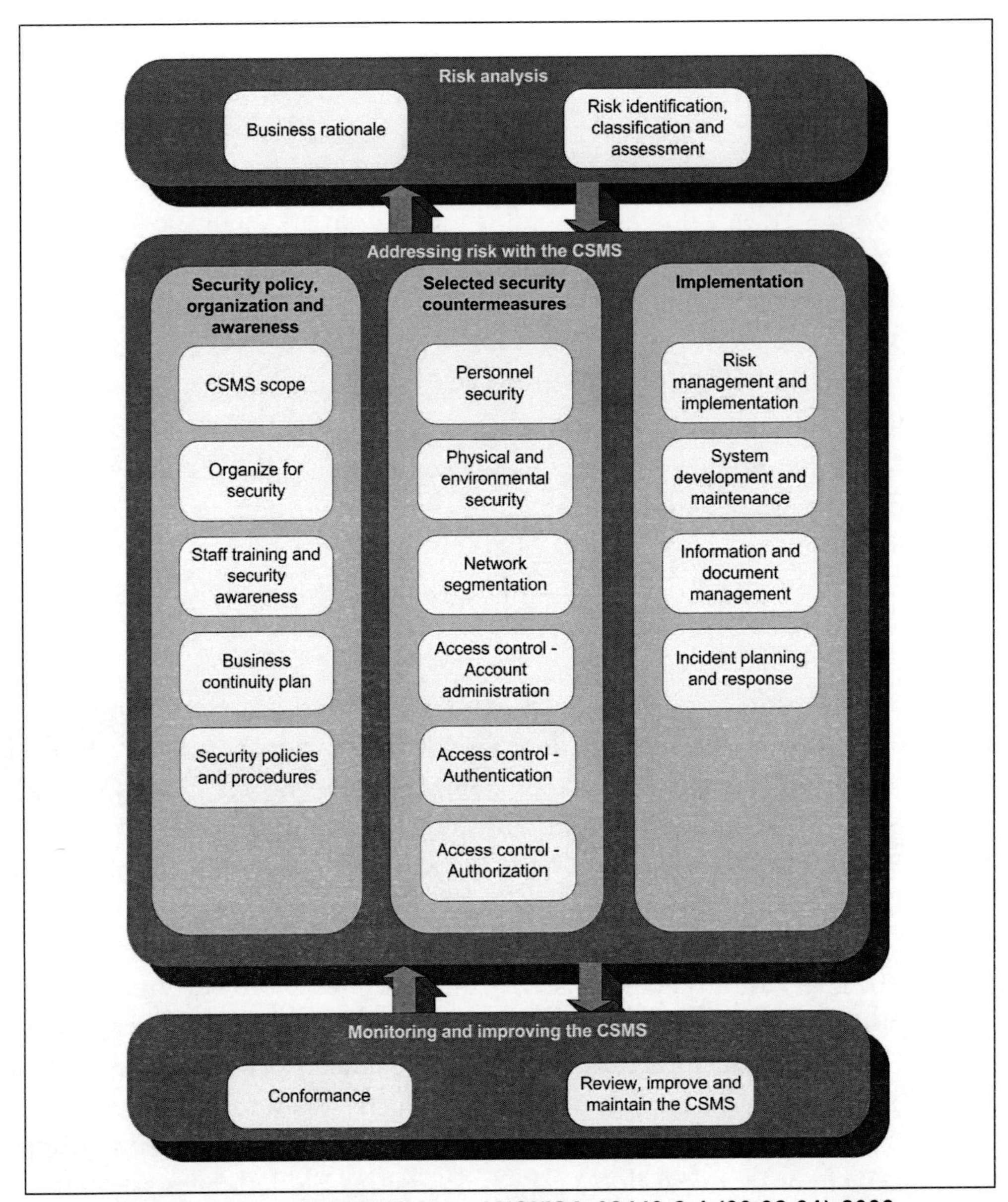

Figure 9-21. Elements of a CSMS from ANSI/ISA-62443-2-1 (99.02.01)-2009

Risk Analysis

Before we can begin to protect our control systems, we must understand what we are dealing with. Risk analysis was described in some detail previously in this chapter; this is a summary of the process in accordance with the tenets of the ISA-62443 standards.

In a security setting, risk analysis is a process that identifies:

- Assets
- Threats
- Vulnerabilities
- Consequences
- Likelihood of a successful attack
- Viable counter measures
  - Technical
  - Administrative
  - Procedural

Why Perform Risk Analysis?

There are many reasons for performing risk analysis, some of which were discussed earlier in this chapter. Below is a summary of the reasons that a formal, documented risk analysis on security should be performed:

- To determine which assets to protect
- To determine credible threats to those assets
- To determine vulnerabilities that currently exist
- To identify the risks posed with regard to the assets
- To recommend changes to current practice (current countermeasures) that mitigate risks to an acceptable level
- To determine implementation priorities
- To provide a foundation for building the security policy and plan
- To provide financial justification for the business case

Risk Analysis: Safety and Security in Industrial Systems

You may have noticed that security is following the same plans of action that safety has in the past: defense in depth, risk analysis, etc. This is because the two are related and can be treated approximately the same way, using identical methodology. Below are ISA definitions of risk as they pertain to safety and security:

- **Safety:** Risk is a measure of human injury, environmental damage, or economic loss in terms of both the incident likelihood and the magnitude of the loss or injury.
- **Security:** Risk is an expression of the likelihood that a defined threat will exploit a specific vulnerability of a particular attractive target or combination of targets to cause a given set of consequences.

Note that both definitions address consequences, which are defined as: the amount of loss or damage that can be expected from a successful attack against an asset. Loss or damage may be monetary but may also include political, morale, operational effectiveness, or other business impacts. Some of the consequences of a cyberattack are outlined in table 9-5.

Table 9-5. Consequences of a Cyberattack

Consequences of a Cyberattack
Reduction or loss of production at one site or multiple sites simultaneously
Injury or death of employees
Injury or death of persons in the community
Damage to equipment
Environmental damage
Violation of regulatory requirements
Product contamination
Criminal or civil legal liabilities
Loss of proprietary or confidential information
Loss of brand image or customer confidence
Economic loss

Summary of Part 2, Annex A, of ISA-62443

Annex A addresses the countermeasures and architecture practices that should be in place in industrial systems. Below are some of the requirements:

1. All high- and medium-risk manufacturing and control networks must be firewalled or disconnected from any external networks (site, corporate, and/or public networks).

2. High-risk manufacturing and control network installations are to be completed with haste.
3. Medium-risk manufacturing and control network installations are to be completed promptly after securing the high-risk installations.
4. All manufacturing and control network firewalls must be:
    - Configured according to a set of published standards established company-wide
    - Centrally monitored in accordance with corporate firewall monitoring guidelines for health and security by the corporate firewall monitoring/support entity
    - Centrally backed up by the corporate firewall monitoring/support entity and have:
        - A documented, viable disaster recovery process
        - A documented escalation, reporting, and change management process in place

Table 9-6 is a summary of countermeasures and architecture practices from Part 2, Annex A.

Table 9-6. Summary of Countermeasures

Countermeasure and Architecture Practices	High-risk IACS	Medium-risk IACS	Low-risk IACS
Two-factor authentication to control access to the device	Required	Required	Optional
Hardening of the operating system	Required	Recommended	Optional
Employ network segmentation	Required	Required	Optional
Employ antivirus application	Required	Required	Required
Use of WLAN	Not Allowed	Not Allowed	Allowed
Strong password authentication at the application level	Required	Recommended	Recommended
Other countermeasures, etc.	Etc.	Etc.	Etc.

As you may surmise, having a working knowledge of the ISA-62443 standards is essential for establishing a cybersecurity management program within industrial settings. In this book, we did not go through the implementation of a CSMS, as that is what the ISA-62443 standards are all about. ISA offers several courses that address implementation. Since the security effort started at the U.S. Department of Homeland Security, you may rest assured it will be implemented (in various forms of conformance) industry wide.

Cyber Emergency Response Team

In the context of cybersecurity, including an IACS, an incident typically entails unauthorized access to computer networks and equipment with actions resulting in some form of negative consequence to the asset owner. The economic and social consequences of a breach could be severe when negative publicity, loss of customer confidence, potential lawsuits, and direct financial loss caused by interruptions in production operations or equipment replacement and repair are considered.

A Cyber Emergency Response Team (CERT) must be in place and a cyber-incident response capability must include proactive elements for incident prevention and response elements for managing an incident after it occurs. The elements of accident prevention include: planning, incident prevention, and post-incident analysis/forensics. The elements that center on detecting and managing an incident are reactive and are typically carried out under severe time constraints and high visibility. These elements include detection, containment, remediation, recovery, and restoration.

The elements of a cyber-incident response are illustrated in figure 9-22. Note that this is a continuous improvement model.

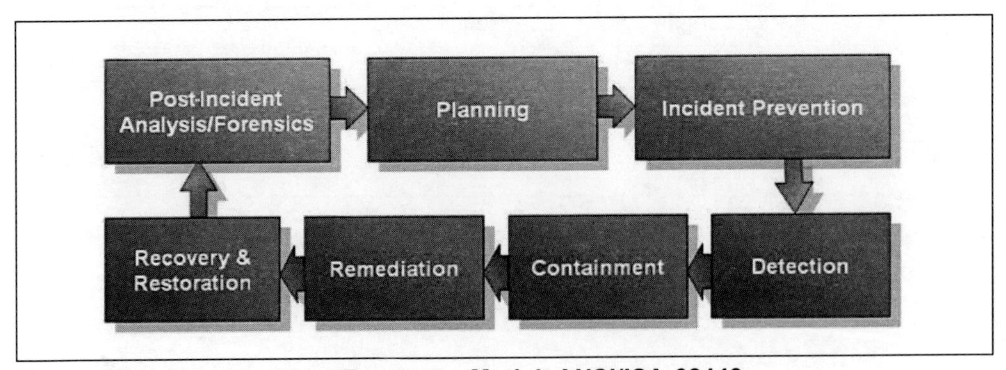

Figure 9-22. Cyber Incident Response Model, ANSI/ISA-62443

Impetus

Where does the impetus to develop policies and procedures for a cybersecurity management model issue from?

The U.S. Department of Homeland Security (DHS): The Office of Cybersecurity and Communications (CS&C) is responsible for enhancing the security, resilience, and reliability of the nation's cyber and communications infrastructure. (The CS&C is a component of the National Protection and Programs Directorate [NPPD], which is charged with protecting the nation's cyber infrastructure.) A cyber infrastructure (as applied to an IACS) is the network of programmable, intelligent systems used to control processes, be they manufacturing or services. A communications infrastructure is the wiring, routers, and switches that make up the communications portion of the cyber infrastructure.

Control Systems Security Program (CSSP): The CSSP has developed the Industrial Control Systems-Cyber Emergency Response Team (ICS-CERT). US-CERT and DOE-CIRC are members of the Government Forum of Incident Response and Security Teams (GFIRST).You (assuming you sign up) may join a variety of listservs (such as US-Cert) that publish vulnerability lists as they are discovered.

ISASecure Certification Program

Cybersecurity is going to become even more complex than it is today. In an effort to assist industrial companies, ISA has set up a third-party certifications system for secure industrial equipment. The ISASecure certification program has been developed by an industry consortium called the *ISA Security Compliance Institute* (ISCI) with the primary goal of accelerating industry-wide improvement of cybersecurity for IACSs. The following are definitions used by the ISA in the certification program.

Embedded Device: A special purpose device that runs embedded software and is designed to directly monitor, control, or actuate an industrial process. The attributes of an embedded device are: no rotating media, limited number of exposed services, programmed through an external interface, embedded OS or firmware equivalent, real-time scheduler, may have an attached control panel, and may have a communications interface. The following are examples:

PLC, field sensor devices, safety instrumented system (SIS) controller, and DCS controller.

Certifier: A chartered laboratory, which is an organization that is qualified to certify embedded devices as ISASecure.

Certification: Third-party attestation related to products, processes, or persons that conveys assurance that specified requirements have been demonstrated. For ISASecure EDSA (embedded device security assurance), this is an authorized evaluation of an embedded device to the ISASecure EDSA criteria, which, when successful, permits the device vendor to advertise this achievement in accordance with certification program guidelines.

Version (of ISASecure Certification): The ISASecure certification criteria in force at a particular point in time, defined by the set of document versions that define the certification program and identified by a year and release number, such as ISASecure EDSA 2010.2.

ISASecure Assessment Criteria: ISASecure assesses devices to three criteria: Communication Robustness Testing (CRT), Functional Security Assessment (FSA), and Software Development Security Assessment (SDSA). The program offers three certification levels for a device, offering increasing levels of device security assurance. The levels correspond to the levels outlined in Section 5.10.1 (table 8) of ANSI/ISA-62443.00.01-2009.

The levels of assessment (illustrated in figure 9-23) are in increasing order of security; suffice it to say that Level 1 does not include the necessity of device authentication. Further information is available at www.isasecure.org.

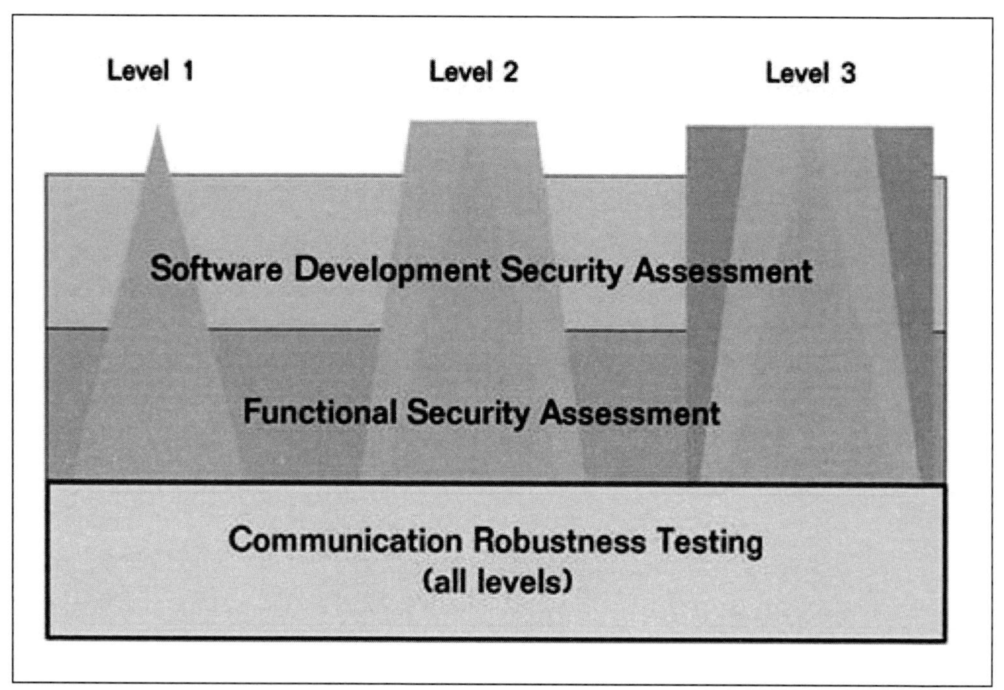

Figure 9-23. ISASecure Levels

Conclusion

This chapter has covered a wide gamut of network security knowledge, from basic security to network management software to encompassing standards. All the topics discussed here will change on a daily basis as technology moves forward, so each topic in this chapter will require that you to give it further study and that you continually update and upgrade your knowledge.

Bibliography

Gilbert S. Vernam. Secret signaling system. U.S. Patent 1,310,719, issued July 22, 1919.

International Society of Automation. ISA Course IC32: "Cyber Security for Automation, Control and SCADA Systems." Research Triangle Park: ISA, 2005.

———. ISA Course TS13: "Industrial Automation Cybersecurity: Principles & Application." Research Triangle Park: ISA, 2012.

———. ISA Course TS14: "IT Survival Basics for I&C." Research Triangle Park: ISA, 2013.

Moore, D. "CSPP Guidelines for Analyzing and Managing the Security Vulnerabilities of Fixed Chemical Sites." AIChE, August 2002.

National Infrastructure Security Coordination Centre. "The Electronic Attack Threat to Supervisory Control and Data Acquisition (SCADA) Control & Automation Systems." London: NISCC, 2003.

———. "The NISCC Good Practice Guide on Firewall Deployment for SCADA and Process Control Networks." London: NISCC, 2004.

"Security for Industrial Automation and Control Systems Part 1-1: Terminology, Concepts, and Models." ANSI/ISA-62443-1-1 (99.01.01)-2007. Research Triangle Park, NC: International Society of Automation, 2009.

"Security for Industrial Automation and Control Systems Part 2-1: Establishing an Industrial Automation and Control Systems Security Program." ANSI/ISA-62443-2-1 (99.02.01)-2009. Research Triangle Park, NC: International Society of Automation, 2009.

Appendix A
Number Systems Review

For this review of number systems, it is important that the emphasis be on the use of the number system as a means of pattern recognition. The number systems addressed in this appendix are the ones currently used in computer-based systems and data communications. Binary is the number system used by computers, decimal is used to represent binary values so they make sense to humans, and hexadecimal (hex) is used to reduce binary streams to recognizable patterns. Computers only work with binary patterns, while humans only really understand decimal.

The Decimal System

All number systems follow the same rules. You are doubtlessly quite familiar with one number system, namely the decimal system. *Decimal* means that the number system is based on 10 digits; therefore, its *base* is 10. The only numbers allowed in the decimal system are 0, 1, 2, 3, 4, 5, 6, 7, 8, and 9. All the numbers that we can use to describe numerical quantities in decimal are made up of those 10 numbers and no others.

Figure A-2 illustrates the powers of 10. Notice that a number such as 4,302.63 is really 4000 + 300 + 2 +.6 + .03, or more correctly, 4×1000 (10 to the 3rd power) + 3×100 (10 to the 2nd power) + 0×10 (10 to the 1st power) + 2×1 (10 to the 0 power or 1) + $6 \times 1 \times 1/10$ (10 to the -1 power) + $3 \times 1 \times 1/100$ (10 to the -2 power).

Binary	Octal	Decimal	Hexadecimal
0000	0	0	0
0001	1	1	1
0010	2	2	2
0011	3	3	3
0100	4	4	4
0101	5	5	5
0110	6	6	6
0111	7	7	7
1000		8	8
1001		9	9
1010			A
1011			B
1100			C
1101			D
1110			E
1111			F

Figure A-1. Decimal Numbers 0 through 9

1 × 1000	1 × 100	1 × 10	1 × 1	Decimal Point	1 × 1/10	1 × 1/100
4	3	0	2	.	6	3

4302.63

Figure A-2. Powers of 10 Example

All other number systems are constructed the same way, except the base is a number other than 10. For the example, the base is 2 (consisting of digits 0 and 1) in the binary system, and the base is 16 (consisting of digits 0, 1, 2, 3, 4, 5, 6, 7, 8, 9, A, B, C, D, E, and F) in the hexadecimal system.

The arithmetic functions for the binary system will be explained next.

The Binary System

To understand digital systems, an acquaintance with the binary number system is necessary, as the binary system is the foundation on which digital technology is built. Computers perform all operations on 4-bit (or some multiple of 4-bit) patterns that are comprised of binary system values 1 and 0. These 4-bit patterns are called *binary coded decimals (BCDs)* and they represent decimal and hexadecimal values that symbolize codes and data. Figure A-3 illustrates the binary values for decimal values 10 through 15 and hex values 0 through F.

Binary	Octal	Decimal	Hexadecimal
0000	0	0	0
0001	1	1	1
0010	2	2	2
0011	3	3	3
0100	4	4	4
0101	5	5	5
0110	6	6	6
0111	7	7	7
1000		8	8
1001		9	9
1010		10	A
1011		11	B
1100		12	C
1101		13	D
1110		14	E
1111		15	F

Figure A-3. Binary Numbers 0 through 15

Below is an example of decimal numbers represented by four binary digits (BCDs).

1 3 0 2	DECIMAL
0001 0011 0000 0010	BCD

The Octal System

Early computers used the octal system to represent computer data and memory address values.

The octal system used 3 bits, arranged in eight unique patterns, to represent digital values 0 through 7. When computers evolved to using an 8-bit byte, 4 bit patterns were necessary to represent digital and hex values.

Conversion from binary to decimal is performed by separating the binary number into groups of three, starting at the binary point. Assign the BCD value for the 3-bit group and you have performed the conversion.

NOTE The binary point is the radix point for base 2. The radix point (the decimal point in base 10) separates the integer part of the number (to the left of the radix point) from the fractional part of the number (to the right of the radix point). In the example below, the binary point is located between the third digit from the right (1) and the fourth digit (0) from the right.

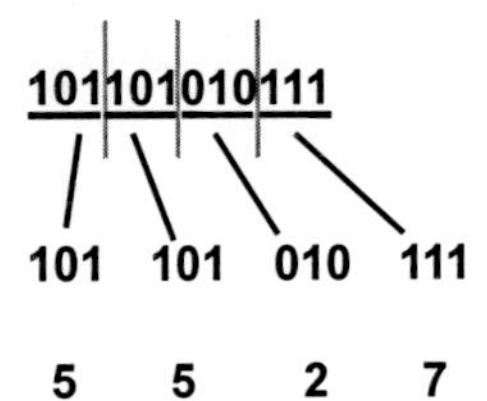

The Hexadecimal System

The hexadecimal (also referred to as *hex* or *HEX*) number system is based on 16 digits: 0 through 9 and A through F. Figure A-3 below shows the relation between the binary coded decimal (BCD) patterns and the corresponding hex numbers.

Binary	Octal	Decimal	Hexadecimal
0000	0	0	0
0001	1	1	1
0010	2	2	2
0011	3	3	3
0100	4	4	4
0101	5	5	5
0110	6	6	6
0111	7	7	7
1000		8	8
1001		9	9
1010		10	A
1011		11	B
1100		12	C
1101		13	D
1110		14	E
1111		15	F

Figure A-4. (Repeat of Figure A-3)

Note that this is the same illustration used to illustrate the binary system. These 16 unique 4-bit binary patterns provide the hexadecimal representation utilized by most modern computers that perform all operations on 4-bit or some multiple of 4-bit patterns.

Convert a Binary Pattern to Hex

For this conversion, you will need to separate the binary pattern into 4-bit groups *starting from the right-most digit,* and then assign the hex representation to each group.

Example A-1

Pattern: 0111010010100101

Solution: Working from right to left, separate the pattern into groups of four bits. Then refer to figure A-4 to determine the correct hexadecimal number for each pattern.

```
0111  0100  1010  0101
  7     4     A     5
```

Convert a Hexadecimal Number to a Binary Pattern

Converting from hex to binary is done simply by writing out the 4-bit binary patterns for each hexadecimal number.

Example A-2

Pattern: 0FF13H (FF13 is hex representation)

NOTE Hexadecimal numbers are generally written as 0hexnumbersH; with a leading 0 and a following H.

Solution:

```
  F     F     1     3
  |     |     |     |
1111  1111  0001  0011

  1111111100010011
```

Convert Binary to Decimal and Decimal to Hex

NOTE If frequent conversions are required, it is recommended that you use an inexpensive calculator.

There are standard number patterns that can be used to facilitate converting binary numbers to decimal and decimal numbers to hex. It is recommended that some patterns should be memorized, both decimal and hex, because they are among the most common patterns encountered. These are shown in table A-1. It would also be useful to know the binary and octal patterns for decimal numbers 0 to 15 and hexadecimal numbers 0 to F listed previously in figure A-3.

Table A-1. Popular Conversion Numbers

Binary Pattern	Decimal	Power of 2	Hex
1 0000	16	4	10
10 0000	32	5	20
100 0000	64	6	40
1000 0000	128	7	80
1111 1111	255	$2^8 - 1$	FF
1 0000 0000	256	8	100
10 0000 0000	512	9	200
100 0000 0000	1024	10	400
1000 0000 0000	2048	11	800
1 0000 0000 0000	4096	12	1000
10 0000 0000 0000	8192	13	2000
100 0000 0000 0000	16384	14	4000
1000 0000 0000 0000	32768	15	8000
1111 1111 1111 1111	65535	16	FFFF

Example A-3 illustrates the process for converting a binary number to decimal using the Popular Conversion Numbers in table A-1.

Example A-3. Converting Binary to Decimal

Problem: Convert 10101010010 to decimal.

Solution: The table below contains standard power of 2 and decimal values and a section for the binary number. Enter the binary number in the Sample Binary Number section. Locate the binary columns that contain a 1 and note the corresponding decimal value for each column. Then, add up the decimal values you noted. The result is the decimal equivalent of the binary number.

Power of 2															
15	14	13	12	11	10	9	8	7	6	5	4	3	2	1	0
Decimal Value															
32768	16384	8192	4096	2048	1024	512	256	128	64	32	16	8	4	2	1
Sample Binary Number															
0	0	0	0	0	1	0	1	0	1	0	1	0	0	1	0

1024
256
64
16
2
1362 is the decimal equivalent of 10101010010

Converting a decimal number to a hexadecimal number is a two-step process: first, you must convert the decimal number to binary (which is much like long division), and then you convert the binary pattern to hex (see Example A-4).

Example A-4. Converting Decimal to Binary and then to Hex

Problem: Convert 953 in decimal to binary and then to hex.

Solution: Using the Power of 2 section of the chart below, determine the largest power of 2 that will go into 953. It is 512. Place a 1 in the 512 column in the Binary Number section of the chart. Because it is a 12-digit binary number and all the digits must be used, enter 0s in the columns for the bits higher than 512.

Then subtract 512 from 953 to determine the remaining amount. Continue determining the largest power of 2 that will fit in the remaining amount and subtracting that amount until you reach 0, as illustrated below. Enter a 1 in the binary columns that correspond to decimal numbers you can use and a 0 in the binary columns for numbers that you cannot use.

Power of 2											
11	10	9	8	7	6	5	4	3	2	1	0
Decimal Equivalent											
2048	1024	512	256	128	64	32	16	8	4	2	1
Binary Number											
0	0	1	1	1	1	0	0	0	0	1	1

512 + 256 = 768. This is less than 953, so enter a 1 in the 256 column.
512 + 256 + 128 = 896, less than 953, enter a 1 in the 128 column.
512 + 256 + 128 + 64 = 950, so enter a 1 in the 64 column.
512 + 256 + 128 + 64 + 32 = 982 so enter a zero in the 32 column.
512 + 256 + 128 + 64 + 16 = 966 so enter a zero in the 16 column.
512 + 256 + 128 + 64 + 8 = 958 so enter a zero in the 8 column
512 + 256 + 128 + 64 + 4 = 954 so enter a zero in the 4 column
512 + 256 + 128 + 64 + 2 = 952 so enter 1 in the 2 column
512 + 256 + 128 + 64 + 2 + 1 = 953 so enter a 1 in the 1 column

Working from right to left, separate the pattern into groups of four bits. Then refer to figure A-3 to determine the correct hex number for each pattern.

```
0011 1100 0011
  3    C    3
```

You will find it necessary to use these conversions when in:

- Learning situations (trying to understand equipment operation, etc.)
- Diagnostics (problem location with differing equipment of different manufacture)
- Software/programming

If you find yourself frequently having to convert numbers between decimal and binary (or hex), it is best to use a calculator that does this, as it is far less time consuming and is usually a great deal more accurate. Converting from hex to binary (and vice versa) only requires knowledge of (or memorization of) 16 patterns, two of which are the same (for both 0 and 1).

While this is not an extensive investigation into number systems, or even one number system, it should provide a sufficient basis for moving between the most commonly used number systems.

Appendix B
Historical Aspects of Industrial Data Communications

Introduction

Modern industrial data communications is a pragmatic discipline: whatever works, regardless of the source, keep it. Due to the changes in technology over the past several centuries, data communications has its origins in many sources. Much of the jargon associated with industrial data communications owes its uniqueness to this mixed parentage. This appendix will examine the major sources of technology that have developed into the modern practice called *industrial data communications*.

To paraphrase the philosopher George Santayana (Jorge Agustín Nicolás Ruiz de Santayana, not to be confused with the contemporary Carlos Santana, a rock musician), "Those who do not study history are doomed to repeat it." Whether this was originally applied to students in a history class or to world leaders, it is particularly true in data communications. The concepts once tried and found to work in one technology keep coming back in the newer versions.

Instrumentation Sources

Closed-loop automatic process control uses the feedback path form of data communications to accomplish automatic control. Figure B-1 illustrates a feedback control loop. Up to the 1970's, the means of transmitting data to the controller and to the indicators was labeled *signal transmission*.

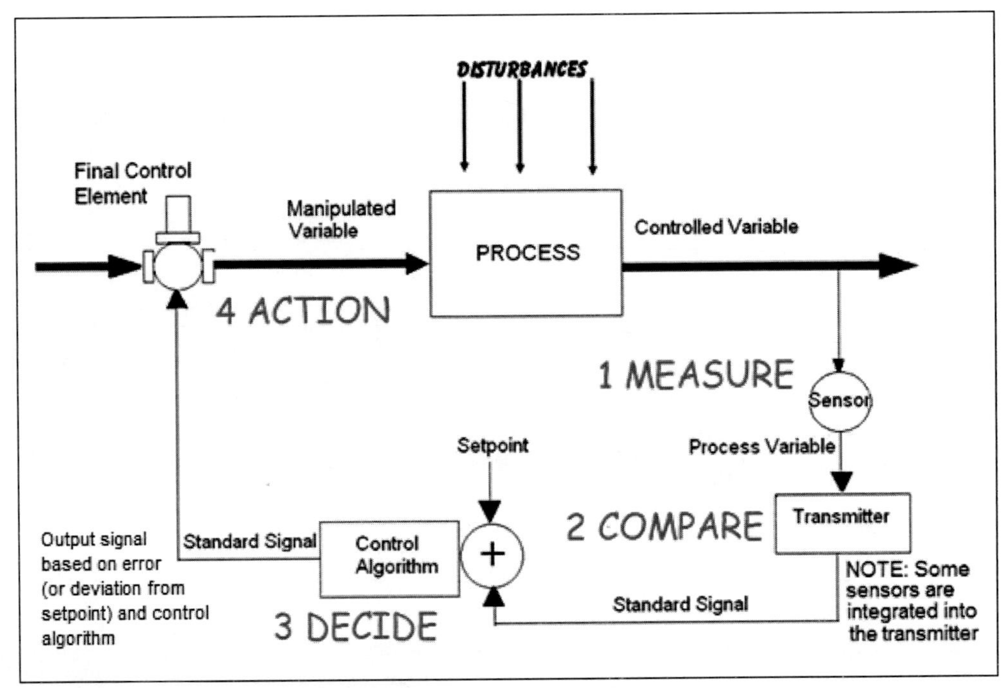

Figure B-1. Feedback Control Loop

In the early 1800s, the precursor to many automatic machines was developed: the Jacquard loom (figure B-2). Its control was via a punched card, actually a punched roll, like that used on an old-fashioned player piano. The punched roll directing the loom weave was an early form of machine control. The use of a punched card for information storage was one precept to form the beginnings of data communications—a century and a half later.

In the past, operators generally were employed to watch gauges and make manual valve changes. This took a large number of operators. It would have been more efficient if there was a means of sending the information to a central location, where operators could scan many more gauges and operate many more valves, as opposed to being in the field and only able to monitor a few gauges—never cognizant of the full impact of adjustments on the whole process (until it was too late). In order to allow fewer operators to have a broader perspective on the process and control more loops, the concept of remote transmitters and remote operation of valves was put into practice. The controllers were generally in a central location—the control room—with connections to the process through pneumatic signaling.

Figure B-2. Jacquard Loom (Photo: David Monniaux, 2006)

While instrumentation brought technical sophistication to pneumatic signaling, it was the industrial requirements pneumatics couldn't meet that led to today's electric industrial communications. Pneumatics lack speed. Pneumatic signal changes traveling at the speed of sound are no match for electric signals traveling at nearly half the speed of light. Electric signals are fine for communication, but they lacked (up through the 1980's) a cost-effective means of moving large objects like valves, so much of the electric instrumentation was developed around electric signaling combined with pneumatic actuation, particularly in the petrochemical industry.

Electric signal transmission never totally displaced pneumatics in signaling; it took a small piece of silicon to do that. In industrial applications, it seemed that one of the most cost-effective uses of a large computer would be in integrating an entire plant's controls. This was tried in the 1960s; it was called *direct digital control (DDC)*. Figure B-3 is a diagram of a DDC system.

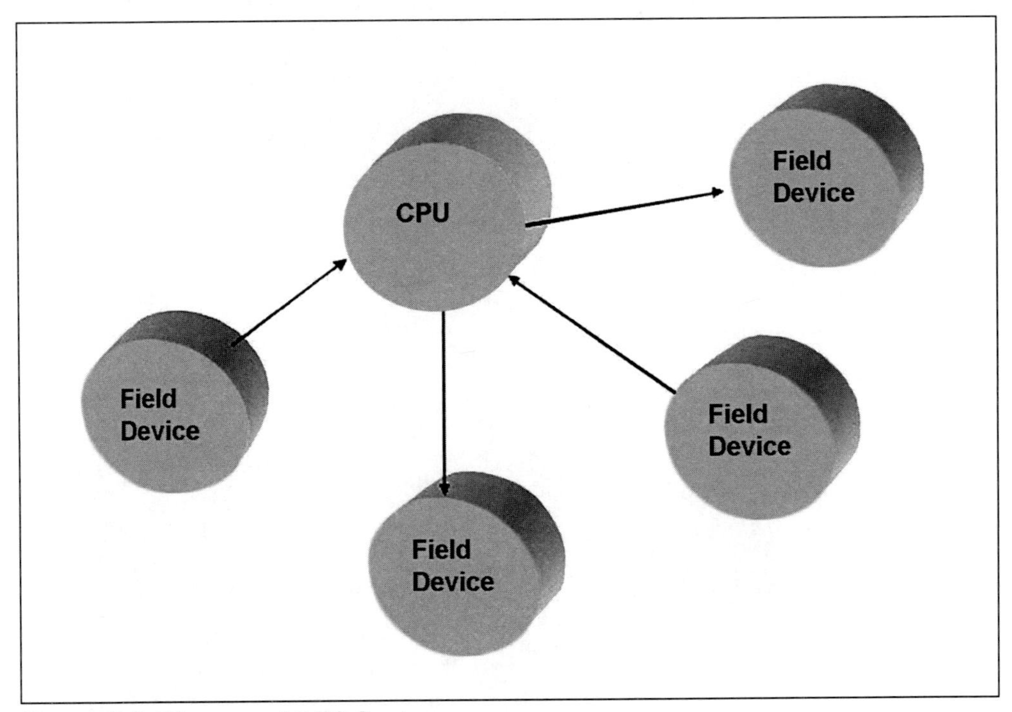

Figure B-3. Diagram of DDC System

DDC systems were proprietary and had several major drawbacks. While the older mainframes had enough computing power to manage a process, from 1960 to 1975 they lacked reliability. The mean time between failures was not long enough to ensure reliable and error-free performance for any extended period of time, unless, of course, an analog (electric/pneumatic) backup was in place, or enough financial resources were available to permit redundant computers. DDC wasn't a bad idea, however, industrial processes are typically not very tolerant of control failures, particularly if all inputs and outputs are controlled by one point and that point stops working.

One way to avoid complete computer system failure (even today) is to have the computer control only set points and to perform data acquisition for optimization; this is called *supervisory control*. Then, if the main computer goes down, the controllers continue to operate at their last set point and only the optimization provided by the large central computer is lost. A supervisory control system (a familiar example is SCADA; supervisory control and data acquisition) uses the same configuration as the DDC system (figure B-3).

Another solution is to use multiple computers with overlapping control responsibilities; if one computer goes down, the entire process is not taken down—just that part of the process. This is the idea behind *distributed process control*. If the other computers can take up a portion of the downed computer's functions so that operation is still possible, the system is (in theory) fault tolerant.

Note in figure B-4 that an essential part of a distributed control system is the communications between the computers. Just when this evolved into a local area network (LAN) depends on one's definition and the point in history being looked at. This was a prohibitively expensive scheme in 1960. Even in the 1970s it was extremely expensive to implement. However, the aforementioned little piece of silicon, the microprocessor, has allowed this arrangement to become the arrangement of choice since the 1980s. Again, note that (by definition) data communications are essential to the operation of distributed process control.

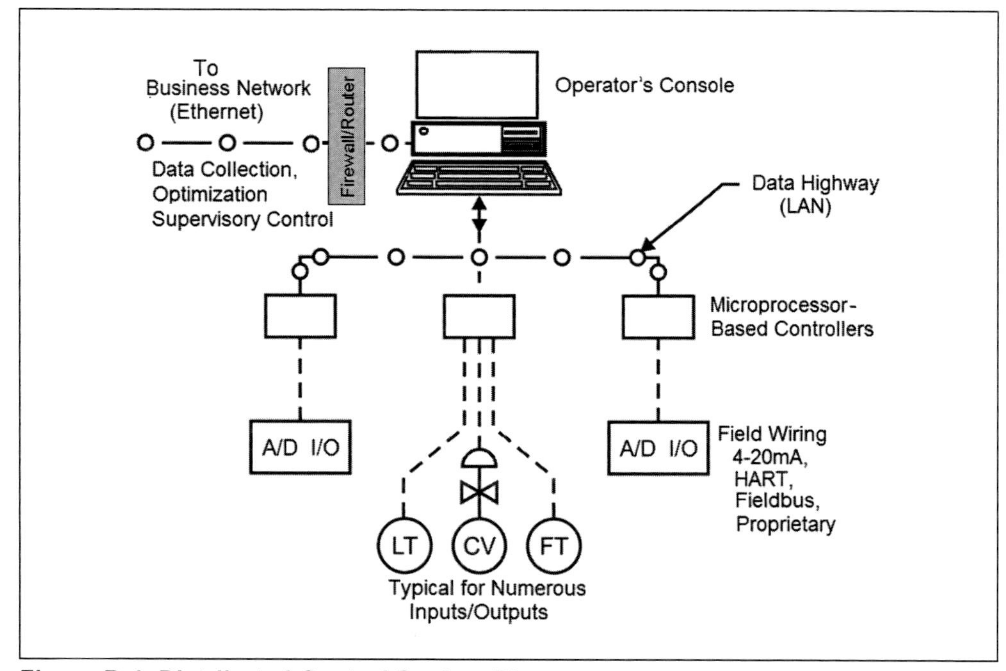

Figure B-4. Distributed Control System Diagram

Progenitors of Modern Industrial Data Communications

One of the primary progenitors of modern industrial data communications was the early telecommunications industry, which was based on the telegraph and the teletypewriter.

Morse Code and the Telegraph

The first progenitor, for all practical purposes, came to be in 1843 with the advent of the first effective binary digital communications system: Samuel Morse's telegraph. While Morse didn't invent the telegraph (he improved it), he and his aide did invent the Morse code that made the telegraph a much more viable system.

Figure B-5 illustrates a simple schematic of the Morse-type telegraph system. Morse's original machines used a mechanical stylus that marked a paper tape, not a sounder. Since telegraphs followed the railroad right of way, they were networked. Operators soon noticed that they could tell when their station was being addressed by the sound of the stylus, so sounders began to be used instead of styluses. Sounders were also introduced because it was easier for the operator to translate an audible signal to the written word than to translate marks on a ticker tape to the written word.

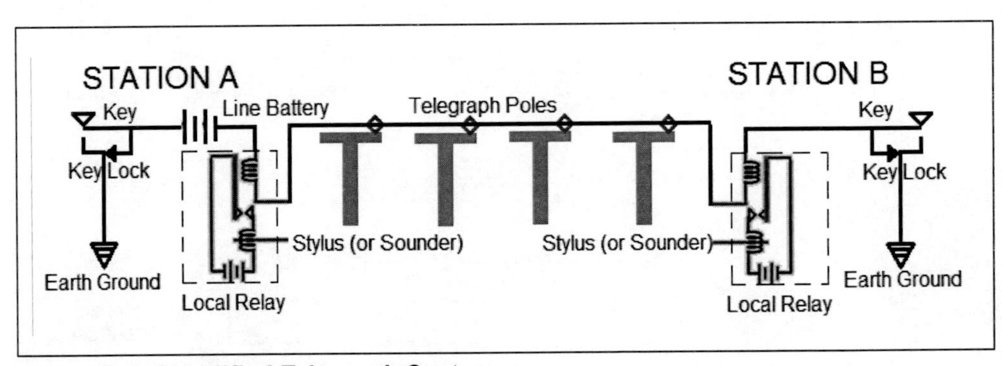

Figure B-5. Simplified Telegraph System

The practical telegraph has significant features that influenced many years of telecommunications design. Note that the system is a series circuit with the locks closed at Station A and Station B. This is the protocol used when neither station has any data to transmit. The schematic shows that there is current on the line with both locks closed. This is necessary for the operator at the receiving end of the communications to be aware of the sending operator's desire to transmit. To transmit, the sending station operator opens the key lock. This puts a space (no current) condition on the line, which opens the sounder at the distant end and indicates a "request to send."

This is an example of a half-duplex circuit: either end may transmit but not simultaneously. If one end only had a receiver and no transmit capability, the system would be a unidirectional (or "pony") circuit in which transmission is possible in only one direction.

If Station A was sending to Station B and Station B developed a high priority need to send, a simple protocol was used. (Recall that a protocol is nothing more than an agreed-upon set of actions for a set of conditions.) To interrupt Station A, the Station B operator opened his lock. This caused the line to go dead, just as if there had been a break in the line. When this happened, Station A knew through the protocol to stop sending, close the lock, and wait to receive Station B's message; that is, of course, if this was an intended "line break" and not a real one somewhere between the two stations.

"Line break" has been shortened over the years to "break." Look at a modern computer keyboard: it will, in most cases, contain a key labeled "break." It is there for the same reason as the telegraph line break. Today, however, it must be recognized by software in order to perform its protocol steps.

Morse's greatest contribution wasn't in hardware; it was in his code. A modern version of Morse code is not much changed from the original version, which is illustrated in figure B-6. The letters most frequently used in the English language have shorter code combinations, as illustrated in the original Morse code. Hence, "E" (the most often used letter) is simply a short current pulse. This consideration of letter frequencies was a factor in older codes and is an early form of data compression.

Signaling devices have used many means over the years other than sounders. As mentioned, Morse's original setup used a device similar to a ticker tape, in which an electromagnet pulled a (non-writing) stylus onto a ribbon of paper to make a mark. A "no current" condition leaves a space. This is the most likely origin of the terms "MARK" and "SPACE," which were used for many years in the teletypewriter field and are still found in technical books.

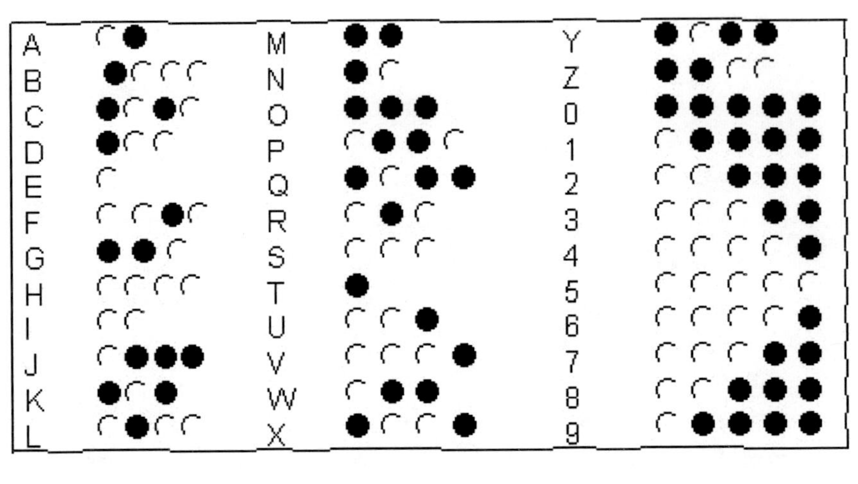

Figure B-6. Morse Code

The Teletypewriter

In the late 1800s, a new office machine made its debut: the typewriter. Although the first versions bear little resemblance to the typewriter's final design, they served the same purpose and freed the office from the need for manual transcription. Until that time, good handwriting was, of course, a valued asset. With the advent of the typewriter, more women begin to enter the office administrative fields, a province once held almost exclusively by men.

As it was a popular device almost instantly, it wasn't long before many people were thinking of ways to apply the typewriter at a distance (i.e., as the *teletypewriter*). This device continued the trend toward the modern communications networks that was started by the telegraph. The teletypewriter contributed to a great extent toward the technological foundations of modern industrial data communications.

Many teletypewriter designs were tried. Basically, the device simply had to take a keystroke and transmit information about that keystroke to a distant device that would print the keystroke. A few early designs used a common line, plus 26 (one for each letter) or more lines. That approach, of course, was not cost effective, but it is an example of parallel transmission. In parallel transmission, a single letter is a character and all character information is sent at once.

In order to use a single pair of conductors (instead of 26 or more) to transmit data, a character had to be coded into a series of on and off pulses (the basis of Morse code), which were called *elements* (the smallest amount of information [time-wise]), and transmitted one at a time. This is known as *serial transmission*, where one element follows another. To consider the magnitude of the technical problems involved in transmitting a series of electrical impulses over a relatively long distance, it would be useful to recall the technical environment of the time—the teletypewriter circuit illustrated in figure B-7. Notice the similarity to the telegraph circuit.

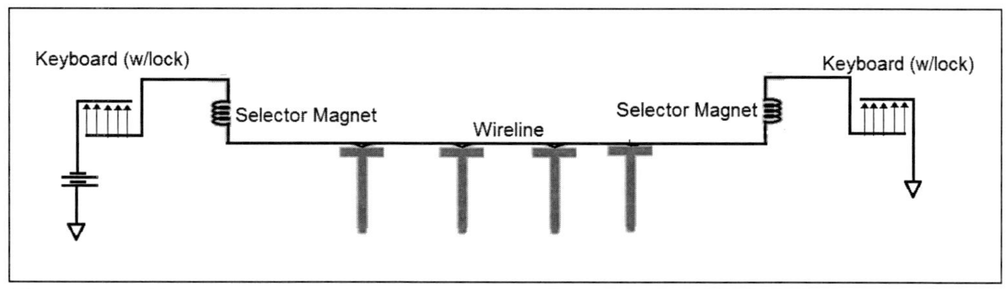

Figure B-7. Teletypewriter Line Diagram

As mentioned above, the telegraph lines followed the great railroad networks, making it a complex network. Switching and routing, necessary functions for railroads, were governed by protocols and routing instructions included in the preamble of a telegraph message. The teletypewriter network was patterned after the telegraph network, which was more advanced than the rudimentary telephone networks that were developing as a form of telecommunications. Yet, networks of these speaking devices were appearing with increasing frequency.

In the early teletypewriter designs, there were no amplifying devices and there was no way to perform electronic synchronizing, so mechanical means were required to synchronize two connected teletypewriters for both character timing and motor speed. There was a code for character timing to convert the parallel action of a keystroke into a series pattern of on and off that would be recognized at the opposite end. Motor speed was determined by markings on the teletypewriter motor flywheel and a tuning fork with an eyepiece used as a strobe, adjustments were made to the centrifugal motor speed switch, which is mounted on the motor shaft and controls the speed of the motor.

How many elements are needed to encode the English language? There are 26 letters, 10 numerals, and various items of punctuation. In addition, since this was an electric typewriter, so to speak, some code patterns must be used to return the carriage, advance the paper, and add spaces between words. While all this requires more than 32 patterns (2 to the 5th power), or 5 elements, the designers had help. They were transmitting only the uppercase letters, and they had the typewriter as a model. Therefore, they put in a mechanical shift and that allowed 26 patterns for letters and 26 patterns shifted for numbers and punctuation. Only 26 patterns were available in either shift, because six patterns were the same for both: Carriage Return, Line Feed, Space, Shift Up, and Shift Down. (If you have added correctly you might wonder what happened to the sixth element. Remember, this is a current activated, electromechanical device. There is one combination of elements that cannot pull in a clutch magnet: the "all space" [no current] combination. It is known as a *blank*.) The pattern of 1s and 0s, the code, was standardized for English and is known as the *International Telegraphic Alphabet #2 (ITA2)*, also commonly known as the Baudot code (it is sometimes it is called by other names depending on minor code variations). The code was meant primarily for text transmission. It had only uppercase letters and was quite usable with paper tape as the storage medium. Most early teletypewriters used basically the same circuit as the telegraph and the mechanicals of the typewriter.

As with the telegraph, a teletypewriter at one end had to have some way of knowing when the other end wished to transmit, so a MARK signal (current ON) would be sent as a *line idle* indication. Since a MARK was the line idle condition, the first element (nowadays called a *bit*) of any code would have to be a SPACE (current OFF) signal; it is known today as the *start space*. In addition, a current ON condition (mark) had to exist after the code pulses (elements) for each character so that the receive device would know when the character was complete and would separate this character from the next character that was transmitted.

This period of current on the line was known as the *stop mark* and was either 1, 1.42, or 2 elements in duration. In the teletypewriter, the element duration (bit time) was determined by the motor speed. Why the varying stop elements (like 1.42)? To allow the receiver time to finish printing the received character before the next character was transmitted. It would be more correct to call each element a binary digit, or "bit" for short. Even today, various devices must be set for the correct number of data and stop bits, although the stop bits are whole units now, either 1 or 2.

An example of a teletypewriter signal is illustrated in figure B-8, along with the teletypewriter code (ITA2) in figure B-9.

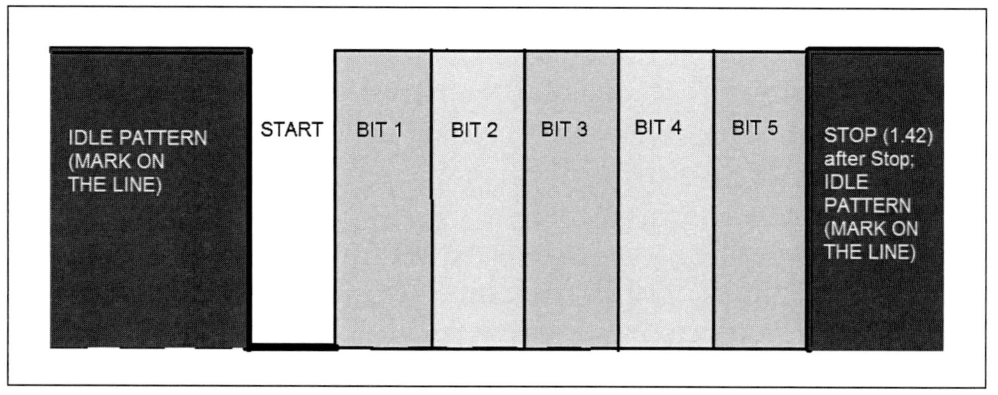

Figure B-8. Teletypewriter Signal

B1 B2 B3 B4 B5	Ltrs	Figs	B1 B2 B3 B4 B5	Ltrs	Figs	B1 B2 B3 B4 B5	Ltrs	Figs
1 1 0 0 0	A	-	1 1 0 1 0	J	'	1 0 1 0 0	S	Bell
1 0 0 1 1	B	?	1 1 1 1 0	K	(	0 0 0 0 1	T	5
0 1 1 1 0	C	:	0 1 0 0 1	L	)	1 1 1 0 0	U	7
1 0 0 1 0	D	$	0 0 1 1 1	M	.	0 1 1 1 1	V	;
1 0 0 0 0	E	3	0 0 1 1 0	N	,	1 1 0 0 1	W	2
1 0 1 1 0	F	!	0 0 0 1 1	O	9	1 0 1 1 1	X	/
0 1 0 1 1	G	&	0 1 1 0 1	P	0	1 0 1 0 1	Y	6
0 0 1 0 1	H	@	1 1 1 0 1	Q	1	1 0 0 0 1	Z	"
0 1 1 0 0	I	8	0 1 0 1 0	R	4			
0 0 0 1 0	Carriage Return		1 1 0 1 1	Shift Figs		0 0 1 0 0	Space	
0 1 0 0 0	Line Feed		1 1 1 1 1	Shift Ltrs		0 0 0 0 0	Blank	

Figure B-9. ITA2 Code

Note that letter frequencies were considered in developing this code, as the most frequently used letters have the least number of elements to eliminate wear on the machine. This ITA2 code is still the most efficient transmission code for narrative text, in terms of transmission overhead, because it requires very little in machine operations or error detection. While no longer widely used, at one time it was the most extensively used digital transmission coding.

In the data processing field, since IBM became a company in the 1920's, much business storage was performed using punched cards. Paper tape was the medium of storage for communications (teletypewriter in particular), and in many industrial environments the first printers were actually teletypewriters, so paper tape was the storage medium. Until the 1970s, paper tape was the most widely used industrial storage medium. When punching a paper tape, if one made an error, it was only necessary to back up the tape to the errored character and over-punch it with the LETTERS key. This made it all 1s (all five holes were punched in the paper tape). The only effect this would have on the receiving machine was to put it in the lower or LETTERS position. It was highly probable that it was already in that position. If the machine was in the upper or FIGURES position, it was a simple matter for the operator to punch the next character as FIGURES and continue on with their work.

To send a teletypewriter message on a network, it had to be formatted correctly. In general, a message would have an identifier, then routing information, and then the message. By keeping a message count and requiring acknowledgments, a primitive form of error detection was performed, with error correction accomplished by retransmitting the message.

Needless to say, many of the teletypewriter conventions have found their way into modern-day data communications. This has come about despite the more significant telecommunications developments that occurred in other fields and the minimal changes in the teletypewriter, which used the same code and basically the same equipment until the late 1970's.

Telephony

From the early 1920s onward, the telephone continued to develop. Radio came into being and radio networks were established. The radio networks usually used both teletypewriter and telephone links as the connection between stations. Just prior to the advent of World War II, the dial telephone network had largely replaced operators in urban areas, but the long distance network had not yet been fully automated.

Various forms of frequency division multiplexing were used on the long lines. (As discussed in the chapters of this book, multiplexing is the term used to represent placing multiple signals on one line. The signals can be separated by frequency or by time.) Frequency division multiplexing (FDM) separates the different signals by frequency, much the same way as you select different channels on your television – in fact, cable TV is an excellent example of FDM. Long distance telephone trunk lines originally used FDM. (In railroad talk, a trunk line is one that carries a lot of traffic, usually between switching centers.) The signals on a telephone trunk line may have many different originations and destinations, but they were carried on this common pathway.

Another method of multiplexing, used sparingly until the late 1960s because its cost-per-channel greatly exceeded FDM, was time division multiplexing or TDM. In the TDM system, the signal is sampled at periodic intervals (more than two times the highest expected input frequency) and the value of that signal is called a *sample*. Each sample is assigned to a particular slot in time; and the total signal is made up of the assigned time slots. This means the slots can be read anywhere a station is in synchrony with the signal. This basic time division multiplexing has evolved into the sophisticated systems used on long lines today. (A point should be made here: TDM did analog samples and sent the analog value in that time slot. TDM is now used digitally by sending a 1 or a 0 in a time slot, so the use of TDM does not automatically mean analog or digital but needs further reference. In the late 1960s, TDM was analog.)

As the price of complex electronics steadily dropped, the employment of TDM systems became more widespread. Anyone making a long distance call on the North American continent today has their voice digitized, multiplexed, de-multiplexed, and re-converted to an analog signal. FDM circuits are no longer used on long distance lines, only all-digital signaling. In fact, in the United States the only place the telephone is analog is from the central office to the subscriber, the so-called "last mile."

Television

One of the technologies that changed both technology and culture is television. A modern television set (and its counterparts the DVD or Blu-Ray for HD) is quite probably the most technically sophisticated device you will ever own. In 1956, a color TV (26" diameter round tube) cost about $525 in 1956 dollars ($5250 in 2014 dollars). It had between 19 and 25 tubes and a minimum of 9 controls on the front, which you had to adjust each time you changed the channel. For those of you not yet born in 1956, this would be the channel selector, fine tune, volume, horizontal hold, vertical hold, brightness, contrast, color intensity, and tint. Nowadays, a 25" LCD TV bought in a furniture store (not a discount house) costs about $425 (2014 dollars) and rarely has to be adjusted, except for channel and volume. If the set flips one time (the signal rolls vertically for several frames or continuously), it is generally assumed that it is time for a new one. The current TV set is absolutely better than its 1956 predecessor in all respects, yet the 2014 dollar is worth about 1/10 of that 1956 dollar. Electronics is one area in which performance has increased while the true costs have decreased.

Cable television is a broadband network; it consists of the same parts as a broadband LAN (cables, modems, connectors, etc.), only it transmits TV signals instead of data. Visual effects demonstrated by television and the ease in conveying information visually have not been lost in either the computer or the automation worlds. Graphical user interfaces (GUIs) have been and are the design goal, and "visually apparent" is a battle cry in manufacturing systems. Television (not just its technology, but its "NOW!" immediacy) has profoundly affected manufacturing systems and will continue to do so as long as manufacturing is a human effort.

Data Communications

In the 1890s U.S. census, a new form of tabulation took place using machines developed by a gentleman named Herman Hollerith. Perhaps he had taken his inspiration from the Jacquard loom and its punch cards—really a string of cards somewhat akin to a piano player roll. Whatever the source of his inspiration, Mr. Hollerith developed a method of punching cards that allowed most characters to be stored using only one or two punches in each column and absolutely no more than three. This avoided the dreaded "wimpy" card (a card with so many holes it that it lacked structural strength and was therefore subject to all forms of physical disaster). He was able to successfully store census data on these cards and tabulate this data mechanically.

Aided by the efforts of several others, Mr. Hollerith went on to form the International Business Machines (IBM) Corporation. IBM stored business data on (of course) punched cards. A Hollerith punched card, circa 1947, is illustrated in figure B-10.

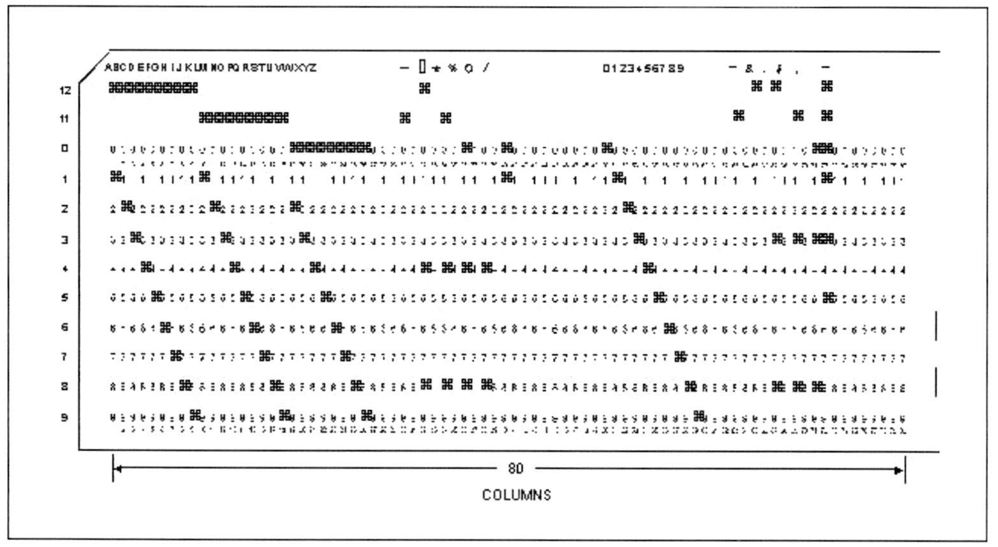

Figure B-10. IBM Data Card

The main means of data transmission at that time was the U.S. mail. After World War II, companies looked for means of electronically transmitting the card data. The Hollerith Card Code, which had 12 elements for each character, would not be efficient.

The teletypewriter code was considered but there was a problem with using it: while quite efficient, it had no means, other than human, of detecting errors. Inventory reference numbers do not have context as narrative text does; therefore, human determination of error was out. IBM settled on a "4 of 8" code, in which 4 of the total 8 bits had to be 1s.

Data communications moved from transporting data over the wide area network (WAN) to local area networks (LANs). (Historically the WAN was the telephone system or special circuits connected by the telephone company, until the Bell System was broken up by the U.S. government in 1984.) Many of the LAN fundamental applications, technologies, and approaches were adopted by the manufacturers of instrumentation systems, primarily distributed control system (DCS) and later programmable logic controller (PLC) "data highways." These LAN fundamental applications, technologies, and approaches are indeed the reason for this text's conception. Data communications terminology and techniques have become widespread throughout instrumentation applications, indeed throughout all forms of automated manufacturing.

Computer Technology

Data communications really started between mainframes. A mainframe was (and is) a large computer, whose performance is such that it may handle many tasks simultaneously (or at least appear to do so). Mainframes have always been large (physically in comparison to lesser performing contemporaries), near state-of-the-art (for a current inventory model), and expensive. They were, and are, a centralized approach to computing. In the early days, mainframe computing made things comfortable for the administrative types (for such things as upgrades, security, and control), but they also represented a single point of failure. In those days, computers were down as much as they were up, due to the technology of the times, so they also constituted a single point of frustration.

Mainframes were the only real computers around for the first decade of computing (1947 through 1957). Limited performance, lower cost computers (known as *mini-computers*) came into being and held sway as the lower-cost alternative until the middle 1970s when they gave birth to the computerization of process control. They were really smaller versions of the mainframe and usually adhered to the star topology, with the computer as the central device. When connected to peripherals or even another computer, they tended to use proprietary architectures and codes.

In the late 1970's, the microprocessor and its progeny, the microcomputer, entered the scene. First thought of by serious computer types and most Information Systems managers as a toy, it quickly gained (10 years is quick in the industrial world) respectability. The price/performance ratio has changed so drastically that many large computer systems are now sophisticated arrangements of microcomputers. As it became cost-effective to put micros at every controller and indeed at every part of an instrument loop, it became painfully obvious that they had to communicate in order to control. Methods that had been used in the past were first applied, and those methods have evolved into our modern industrial data networks.

Current Status

The basis and rationale behind most systems comes from experience, not only in the three technological areas described above but in many other areas that have contributed this principle or that technique. But what led to the necessity to connect our "islands of automation" with corporate administrative and management systems? Competitiveness, economic constraints, proper management of resources, better control of the process, customer directed manufacturing... all of these reasons and more. Production does not stand by itself. In any manufacturing company, there are some common threads. Orders must be taken, the resources to fabricate or produce the orders must be procured, the correct operational staff (maintenance, too) must be in place, and the correct process must be followed. The manufactured item must be shipped and then the customer must be billed.

It is extremely important that on-time shipments be made to the customer, that no excess inventory or work-in-process accrues, and that no time in the schedule is wasted waiting for a resource, such as the inventory in a storage tank or the flow rate of sub components to a product. Small lot sizes and short cycle times are a hallmark of the "world class manufacturer." Small lot sizes usually means the loss of the economy of scale (manufacturing 100,000 widgets or 100,000 tons of product will provide more discounts on supplies and materials than 50 widgets or 100 tons of product). Short cycle times mean frequent tool or recipe changes, resulting in a higher ratio of overhead to production. Without industrial data communications, it would not be possible to meet these conflicting requirements and still be competitive and profitable.

Add in the Internet (which provides fast, inexpensive digital communications) and it becomes possible to tie all of a company's departments together around a website that customers and company staff can use for ordering, shipping, and all the functions a business must perform.

The current status of technology in the process industry is that the industry is proceeding toward the goal of relatively open systems, non-proprietary network standards, interconnection between disparate devices, and the inclusion of control and measurement systems into a master integrated network. The goals are closer than ever to a reality today, yet the implementation will still take time, as vendors and users sort out which systems will be sustainable, reliable, and have technologically correct interfaces; which implementations are cost-effective; which systems meet their needs; and what must be done to make these interoperable systems secure, yet usable. Figure B-11 illustrates a block diagram view of such an integrated system.

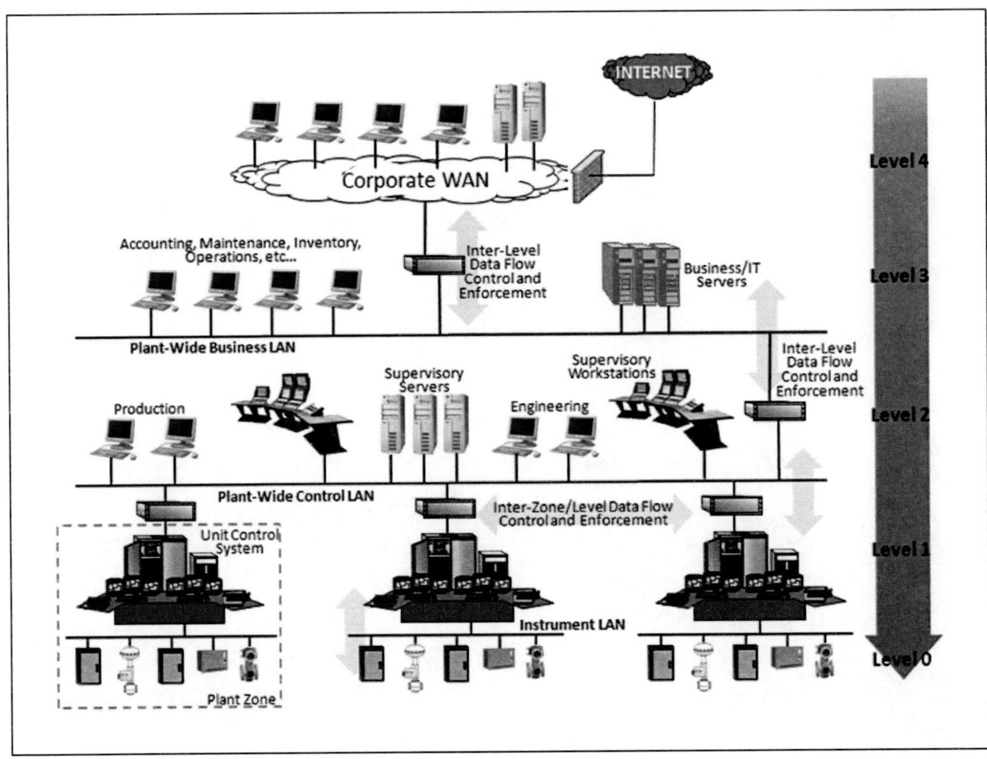

Figure B-11. Hierarchy of Systems (from ISA-62443)

In Closing

This appendix has listed many, but certainly not all, of the contributors to modern industrial data communications. It has been said that human progress is a result of standing on the shoulders of giants. A cursory review of this appendix shows that saying to be particularly applicable to data communications. The Internet is a valuable resource for researching the impact the early practitioners had, not only in design and development, but on the role that the ebb and flow of serendipity played in technological development.

Appendix C
Media

In the chapters of this text, we looked at the way data is represented as a set of patterns consisting of 1s and 0s, generally in the form of a set of these patterns, such as an octet or a byte (or sometimes a character when referring to text). We also looked at the functions required to move these data representations from the source (end user) to the destination (end user). We will now look at another of the fundamentals of communications, the media over which we transmit those data representations: the very bottom of the OSI model. This is the location where the Physical layer (the conduit for data) connects to external circuitry (i.e., other networks, WAN, etc.).

All things in data communications are related, and if they are not all correct, it is not possible to communicate very well. Sometimes called *pipes*, the media are the main constraint on data transfer speed. There are three general types of media in common use in data transmission that we will discuss, these are: copper, fiber optic, and wireless.

Copper Media

Of the copper types of media, three generally come to mind in the LAN and industrial areas: unshielded twisted pair (UTP), shielded twisted pair (STP), and coaxial cable.

UTP

Unshielded twisted pair is just as the name implies. It is two conductors twisted around each other. In theory, twisting the two conductors will cause an induced electrical field (such as spark noise or other electromagnetic radiation) to be equal in both conductors because the conductors are of opposite polarity (the induced voltage is across the pair, not across one conductor and the 0-volt DC reference known as *ground*) and the induced fields will cancel out. The reduction in noise is dependent on the number of turns per inch and the uniformity of the twist symmetry. In UTP, the larger the number of turns per inch and the more uniform the turns, the more significant the noise reduction and the higher the data rate. Typically, the higher the data rate, the higher the cost of the cable (or any media), at least for initial deployment before volume pricing sets in. In a standard UTP cable, there are normally either two or four pairs (four or eight conductors) of wires surrounded by a non-metallic insulating sheath that keeps them all together. The wire pairs are color coded; that is, one cable will be a solid color and its partner will (usually) be white with a trace of the partner's cable color, typically a stripe, so that it may be identified. According to the EIA/TIA-568 cabling standard, UTP has six Category levels, 1 through 8, with 1 having the widest tolerances (for impedance, noise, etc.) and providing the lowest data rate capacity and 8 (as of 2014) having the highest data rate and tightest tolerances, as shown in table C-1. Draft versions of ISO/IEC TR 11801-99-1 and ANSI/TIA-568-C.2-1 have been reconciled to reduce the technical differences in the standards for Category 8 (2014).

Table C-1. UTP Cable Categories

Category Level	Typical Data Rate	Some Applications
1	.3 to 3 kHz	Telephone lines
2	4 Mbps	4-Mbps Token Ring
3	10 Mbps	10BASE-T Ethernet (802.3)
4	16 Mbps	16-Mbps Token Ring
5	100 Mbps	100BASE-T Ethernet
5e	100/1000 Mbps 100/1000BASE-T Ethernet	(100 MHz cable)
6a	1000 Mbps	1 GbE (250 MHz cable)
7	10000 Mbps	10 GbE (600 MHz cable)
8	40000 Mbps	40 GbE (1.2 GHz cable)

UTP has a characteristic impedance of 100 ohms; theoretically, the line exhibits this impedance regardless of length at the frequency of interest. High conformance to the characteristic impedance (close tolerances) is necessary because maximum power is transferred between matched impedances. If the source (generator) has an output impedance of 100 ohms, and the receiver (load) has an input impedance of 100 ohms, and they are connected using UTP, there should be no wasted energy in the form of heat or reflected power.

In 2015, it is possible to purchase Category 6a cable for substantially less than Category 5e. In an industrial setting, end user connections are typically at 100 Mbps (megabits per second) with inter-switch connections at 1 GbE (gigabits Ethernet), therefore, Category 5e or 6a will normally be installed. Note that 6a has a slightly larger diameter than 5e and different termination pieces are required.

UTP cables are available in two varieties: plenum and non-plenum. The difference is in the insulation materials. Plenum cables will not exude toxic smoke when they are burning; therefore, most local code agencies, as well as the National Electric Code (NEC), require the use of plenum-type cable in crawl and air spaces.

EIA/TIA-568 Wiring for UTP

The EIA/TIA-568 (Electronic Industries Alliance [EIA] and Telecommunications Industry Association [TIA]) Commercial Building Telecommunications Cabling Standard specifies which color pairs are be used when wiring terminations. The terminations are commonly referred to as *RJ45 jacks and plugs,* however, there is technically no such thing as an RJ45 plug, just a jack that appears similar to the Bell Labs RJ45 jack (RJ = registered jack). An 8p8c (8 pole, 8 conductor) jack and plug is typically used. For reduced crosstalk (signals induced from one pair to another) on cables with a data rate of 100 Mbps and above, it is essential that pairs 1 and 2 be reversed in relation to each other. There are two wiring schemes for doing this, 568-A and 568-B (figure C-1). Both work fine for direct cables as long as both ends are wired the same. If one end is 568-A and one end is 568-B, it is necessary to cross-connect the cables.

Shielded Twisted Pair (STP)

This is the type of cable most persons with any industrial electrical experience know as signal cable. It has one or more twisted pairs enclosed within a metal-

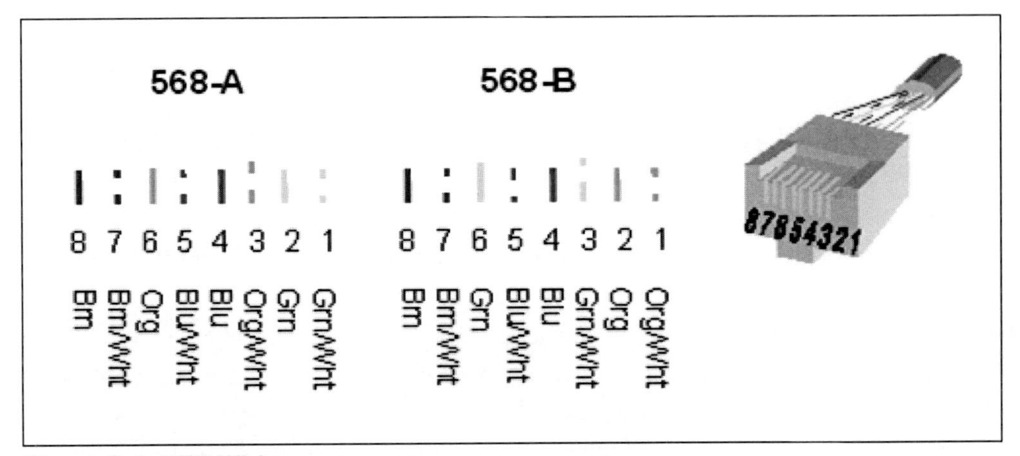

Figure C-1. UTP Wiring

lic sheath called a *shield*. In turn, this shield is enclosed in a non-conducting sheath. The size of the conductors (based on whether each pair has a shield, all pairs have an overall shield, or all pairs are individually shielded and there is an overall shield) and the type of shield (foil or braid) determines the quality and suitability for the application.

Higher data rates could (in theory) be achieved over STP cable due to decreased noise levels from the shielding; however, the shield results in a higher capacitance to ground and cannot be run for as long a distance as UTP (25 m for STP, 100 m for UTP at 100 Mbps). UTP Category 6a cable can transmit 1000 Mbps for 100 m.

Coaxial Cable

Coaxial cable, known as *coax* (*coe*-ax), consists of two concentric conductors (see figure C-2) separated by an insulator with uniform physical dimensions and dielectric consistency, sheathed in an insulating cover. Coax has been used for years to transmit large bandwidths (high frequencies). The reason it may do so is that its characteristic impedance is very closely controlled, due in large part to its physical construction. Table C-2 lists some commonly used coax types.

Good engineering practice, as based on many observations as well as on theoretical proofs, states that one should never use the signal return as a shield. In order to have a complete circuit, there must be one conductor from the source potential to the destination load and one conductor from the destination load back to the source potential [signal return]. The two conductors are generally

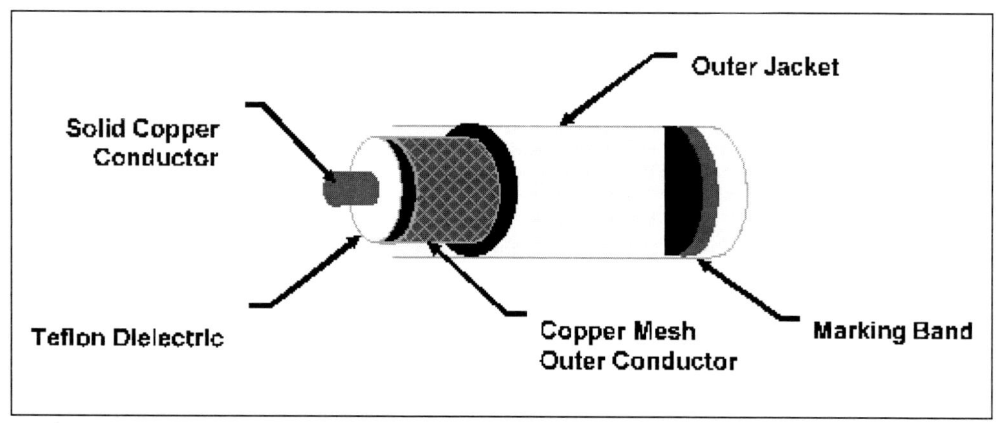

Figure C-2. Coaxial Cable Construction

Table C-2. Some Commonly Used Coaxial Cable Types

Type	Impedance	Comments
RG-8A/U	50	Replaced by RG-212/U
RG-58/U	50	Replaced by RG-213/U
RG-59B/U	75	
RG-62B/U	93	
RG-148/U	72	RG-8/U with spiral armor

referred to as a *signal pair*. So do not consider a coax cable to be just a big shielded conductor: it is not. The outer conductor is made of a flexible mesh so that the cable may be more easily bent, otherwise a solid conductor would be much like a conduit. If the application calls for a shielded coaxial cable, use triax or shielded coax.

Coax is more expensive than twisted pair. In the larger diameters, it is difficult to install and work with—it is just not as physically flexible as UTP. The higher the quality coax, the larger diameter with less signal drop per unit length.

As you may note from table C-2, there are different impedances for coax, the two most widely used being 50 and 75 ohms. Design of a system with a specific characteristic impedance has a lot to do with transmission line theory – a fascinating subject, but well beyond the scope of this book. As far as we are concerned here, if the system calls for 50 ohm (or another specific impedance specification) coax, then that is what should be used.

Fiber Optic Media

Transmission of a signal using light is an astoundingly simple concept and, since light is immune to all but the strongest electrical environment interference, it would be ideal in industrial facilities. If this is so, why isn't it used more than all other media? The reason is that simple concepts are sometimes difficult and/or expensive to implement. We will provide a cursory explanation of fiber optic transmission in order to explain the difficulties with the media itself. It should be stated, however, that the high costs formerly associated with fiber optic transmission had more to do with the interface electronics (i.e., how electronic devices are connected to the cable) and the high costs of maintenance. These costs have dropped dramatically in the last 2 decades, so much so that fiber is competitive with copper in many applications.

Fiber Optic Operation Principles

The principles of physics that allow fiber optics to work are refraction and reflection. Refraction occurs because the speed at which light travels varies with the media. As an example, light travels faster in air than in glass; light travels faster in some compositions of glass (fast glass) and slower in other compositions (slow glass). Reflection occurs when light approaches above the critical angle. Figure C-3 illustrates the optical principles involved.

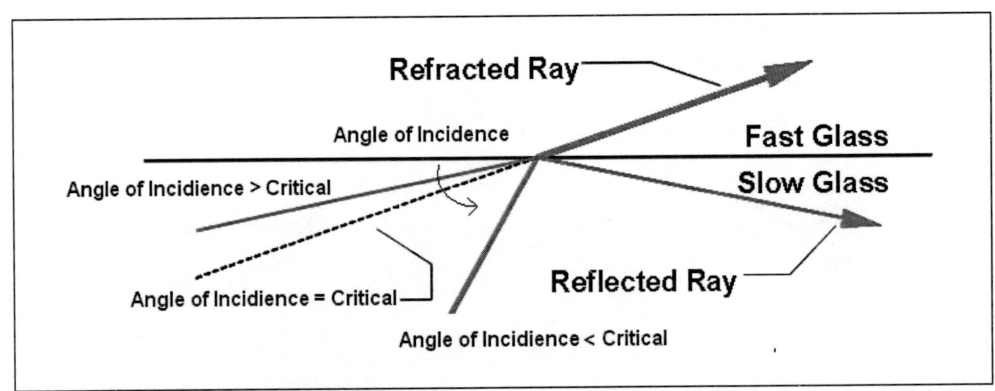

Figure C-3. Angle of Incidence, Refraction, Reflection

Note that there is a critical angle of incidence. When light approaches the interface between the fast glass and the slow glass at less than the critical angle, the light leaves the slow glass but it is not reflected, it is refracted. At the critical angle, the light never leaves the surface of the interface but travels parallel to it. In summary: light approaching a surface at less than the critical angle is refracted; light approaching a surface at the critical angle is refracted parallel to the surface; and light approaching the surface at an angle greater than the critical angle is reflected.

The angle of incidence (i) and the angle of refraction (r) are related by the ratio of their sines. At the critical angle, the angle of incidence is 90 degrees from the angle of refraction (they form a right triangle). The critical angle varies for different interfaces, such as fast glass and slow glass. The relationship between the angle of refraction and the angle of incidence is the ratio of their sines. The "index of refraction" is expressed as:

$$\eta = \text{sine angle i} \ / \ \text{sine angle r}$$

This is one of the more important parameters behind the operation of a fiber optic cable. Since the critical angle is the point between refraction and reflection, all light approaching a fiber optic cable must be at some angle greater than the critical angle.

Figure C-4 shows the makeup of a fiber optic cable.

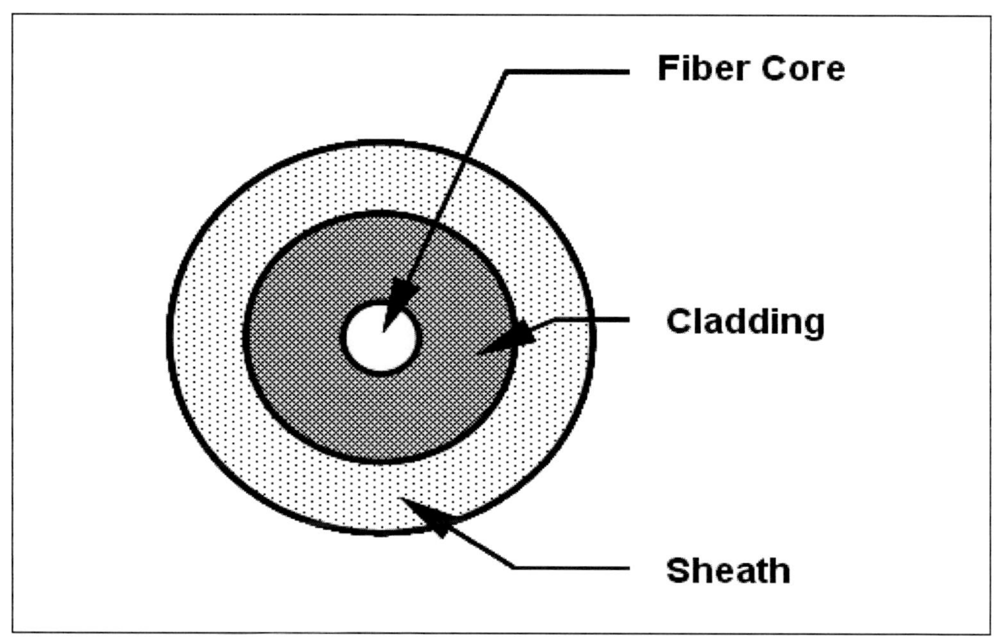

Figure C-4. Fiber Optic Cable

The cladding and the core material are such that the index of refraction for the core material is greater than the index of refraction for the cladding. Since light must approach at a greater than critical angle, if a line is drawn that extends from the center of the fiber optic cable and another line is drawn from the surface at the critical angle, the angle of acceptance, φ, is produced. If this angle is rotated a cone is generated, called the *cone of acceptance*. Light must fall within the cone to be transmitted through the fiber. Figure C-5 illustrates the cone of acceptance.

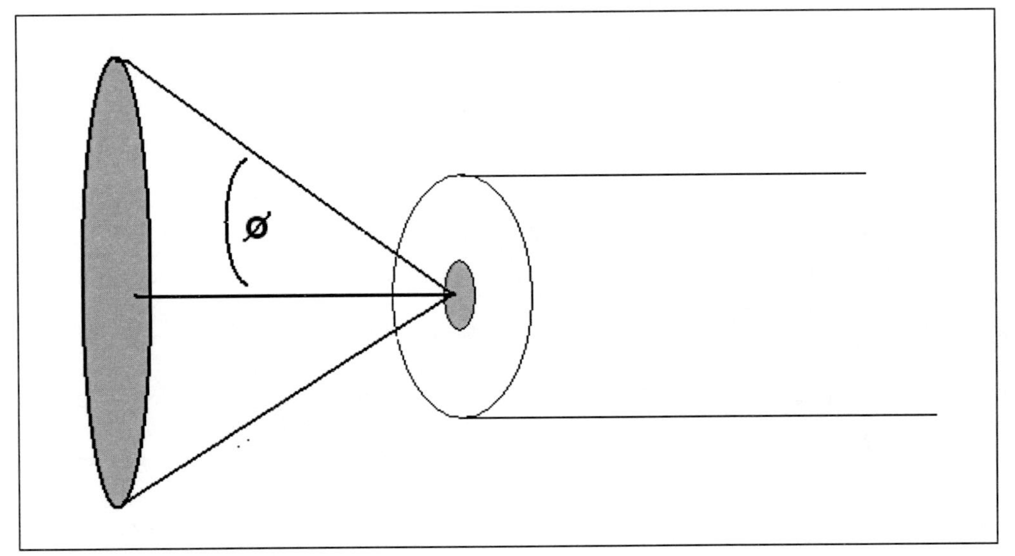

Figure C-5. Cone of Acceptance

Losses

Fiber optic cables have some specific losses. One is scattering, another is absorption. Scattering is the act of light being spread out by the molecular structure of the fiber material. Scattering is also caused by impurities in the fiber as well as bubbles, scratches, and any other imperfections in the fiber surface. Manufacturing techniques and good installation practices can eliminate all the scattering losses except those caused by the molecular structure. This is a fundamental loss that cannot be reduced further. Absorption losses are also caused by impurities in the fiber material. An almost semiconductor level of purity must be obtained if absorption loss is to be limited in a fiber.

The size of a fiber optic cable determines the mode of operation and, to a great extent, its losses. Single-mode glass fibers have a very small diameter fiber in relation to the cladding. Multi-mode fibers have a larger core to cladding ratio. *Single-mode* and *multi-mode* refer to the number of paths and hops that light can take to reach the end. *Single mode* has a core of such small diameter that there is only one mode (path) for light to reflect back and forth (from cladding to cladding in the core) to the destination end; *multi-mode* has a wider (in diameter) core and provides multiple reflective paths to the destination end. Single-mode fibers have good transmission throughput and the cable is easier to handle, but single-mode cable requires far more precise mechanical alignment of terminations and splices than multi-mode types (single-mode cable also costs more). Table C-3 is an illustration of some fiber optic sizes.

Table C-3. Selected Fiber Optic Sizes

Mode	Core (mm)	Cladding (mm)
Single-mode	8	125
Multi-mode	50	125
Multi-mode	62.5	125
Multi-mode	100	140

Most of the losses that occur in a fiber optic application happen at connections and termination points. In practice, fiber optic cable is not run directly from device to device; instead, a multi-fiber trunk cable is run from area to area and is permanently installed into a fiber patch panel (a system of fiber optic jacks) at each end, then fiber patch cords are used to make the connections from the patch panel to the end devices. Figure C-6 shows a diagram of how this is typically done.

To ensure good termination, each fiber has to be finished and polished and then inserted into a connector whose job is to align this fiber and the one in the connector on the other side of the joint. Any number of problems, such as misalignment, contamination, or air gaps can create loss at the connection. Figure C-7 shows perfect termination along with some common fiber connection problems.

In the discussion of the physics of refraction and reflection above, the idea of a critical angle of incidence was presented. When a fiber optic cable has to make a sharp bend, it is likely that some of the light traveling down the cable will exceed the critical angle and be absorbed into the cladding, causing additional loss. The requirement is that the total loss of light energy from all causes must

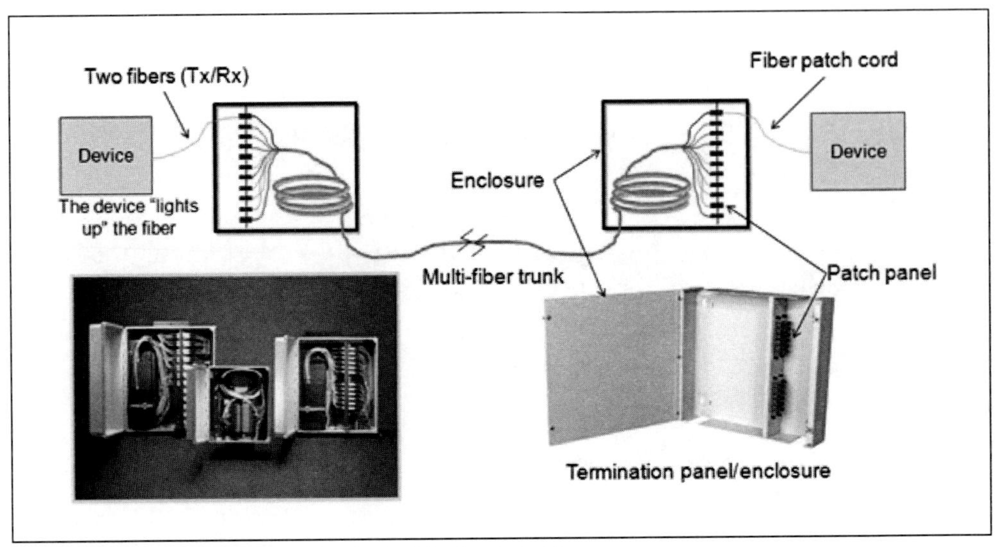

Figure C-6. Typical Fiber Optic Cable Installation

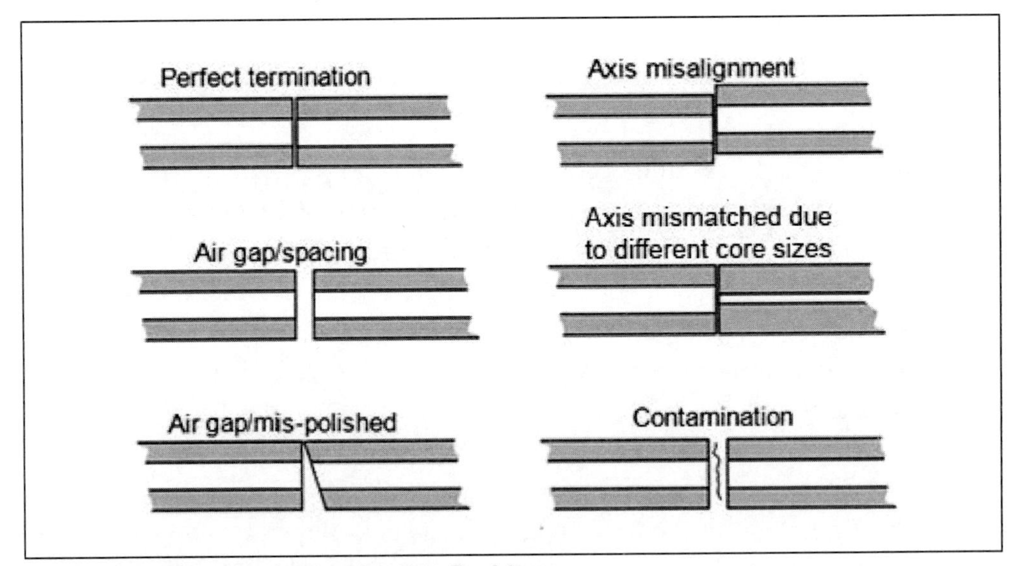

Figure C-7. Fiber Optic Termination Problems

not be to the level where the receiving light sensor cannot accurately detect the light pulses being transmitted.

Single vs. Multi-Mode Fiber Cable

Light travels at a constant speed through a fiber optic cable. Light is composed of discrete photons and these photons, based on the angle at which they enter the fiber, may take a direct path or a much longer path based on numerous reflections. Since all photons travel at the same speed, some will arrive at the receiver sooner than others, even though they all started at the same time. Figure C-8 illustrates this concept.

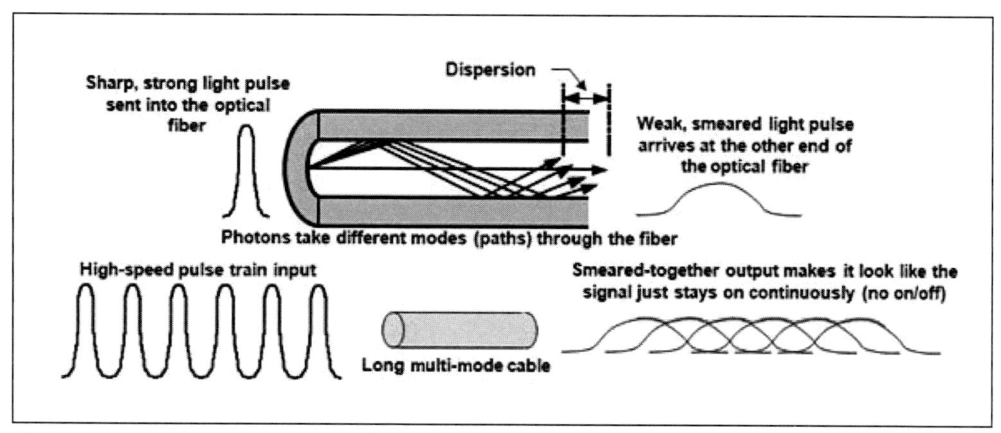

Figure C-8. Multi-Mode Fiber Optic Cable

What this means is that there needs to be sufficient time between transmitted light pulses to allow the dispersed ("smeared") pulses at the receiving end to be discerned and differentiated. This limits the maximum signaling rate possible; as the cable gets longer, the signaling rate has to drop. Single-mode fiber has a very small core to ensure that photons do very little bouncing about and all arrive at the receiving end at about the same time (minimal dispersion.) Single-mode fiber permits much higher signaling rates than does multi-mode fiber. Even so, multi-mode fiber is more than adequate for most industrial uses (it can easily support 100 Mbps and even 1 Gbps Ethernet speeds) and it is somewhat easier to terminate, polish, and "connectorize" due to the much larger core size.

As mentioned above, fiber optics would appear to be the answer to most industrial applications. The reason that they have not been utilized is the high cost, not necessarily of the fiber optic material but of the electronics required for getting a signal on and off of the cable. However, due to fiber's capability of much higher bandwidth; noise immunity and electrical isolation; and the significant reduction in cost in the last few years; fiber has come to be the recommended medium for LAN backbone cables (the main trunk cable serving many networks) by many automation vendors. History has taught where there is a want (or need), eventually the answer will be cost-effective. Look for fiber optic media to continue to make inroads into industrial areas for interconnection; however, it is restricted in tying field devices together by its inability to supply power to the devices using the signal media, as is done in the present two-wire loop or even power over Ethernet (PoE).

Wireless Media

Any discussion about wireless media should include a discussion on the properties of the atmosphere, which is where the energy is propagated. Wireless *is* radio communications. But this is not a simple amplitude modulated (AM) or even a frequency modulated (FM) broadcast-type signal. When referring to wireless data communications, reference is made to a method of propagating electrical signals by impressing these signals (the modulating frequencies) on a much higher frequency carrier signal. For industrial applications it is ultra-high frequency (UHF, between 800 MHz and 5.2 GHz).

Wireless (or if you will, radio) has been around for some time but was not used in industry for a number of reasons, the primary one being the high ambient electrical noise in an industrial facility. Other interference factors from external sources, such as walkie-talkies, fixed radio and TV, plus multipath reception problems have given this medium a less than reliable reputation in such an environment.

A signaling method designed to overcome jamming (where a high energy radio signal is used to overwhelm the intended signal) is proving effective in the industrial area. It is known as *spread spectrum.* In the frequency hopping type of spread spectrum (FHSS), a predetermined algorithm selects the transmit frequencies throughout the authorized spectrum, never sending more than a packet or so at any one frequency. The receiver has the same algorithm so it tracks along with the transmitter.

A second type of spread spectrum signaling is direct sequence spread spectrum (DSSS). A direct sequence spread spectrum transmission takes the data being transmitted and multiplies it by a deliberately generated noise signal. This noise signal is a random appearing sequence of 1 and –1 values, whose frequency is much higher than that of the original signal. The result of the multiplication produces a signal that resembles white (random) noise. This apparent noise signal is multiplied by the same noise signal at the receiving end to reconstruct the original data. This process reproduces the transmitted sequence (before the noise multiplication) because the receiver knows the exact noise sequence the transmitter is using.

With either type, the signal is spread over a wide band of frequencies and the algorithm used, whether frequency hopping or direct sequence spread spectrum is used only within the system. These two are identified in IEEE 802.11.

A second method, used to more efficiently use the assigned radio spectrum, is orthogonal frequency division multiplexing (OFDM), also called discrete multi-tone modulation (DMT). This method was originally used in many wireline modems and it is now in use in wireless modems. Spread spectrum and OFDM are superior to other methods in noise suppression, and they represent a more efficient use of assigned bandwidth, as opposed to assigning each sender a transmit frequency for all transmissions or even for a single transmission.

At present, wireless applications in industrial networks are not widespread, but their incidence of use is rising. When one realizes just how much money is "capital at rest" in a conventional wiring system and the number of times processes or parts of processes are moved or re-engineered, a wireless alternative may seem like a cost-effective approach. Nonetheless, wireless as an industrial medium will have to prove itself in more than support applications before there is widespread installation. Wireless LANS were discussed at some length in chapter 4 and wireless WANS were discussed in chapter 7 of this text.

Media Summary

The three major categories of media used in the industrial area were discussed: copper, fiber optic, and wireless. Of the three, copper is by far the most commonly installed, but it has limitations in costs (both media and infrastructure), distance, bandwidth, and data rate capability. Both wireless and fiber optic are efforts to overcome some of copper media's limitations. Each of the alternatives to copper has limitations of its own, which so far have limited their acceptance in the industrial area; however, fiber optic is becoming the standard for the transmission of data over distance in a facility or between facilities.

Glossary

Abort—To terminate or exit a program or communication hastily without any data or states being saved, not a desirable end to a program.

ACK—acknowledgment. A message from the receiver that a transmission or communication has been received.

A/D—analog to digital converter.

Adapter—A device to connect different parts of a system or provide an interface. An example would be an EIA-485 adapter that plugs into a PC and provides 485 termination in place of the PC COM port.

ADCCP—Advanced Data Communications Control Procedure. ANSI version of LAP-B (see LAP-B).

Address—The identifier for a logical or physical element.

Agent—A software function that responds to network management requests.

AIX—IBM's version of UNIX.

Algorithm—A set of steps (procedure) to a solution; it may be mathematical, but does not have to be.

Analog—Any value between the lower and upper range values, a continuum of values.

ANSI—American National Standards Institute.

ANSI/EIA-568—A telecommunications wiring standard for commercial buildings.

ANSI/ISA-100.11a-2011—A standard for wireless systems for industrial automation.

API—application programming interface. A standard format for connectivity to a program or function.

ARCnet—Attached Resource Computer (registered trademark of Datapoint Corporation). A data-point-derived token-passing bus network.

ARPANET—Advanced Research Projects Agency Network. A packet switching network sponsored by the U.S. Department of Defense, the forerunner of the Internet.

ARQ—automatic retransmission (repeat) query (reQuest). An error correction strategy.

ASCII—American Standard Code for Information Interchange. A 7-bit character set, often confused with the 8-bit extended ASCII.

ATM—Asynchronous Transfer Mode. A connection-oriented, cell-oriented, high-speed transmission protocol.

AUX—Apple Corporation's version of UNIX.

AWG—American wire gauge. Standard for wire sizes.

B-channel—The bearer channel in ISDN (see ISDN).

backbone—The main connectivity media for distributed systems.

BASIC—beginners all-purpose symbolic instruction code. An interpreted or compiled programming language.

baud—A unit of line modulation rate. The baud rate is determined by dividing the time of the smallest occurring element into 1. It is the decision rate the media will support.

BER—bit error rate. The ratio of errored bits to bits transmitted.

binary—A base 2 digital system.

BIOS—basic input/output system. Firmware that specifies certain results regardless of physical architecture for predetermined inputs.

bit rate—Bits per second.

block—A unit of data usually stored and perhaps transmitted as a unit (see packet and frame).

bridge—A device that connects two segments of a LAN. Contains twin sets of Layer 1 and Layer 2 functions.

CAN—controller area network.

category cable —Cable that complies with EIA/TIA TSB-36.

CCITT—Comité Consultatif International Téléphonique et Télégraphique standards setting body replaced by ITU-ITS.

client/server—An arrangement in which the client shares server resources.

CIFS—Common Internet File System.

CLNP—Connectionless-mode Network Protocol. Part of OSI Standards, referred to as OSI-IP.

CLNS—Connectionless-mode Network Service. A connectionless network protocol. One of the two options in the OSI Standards.

COM—Component Object Model.

CONS—connection-oriented network service. The OSI Standard for connection-oriented service.

CRCC—cyclic redundancy check character. An error detection scheme.

CSMA/CD—Carrier Sense Multiple Access with Collision Detection. A media access strategy.

CSU—channel service unit. Used to terminate a digital line at the customer site.

D/A—digital to analog.

DBMS—database management system.

DCE—data communications equipment. The interface at both ends of the channel that provides the necessary connectivity for the communications infrastructure to create an end-to-end channel (located between the DTE and the communications channel).

DCOM—Distributed Component Object Model.

DCS—distributed control system. A networked system of client-server devices, in which the intelligence is distributed throughout the system to perform parts of the whole task.

DDD—direct distance dialing.

DDS—digital data service.

decimal—A base-10 number system.

DES—data encryption standards. A cryptographic block algorithm designed by NBS (NIST).

DIX Ethernet—Digital Intel Xerox Ethernet (also known as Ethernet 2.0).

DLL—dynamic link library.

DSAP—destination service access point. User destination address for a service.

DSL—digital subscriber line.

DSU—digital service unit. Terminates the data circuit to the CSU.

DTE—data terminal equipment. The end device (either data source or destination) that sends (originates) data to or receives data from another DTE (e.g., a computer, a printer, or a PLC).

duplex—Formerly full duplex. Data is transmitted simultaneously in both directions.

EBCDIC—extended binary coded decimal interchange code. IBM's proprietary 8-bit character set.

end user—The source/destination of data sent through a communications system.

EIA—Electronic Industries Association.

EIA-232—EIA standard for an unbalanced-to-ground digital interface, up to 20 Kbps.

EIA-422—EIA standard for a balanced-to-ground transceiver, up to 10 Mbps.

EIA-423—EIA standard for an unbalanced-to-ground transceiver, up to 20 Kbps.

EIA-485—EIA standard for a multidrop balanced-to-ground electrical specification.

ES—end system. Defined by OSI.

ES-IS—end system to intermediate system. Defined by OSI.

Ethernet—Term applied to both IEEE 802.3 and Ethernet 2.0 systems that use CSMA/CD and differ only in minor structure.

FDDI—fiber distributed data interface. A token-passing, fiber-optic ring capable of 100 Mbps transmission.

FEC—forward error correction. The use of delay- and error-detection algorithms to protect data and correct all errors within the protection of the algorithm.

fiber-optics—A transmission media using light as the carrier.

Fieldbus—A network connected to field devices; also used for the SP50 Fieldbus, a standard.

frame—A sequence of octets with a header and a trailer; usually a Layer 2 contrivance, similar to the packet or the older block in Layers 3 and above.

frame relay—A connectionless packet system.

FTAM—file transfer, access control, and management. Programs that interface to the Applications layer of the OSI Model.

FTP—File Transfer Protocol.

gateway—A twin seven-layer device for connectivity between dissimilar systems.

Gbps—gigabits per second.

GOSIP—Government Open System Interconnection Profile. The U.S. Government version of OSI.

GUI—graphical user interface.

half duplex—Data can be transmitted in either direction on a carrier, from A to B or from B to A, but not at the same time.

HART—Highway Addressable Remote Transducers. A method of communicating with smart instruments.

HDLC—high-level data link control. Similar to ADCCP and LAP-B.

IEC/ANSI/ISA-62443 a series of published standards for implementing a cybersecurity program (formerly ISA99).

IEEE—Institute of Electrical and Electronic Engineers. A professional society that sets standards.

Internet—An international, public, packet-switched network consisting of numbers of backbones and many nodes using TCP/IP.

Intranet—Corporate networks (enterprise) connected using Internet technologies.

IP—Internet Protocol. A Layer 3 routing protocol.

IPX—Internetwork Packet Exchange. A modification of XNS for LANs.

IS-IS—intermediate system to intermediate system. Defined by OSI.

ISA—International Society of Automation. Professional society that establishes standards for process automation.

ISDN—integrated services digital network. An older technology for bringing digital to the desktop.

ISO—International Standards Organization. An international standards-setting organization.

jabber—To continuously transmit. Indicates a failed adapter.

Kbps—kilobits per second.

LAN—local area network. More than two nodes connected together serving a function; the media and distribution is privately owned.

LAP-B—Link Access Protocol-Balanced. A duplex version of a bit-oriented protocol similar to HDLC, SDLC, and ADCCP.

LATA—local access and transport area. An area or region in which a telephone provider company is permitted to offer services.

LLC—Logical Link Control. The upper sub-layer of Layer 2 (Data Link layer).

MAC—Media Access Control. The lower sub-layer of Layer 2 (Data Link layer).

MAN—metropolitan area network. Defined in IEEE 802.6.

Mbps—megabits per second.

media—The plural of medium. The entity (e.g., cable and fiber-optics) used to propagate the communications signal.

MIB—management information base. Database used in SNMP.

MIS—management information systems. Organization providing computing and support.

MNP—Microcom Networking Protocol. A set of error detection and compression protocols, MNP-1 through 10.

modem—MODulator/DEModulator. Converts digital data to analog signals for transmission, then converts the signals back to digital data upon reception.

multi-tasking—Two or more program segments running simultaneously.

multi-threaded—A method of segmenting instructions.

multi-user—The system supports multiple users simultaneously.

multiplexor—A device used to place more than one signal at a time on a single line.

NAK—negative acknowledgment. Signifies that the previous transmission was errored.

.NET—A system based in part on XML and SOAP.

NetBEUI—A NetBIOS-enhanced user interface. NetBIOS is an interface between Layer 2 and all intervening layers.

NetBIOS—network basic input/output system. A set of commands and protocols to allow network traffic between a workstation and server.

NIC—network interface card. The network adapter that plugs into a PC.

NIST—National Institute for Standards and Technology. Previously, National Bureau of Standards.

node—An addressable entity connected to a network.

OOP—object-oriented programming.

OPC—Open Platform Communications, formerly: object linking and embedding (OLE) process control.

OSI—Open Systems Interconnection. An open set of standards for networking.

packet—A unit of data, usually from Layer 3 or higher (see block and frame).

parity—An error-checking technique dependent upon the number of 1 states in a character.

Physical layer—Layer 1 of the OSI Model.

PLC—programmable logic controller.

polling—A form of master-slave in which one station (hub or master) queries all connected stations.

RIP—Routing Information Protocol. A method for determining addresses on a multi-server network.

SAP—service access point. The interface address between layers.

SATA—serial AT attachment. This connects EIDE drives to PC motherboards.

SCSI—small computer system interfaces. A method of connecting peripherals to small systems.

SDLC—synchronous data link control. IBM's version of ADCCP, HDLC, and LAP-B.

SMTP—Simple Mail Transport Protocol.

SNMP—Simple Network Management Protocol. Performs network management over TCP/IP systems.

SOAP—Simple Object Access Protocol.

SSAP—source service access point. The source address of a service between layers.

STP—shielded twisted pair. A twisted pair of conductors surrounded by their individual shield.

TCP—Transmission Control Protocol. The Layer 4 transport for TCP/IP transmission.

TIA—Telecommunications Industry Association.

token—A binary pattern.

token bus—A bus topology network in which a node must have the token to initiate transmission.

token ring—A ring topology network in which a node must have the token to initiate transmission.

two-wire loop—An SP50 standard used to connect field devices. The loop supplies power to the transmitter and provides signal to the receiver.

USB—universal serial bus. Allows plug-and-play operation for PC peripherals.

UTP—unshielded twisted pair.

WAN—wide area network. The media is leased or rented.

XML—extensible markup language.

zero insertion—In normal transmission, the rule is to insert a 0 any time the protocol detects five 1 bits in the data stream.

Index